T0254910

WATER AT INTERFACES

A Molecular Approach

WATER AT INTERFACES

A Molecular Approach

Jordi Fraxedas

CRC Press
Taylor & Francis Group
Boca Raton London New York

CRC Press is an imprint of the
Taylor & Francis Group, an **informa** business

CRC Press
Taylor & Francis Group
6000 Broken Sound Parkway NW, Suite 300
Boca Raton, FL 33487-2742

First issued in paperback 2018

© 2014 by Taylor & Francis Group, LLC
CRC Press is an imprint of Taylor & Francis Group, an Informa business

No claim to original U.S. Government works

ISBN-13: 978-1-4398-6104-2 (hbk)
ISBN-13: 978-1-138-37450-8 (pbk)

Visit the Taylor & Francis Web site at
http://www.taylorandfrancis.com

and the CRC Press Web site at
http://www.crcpress.com

Dedication

Hay libros que no son de quien los escribe
sino de quien los sufre
Relato de un náufrago, G. García-Márquez

Dedicated to my wife Montse
and my sons Roger and Marc

Contents

Preface

Guárdate del agua mansa, the Spanish translation of the known proverb, *beware of still waters*, is a theater piece of the great Golden Age Spanish writer Pedro Calderón de la Barca based on a story of love and honor. This title is perfectly suited to any work devoted to the understanding of water, this vital, essential, fascinating, surprising, and eccentric small molecule that continues to preserve an aura of mystery around it, as if to remind us that without it, life (as we understand it) is not possible. The water molecule can be considered as a very simple one, given that it is constituted by only one oxygen and two custodian hydrogen atoms with a shape that recalls a very popular cartoon character, but we have to be aware when water molecules condense into liquid and solid states through hydrogen bonding because this simplicity is only apparent. Below such an apparent simplicity a complex and mysterious creature is hidden. It is a sort of complex simplicity fully justifying the term, *Guárdate del agua mansa*. We could find a parallelism between water and Fermat's last theorem, where no integer solution exists for the equation $x^n + y^n = z^n$, for $n > 2$. It appears to be a rather simple problem but it took more than 300 years to solve it!

Water is perhaps the only compound having a biography (P. Ball, *Life's Matrix: A Biography of Water*, University of California Press, 2001) and many books, articles, reviews, and so on, have been and continue to be published focused on diverse aspects. I strongly recommend M. Chaplin's and US Geological Survey webpages http://www.lsbu.ac.uk/water/ and http://www.usgs.gov/, respectively. However, an introductory book on *interfacial* water trying to reach a broad multidisciplinary audience was missing in my view, and this was the pivotal motivation, or excuse, that drove me to conceive the work you are now reading. Among the myriad published works devoted to water and interfacial water, only a partial selection of them emerge in this book, as the tip of an iceberg. The rest, those below the surface, indeed contribute to the general knowledge, but this book is not intended as a review of all published material. That would have little sense for a book because of its intrinsic static nature in printed version in a quite dynamic and active subject. I just mention that more than 1.5×10^6 articles have been published on water since 1900, according to online bibliographic research platforms, and that about 4 and 18% of them contain the terms "interface" and "surface", respectively. Water at interfaces is thus a visible part of the water iceberg (only about 10% of the volume of an iceberg emerges).

This book is conceived to go from basic simplified concepts toward more complex issues, increasing the degree of complexity. We start in Chapter 1 with isolated water molecules, because it is fundamental to know its properties in the absence of intermolecular interactions. Then we allow them to interact through van der Waals interactions, then through hydrogen bonding to build liquids and solids, and we submerge in Chapter 2 into the (pristine) interfaces built by water in a broad sense, including flat liquid/vapor, liquid/oil, and liquid/solid interfaces. Chapters 3 and 5 are devoted to interfaces involving both ideal and real surfaces, respectively, and Chapter 4 discusses the affinity of water to surfaces (hydrophobicity and hydrophilicity). Finally, Chapter 6 deals with the interaction of water with biomolecules. Interfaces are considered in a broad sense and at different length scales: the zero-dimensional case of water sequestering small molecules (clathrate hydrates), water surrounding macromolecules, liquid water wetting (or not) extended flat surfaces, and so on.

Finally, I would like to acknowledge many people who have participated to different extent during the preparation of this book, but to the tip of the iceberg I would first invite E. Canadell and C. Rovira for extended Hückel and DFT calculations, M. M. Conde and C. Vega for MD simulations, M. V. Fernández-Serra for discussions on DFT calculations, and T. Hernández and A. Santos for help on documentation. I would skip my family, my wife Montse and my sons Roger and Marc, from the uncomfortable, cold, and slippery honorary iceberg tip and reserve for them a more hospitable place for their warm permanent support.

Abbreviations and Symbols

A	Acceptor
A_0	Free Oscillation Amplitude
A_c	Critical Free Amplitude
AFM	Atomic Force Microscope
AFP	Antifreeze Proteins
AM-AFM	Amplitude Modulation Atomic Force Microscope
AQP1	Aquaporine-1
ASW	Amorphous Solid Water
AU	Astronomical Unit
B3LYP	Becke Three-Parameter Lee–Yang–Parr
CERN	European Organization for Nuclear Research
CNT	Carbon Nanotube
CPMD	Carr–Parrinello Molecular Dynamics
CVD	Chemical Vapor Deposition
D	Donor
DFT	Density Functional Theory
DNA	Deoxyribonucleic Acid
DOS	Density of States
DPPC	Dipalmitoylphosphatidylcholine
DWNT	Double-Walled Nanotube
e	Electron Charge
E_{ads}	Adsorption Energy
E_{coul}	Coulomb Interaction Energy
E_F	Fermi Level
E_{LJ}	Lennard–Jones Interaction Energy
E_{LJ}^0	Minimum of the Lennard–Jones Interaction Energy
E_{vdW}	van der Waals Interaction Energy
EELS	Electron Energy Loss Spectroscopy
EH	Extended Hückel
EWOD	Electrowetting on Dielectric
EXAFS	Extended X-ray Absorption Fine Structure
F_{cap}	Capillary Force
FIR-VRT	Far-Infrared Vibration-Rotation-Tunneling
FWHM	Full Width at Half Maximum
g	Gravity Constant
G	Gibbs Free Energy
GAXS	Glancing-Angle X-ray Scattering
GGA	Generalized Gradient Approximation
h	One-Electron Hamiltonian
HAS	Helium Atom Scattering
HDA	High-Density Amorphous
HF	Hartree–Fock
HGW	Hyperquenched Glassy Water
HOMO	Highest Occupied Molecular Orbital

HOPG	Highly Oriented Pyrolytic Graphite
HREELS	High-Resolution Electron Energy Loss Spectroscopy
Ih	Hexagonal Ice
INP	Ice Nucleating Protein
IR	Infrared
IUPAC	International Union of Pure and Applied Chemistry
K	Kinetic Energy
k_B	Boltzmann Constant
k_c	Force Constant
k_{GT}	Gibbs–Thomson Coefficient
KPFM	Kelvin Probe Force Microscopy
l_c	Capillary Length
l_v	Latent Heat of Vaporization
LB	Langmuir–Blodgett
LCAO	Linear Combination of Atomic Orbitals
LDA	Low-Density Amorphous
LEED	Low-Energy Electron Diffraction
LJ	Lennard–Jones
LUMO	Lowest Occupied Molecular Orbital
MD	Molecular Dynamics
MEMS	Microelectromechanical System
ML	Monolayer
MO	Molecular Orbital
MP2	Moller–Plesset (second-order perturbation theory)
NAPP	Near Ambient Pressure Photoemission
NASA	National Aeronautics and Space Administration
NEMS	Nanoelectromechanical System
NEXAFS	Near-Edge X-Ray Absorption Fine Structure
NMR	Nuclear Magnetic Resonance
NP	Nanoparticle
p_v	Vapor Partial Pressure
p_v^{sat}	Vapor Partial Pressure at Saturation
PBE	Perdew–Burke–Ernzerhof
PDOS	Partial Density of States
PECVD	Plasma Enhanced Chemical Vapor Deposition
Q	Q-Factor
QLL	Quasiliquid Layer
r	Radius
R	Molar Gas Constant
r^*	Critical Radius
r_d	Droplet Radius
r_{ion}	Ionic Radius
r_{pore}	Pore Radius
r_0	Intermolecular Distance at Minimum LJ Potential Energy
r_μ	Dipole Radius
RAIRS	Reflection Absorption Infrared Spectroscopy
RH	Relative Humidity
RT	Room Temperature

S	Entropy
s_{jk}	Overlap Integral
SAM	Self-Assembled Monolayer
SEM	Scanning Electron Microscopy
SFG	Sum-Frequency Generation
SFM	Scanning Force Microscopy
SPFM	Scanning Polarization Force Microscopy
SPM	Scanning Probe Microscopy
SR	Synchrotron Radiation
STM	Scanning Tunneling Microscopy
SWNT	Single-Walled Nanotube
T_g	Glass Transition Temperature
T_m	Bulk Melting Point
T_s	Onset Temperature
UHV	Ultra High Vacuum
UN	United Nations
UPS	Ultraviolet Photoemission Spectroscopy
UV	Ultraviolet
V	Potential Energy
V_m	Molar Volume
vdW	van der Waals
VHDA	Very-High-Density Amorphous
VUV	Vacuum Ultraviolet
W	Bandwidth
W_N	Weber Number
XAS	X-Ray Absorption Spectroscopy
XES	X-Ray Emission Spectroscopy
XPS	X-Ray Photoemission Spectroscopy
XRS	X-Ray Raman Spectroscopy
z_G	Position of the Gibbs Surface
0D	Zero-dimensional
1D	Two-dimensional
2D	Two-dimensional
3D	Three-dimensional
α	Electronic Polarizability
γ_{lv}	Surface Tension
ϵ	One-Electron Energy
ε	Relative Permittivity
ε_0	Dielectric Constant of Free Space
ζ	Order Parameter
Θ	Surface Coverage
θ_c	Contact Angle
θ_c^a	Advancing Contact Angle
θ_c^r	Receding Contact Angle
θ_{CB}	Cassie–Baxter's Contact Angle
θ_W	Wenzel's Contact Angle
Θ_D	Debye Temperature
κ	Curvature
κ_D^{-1}	Debye Length
λ_K	Kelvin Length

μ	Electrical Dipole
χ_s	Surface Potential
$|\chi^{at}\rangle$	Atomic Orbitals
ρ	Density
$\rho_{ch}(x)$	Spatial Charge Density
ϱ	Roughness Factor
ϕ	Phase
ϕ	Work Function
φ	Electrostatic Potential
$|\psi\rangle$	One-Electron Wave Function
$\hbar$	Normalized Planck Constant
ω_{red}	Reduced Normalized Frequency

1 An Introduction to Water

que las aguas mansas son
de las que hay que fiar menos
P. Calderón de la Barca, Guárdate del agua mansa

From our everyday life experience we all know that water vapor from the air condenses in droplets upon decrease of temperature on distinct surfaces such as windows, leaves, and the like, and that a further decrease of temperature leads to the formation of solid water (ice) below 273 K (0°C) at about standard atmospheric pressure (1013 hPa). Up in the troposphere, the clouds have no large surfaces to condense on and for this reason they can stay there as huge wandering shape-changing beings with a milk-white appearance. Their characteristic color arises from the scattering of visible (white) light with water droplets whose sizes are comparable to the wavelength of such light (~500 nm). Such droplets, if pure, can be cooled down to about 233 K (−40°C) without the formation of ice (Schaefer 1946). Water is then said to be in a supercooled state. But ice in the clouds can be formed at higher temperatures (or not that low) as a result of the presence of foreign bodies such as dust particles, contaminants, nanoparticles, and so on, both from natural and anthropogenic origin, which act as nucleation agents (Langmuir 1950), in a process known in the crystal growth community as heterogeneous nucleation. This is one aspect of the water–surface interactions that is discussed at length in this book. The surface of ice plays a relevant role for atmospheric phenomena because it is able to catalyze chemical reactions such as those involved in polar ozone depletion, and those who love skiing should be grateful for the presence of a thin film of liquid water at the surface of snow below 273 K, where water would be expected to be in the solid state (again a supercooled liquid but this time on top of solid water). When the liquid water film is too thick (above 273 K) the resulting (spring) snow makes skiing a difficult task.

In this book we consider interfaces in a broad sense and at different length scales: water sequestering small molecules, which may lead to burning ice when the kidnapped molecule is methane, surrounding macromolecules such as proteins actively participating in their biological function, wetting or not extended flat surfaces, and the like. In this chapter a general introduction to water is given, describing the basic concepts needed for the rest of the book. Among the many books specifically devoted to water, I make a personal choice of:

- *The Structure and Properties of Water*, by D. Eisenberg and W. Kauzmann (1969)
- *Water: A Matrix of Life*, by F. Franks (2000)

- *Life's Matrix: A Biography of Water*, by P. Ball (2001)
- *Physics of Ice*, by V. Petrenko and R. W. Whitworth (2006)
- *The Hydrogen Bonding and the Water Molecule*, by Y. Maréchal (2007)
- *Interfacial and Confined Water*, by I. Brovchenko and A. Oleinova (2008)

In addition, there are books where water benefits from a privileged although not pivotal position as in *Intermolecular and Surface Forces* of J. Israelachvili (1991; 2011), and in *Physics and Chemistry of Interfaces* of Butt, Graf and Kappl (2013) and a plethora of excellent review articles that are referred to in this book. But before going into detail, just refer to the recent accurate determination of the Avogadro constant, which links the atomic and macroscopic properties of matter (Andreas et al. 2011). Thus, 18.0152 g of water (one mole) has $6.02214078(18) \times 10^{23}$ molecules and lots of surprises.

1.1   WHERE DOES WATER COME FROM?

Water is one of the most abundant constituents of the universe and of our solar system and is composed of hydrogen and oxygen, the first and third most abundant elements in the universe, respectively. It was A. Lavoisier (1743–1794) who gave the name to hydrogen and oxygen, before losing his head. They mean water and acid formed, respectively, from Greek roots. But before they could combine into the magic 2:1 proportion both elements had to be produced and in order to see how they came on the scene we have to go back to the very origin of the universe, when and where it all started. Matter was formed after the Big Bang, the most solid model describing the origin of the universe, when the incipient universe was cooling upon expansion. Hydrogen was generated when the temperature was sufficiently low in order to allow the young protons and electrons to combine, but oxygen had to wait much longer. It had to wait for the formation of stars, the veritable alchemists of the universe, which was made possible thanks to the gravitational force that started to aggregate the existing matter, mostly in the form of gas, up to a point where the increasingly denser matter collapsed, becoming heated up and starting ignition. The fuel was hydrogen and the final product helium (^4He), the second most abundant element in the universe, plus energy in the form of radiation. But the production, nucleosynthesis is the correct word, of heavier elements such as boron (^8B), carbon (^{12}C), oxygen (^{16}O), neon (^{20}Ne), magnesium (^{24}Mg), silicon (^{28}Si), and beyond needed the burning of helium after the gravitational collapse of the stars (Burbridge et al. 1957). So hydrogen had to wait quite some time for the successful alliance with oxygen, one of its great-grandsons.

The next question is that of how water was incorporated to the Earth, our beloved planet. It is assumed that such a process took place both during the formation of the Earth, by the accretion of planetary embryos, also called planetesimals, bringing both absorbed and adsorbed water with them, as well as at a later stage, which includes the late impact of external bodies after the formation of the moon (which would have vaporized the pre-existing water), such as carbonaceous chondrites, meteorites, asteroids, and comets which brought lattice water, that is, water molecules incorporated in their chemical structure, and external ice (Mottl et al. 2007; Hartogh et al. 2011). While writing this book, the journal *Nature* published the detection of water ice on the surface of an asteroid (24 Themis) by spectroscopic methods (Campins et al. 2010; Rivkin and Emery 2010). The observed absorption line at 3.1 μm, in the infrared (IR) region, has been ascribed to water ice in the form of thin films, with a thickness ranging from 10 to 100 nm. Such a finding provides a further example, fallen from heaven, of the importance of interfacial water, in this case as water supply. About 3.8–3.9 billion years ago the conditions on Earth were clement enough to become the scenario for the most important performance, the emergence of life (Luisi 2006; Lynden-Bell et al. 2010).

The search for extraterrestrial life in the solar system goes hand in hand with the detection of water, the well-known *follow the water* principle, hence the efforts mainly coordinated by the National Aeronautics and Space Administration (NASA) to look for water in planets and in their satellites. The Phoenix Mission was sent to Mars to verify the presence of subsurface water ice. It landed on May 25, 2008 in a northern arctic region and has provided evidence of the existence of both ice and atmospheric water (Smith et al. 2009). Water ice has also been observed on the moon, in the south pole crater Cabeus, after impact of a rocket and analysis of the ejected material (Colaprete et al. 2010), but its existence was predicted earlier by Hergé, Tintin's father, in *Explorers on the Moon*. This story was published in 1954, well before the Apollo 11 landing in 1969. On the other hand, Europa, a moon of Jupiter, is believed to host a deep ocean of liquid water beneath an icy shell (Carr et al. 1998).

But water has also been detected in the atmosphere of extrasolar planets (Seager and Deming, 2010). The atmosphere of the exoplanet with the nonmythological given name HD 189733b possesses water vapor, as observed with the powerful NASA Spitzer Space Telescope using an IR array camera (Tinetti et al. 2007). HD 189733b is an enormous gas giant with temperatures around 1,000 K, not the most indicated place for water-based life. Life as we know it needs the presence of liquid water and, for a given solar system, this imposes some constraints in terms of the star–planet distance. If the distance is too small the temperature becomes too high and vice versa. The *habitable zone* is defined as the distance that permits water to be in the liquid state. In our solar system this distance is about 1 astronomical unit (AU), the distance between the sun and the Earth. The GJ 1214b exoplanet has a water ice core surrounded by hydrogen and helium (Charbonneau et al. 2009). It is more massive than the Earth (6.55 times) and larger with a radius 2.68 times Earth's radius and the estimated temperatures lie around 475 K, still too high for liquid water but encouraging further research. Recently, the first confirmed exoplanet orbiting in the middle of the habitable zone of a star similar to our sun, initially termed Kepler–22b, has been discovered. It has a radius about 2.4 times that of the Earth and is about 600 light years away with a 290-day year (Borucki et al. 2012). The growing interest in exoplanets will bring in the forthcoming years a less Earth-centric vision of the universe. But let us come back to Earth after such a short journey to distant parallel worlds and explore what we know about the fascinating H_2O molecule.

FIGURE 1.1 Ball-and-stick model of a water molecule. Oxygen and hydrogen atoms are represented by dark gray and white balls, respectively.

1.2 MOLECULAR STRUCTURE OF ISOLATED WATER MOLECULES

1.2.1 A NONLINEAR POLAR MOLECULE

Let us now start the study of the molecular structure of individual (isolated) water molecules. In Chapter 3 we discuss the experimental visualization of single molecules (monomers) and of small clusters on surfaces but here we consider the case of non-interacting molecules as in the gas phase. Understanding the properties of isolated molecules is essential. When the intermolecular interactions are much weaker than the intramolecular interactions the electronic and vibrational properties of the condensed phases can be rationalized in terms of free molecules, a strategy that has been success-fully applied to molecular organic materials (Fraxedas 2006). In the limit of weak intermolecular interactions physicists like to describe a solid in terms of a weakly interacting gas whereas chemists prefer a supramolecular description.

The isolated water molecule has a V-shape (see Figure 1.1), with an H–O–H angle of $104.52 \pm 0.05°$ and an O–H distance of 0.9572 ± 0.0003 Å (Benedict, Gailer, and Plyler 1956). Given the different electron affinities of oxygen and hydrogen, the molecule is polarized with a permanent dipole $\mu = 1.8546$ D (Clough et al. 1973), where 1 D (Debye) = 3.336×10^{-30} C m, but is only slightly polarizable, with an electronic polarizability $\alpha = 1.64 \times 10^{-40}$ C^2 m^2 J^{-1} arising mostly from the oxygen atom (Murphy 1977; Tsiper 2005). We interchangeably use the term atom or ion (hydrogen atoms or protons) throughout the book.

Figure 1.2 shows a water molecule oriented in a Cartesian coordinate system, where the origin is arbitrarily set at the oxygen atom and both hydrogen atoms are contained in the yz-plane ($z > 0$). With this convention, that is kept throughout the book, the water dipole is defined along the positive z-axis. The water molecule remains invariant under the symmetry operations:

$$E, C_2(z), \sigma_v(xz), \sigma_v(yz)$$

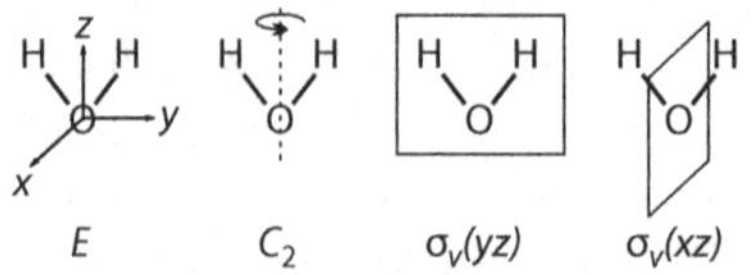

H $\overset{z}{\underset{x}{\overset{|}{\underset{}{\text{H}}}}}$ H $\overset{\leftrightarrow}{\text{H}}$ H   H H $\diagup$ H

E C_2 $\sigma_v(yz)$ $\sigma_v(xz)$

FIGURE 1.2 Symmetry operations for the water molecule corresponding to the C_{2v} point group.

TABLE 1.1
Character Table of the C_{2v} Point Group

C_{2v}	E	C_2	$\sigma_v(xz)$	$\sigma_v(yz)$
A_1	1	1	1	1
A_2	1	1	-1	-1
B_1	1	-1	1	-1
B_2	1	-1	-1	1

where E represents the identity operator (from German *Einheit*), $C_2(z)$ stands for a proper rotation of π degrees (rotation of order 2) with respect to the z-axis, and $\sigma_v(xz)$, $\sigma_v(yz)$ are planes of symmetry with respect to the xz and yz planes, respectively (from German *Spiegel*). The full set of such symmetry operations leads to the C_{2v} point group, using Schönflies symbols. Symmetry is a very important parameter when characterizing molecules (and solids) because it helps in simplifying the resolution of the Schrödinger equation, as discussed later in this chapter.

Table 1.1 shows the character table corresponding to the C_{2v} point group, where every row corresponds to an irreducible representation. Cotton's book *Chemical Applications of Group Theory* is a must for those interested in learning more about group theory applied to chemistry (Cotton 1990).

We are interested in the electronic and vibrational structures of the water molecule but before going into detail let us start first with a simple physical water model consisting of three point charges, $-2q$ for the oxygen ion and $+q$ for both hydrogen ions ($q > 0$) and see what we can learn from it. The model is shown in Figure 1.3a. The total interaction energy $E_{coul}(r)$ is given by the sum of the two O–H and one H–H Coulomb contributions:

$$E_{coul}(r) = \frac{1}{4\pi\varepsilon\varepsilon_0} \left\{ -\frac{4q^2}{r} + \frac{q^2}{2r\sin\theta} \right\} \tag{1.1}$$

where ε and ε_0 represent the relative permittivity and the dielectric constant of free space ($\varepsilon_0 = 8.854 \times 10^{-12}$ C^2 J^{-1} m^{-1}), respectively, r the O–H distance, and θ half the angle between both O–H pairs. The angle that minimizes the total energy is $\theta = \pi/2$, as can be analytically obtained and initially guessed without any calculation; inasmuch as both $+q$ charges repel each other, hence they will tend to stay as far away as possible. This model is clearly unable to describe the known V-shape and thus additional conditions have to be considered.

If we allow, for instance, the oxygen ion to be polarized, then the total interaction energy becomes:

$$E_{coul}(r) = \frac{1}{4\pi\varepsilon\varepsilon_0} \left\{ -\frac{4q^2}{r} + \frac{q^2}{2r\sin\theta} \right\} - \frac{1}{(4\pi\varepsilon\varepsilon_0)^2} \frac{4\alpha q^2}{2r^4} \cos^2\theta \tag{1.2}$$

This improved model is represented in Figure 1.3b. The second term of this expression corresponds to the interaction between an ion and an uncharged polarizable atom or molecule. Minimization of (1.2) gives rise to two solutions: $\cos\theta = 0$ and

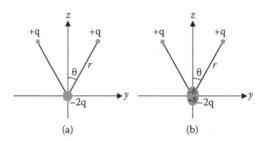

FIGURE 1.3 (a) Unpolarized and (b) polarized point charge model of a water molecule. The O–H distance and half the H–O–H angle are represented by r and θ, respectively.

$\sin^3 \theta = 4\pi\varepsilon\varepsilon_0 r^3/8\alpha$. The former corresponds to the solution found from (1.1) and the latter implies that $\alpha > 4\pi\varepsilon\varepsilon_0 r^3/8$ and that the angle θ becomes more acute when α increases, which can intuitively be understood as caused by the induced positive charge in the oxygen ion which exerts a repulsion in both hydrogen ions. Indeed, with $r = 0.9572$ Å, as mentioned earlier, we obtain $\alpha > 0.12 \times 10^{40}$ C^2 J^{-1} m^{-1}, a condition that is fulfilled because the experimental value is $\alpha = 1.64 \times 10^{-40}$ C^2 m^2 J^{-1}. With this simple model we cannot know much about the electronic structure of the water molecule but it provides us a hint of the importance of the charge distribution opposite the hydrogen ions as discussed later. To go further we need some basic concepts of quantum chemistry and this is exactly what we introduce next. Many introductory textbooks on quantum chemistry can be found in the literature and interested readers are referred to them. Among them I would suggest that from I. N. Levine (2008).

1.2.2 ELECTRONIC STRUCTURE

In what follows we describe the electronic structure of the water molecule or equivalently its molecular orbitals (MOs). Physicists prefer the former term and chemists the latter, although they are referring to the very same concept. A detailed discussion on MOs and their interactions can be found in the book by Albright, Burdett, and Whangbo (1985). Let us start by considering the general many-electron problem of N_e electrons contributing to chemical bonding, and N_{ion} ions, which contain the nuclei and the tightly bound core electrons. This problem can be described quantum-mechanically, in the absence of external fields and ignoring atomic vibrations (we consider the latter in the next section), by the 0-order Hamilton operator:

$$H_0 = H_{ee} + H_{ion-ion} + H_{e-ion} \tag{1.3}$$

where H_{ee}, $H_{ion-ion}$, and H_{e-ion} correspond to the Hamilton operators concerning electron–electron, ion–ion, and electron–ion interactions, respectively. Thus, the Schrödinger equation to be solved is:

$$H_0|\Psi\rangle = E_0|\Psi\rangle \tag{1.4}$$

where $|\Psi\rangle$ and E_0 represent the N_e-electron wave (eigen)function and energy, respectively. In the vast majority of cases (1.4) cannot be solved exactly and approximations are needed. A successful strategy consists of assuming that every electron is subject to an effective interaction potential, the so-called one-electron approximation. Ideally we would like to express H_0 in the form:

$$H_0 \simeq \sum_{i=1}^{N_e} \left\{ -\frac{\hbar^2}{2m_e} \nabla_i^2 + V_i \right\} = \sum_{i=1}^{N_e} h_i \qquad (1.5)$$

where the Hamiltonian H_0 of an N_e-electron system can be expressed as the sum of N_e one-electron Hamiltonians h_i where $h_i = -\hbar^2/2m_e \nabla_i^2 + V_i$, the first and second terms accounting for the kinetic energy and the mean interaction potential, respectively, where $\hbar$ represents the normalized Planck constant (1.054×10^{-34} J s $= 6.582 \times 10^{-16}$ eV s), m_e the electron mass, and the ∇^2 operator is given by $\partial^2/\partial x^2 + \partial^2/\partial y^2 + \partial^2/\partial z^2$.

Within the one-electron approximation, we can obviate the i index and the Schrödinger equation to be solved becomes:

$$h|\psi\rangle = \epsilon|\psi\rangle \qquad (1.6)$$

where h, ϵ, and $|\psi\rangle$ stand for the one-electron Hamiltonian, energy, and wave function, respectively. In the case of molecules $|\psi\rangle$ represents the MOs and can be simply expressed as a linear combination of N_{at} atomic orbitals $|\chi^{at}\rangle$ (LCAO) of the different atoms forming the molecule:

$$|\psi_j\rangle = \sum_{k=1}^{N_{at}} c_{jk} |\chi_k^{at}\rangle \qquad (1.7)$$

where j runs from 1 to N_{at} and the c_{jk} coefficients have to be determined. For most purposes these atomic orbitals can be assumed to be real functions and normalized such that the probability of finding an electron in $|\chi_j^{at}\rangle$ when integrated over all space is one: $\langle \chi_j^{at}|\chi_j^{at}\rangle = 1$. On the other hand the MOs should be orthonormal, that is, orthogonal and normalized: $\langle \psi_j|\psi_k\rangle = \delta_{jk}$, where $\delta_{jk} = 1$ if $j = k$ and $\delta_{jk} = 0$ if $j \neq k$.

Introducing (1.7) into (1.6) we obtain:

$$\sum_{k=1}^{N_{at}} c_{jk} \left\{ h|\chi_k^{at}\rangle - \epsilon|\chi_k^{at}\rangle \right\} = 0 \qquad (1.8)$$

and multiplying (1.8) to the left by $\langle \chi_j^{at}|$, the complex conjugate of $|\chi_j^{at}\rangle$, we obtain:

$$\sum_{k=1}^{N_{at}} c_{jk} \left\{ \langle \chi_j^{at}|h|\chi_k^{at}\rangle - \epsilon \langle \chi_j^{at}|\chi_k^{at}\rangle \right\} = 0 \qquad (1.9)$$

By defining $h_{jk} = \langle \chi_j^{at}|h|\chi_k^{at}\rangle$, the interaction energy, and $s_{jk} = \langle \chi_j^{at}|\chi_k^{at}\rangle$, the overlap integral, where $j, k = 1, \ldots, N_{at}$, (1.9) is simplified to the expression:

$$\sum_{k=1}^{N_{at}} c_{jk} \left\{ h_{jk} - \epsilon s_{jk} \right\} = 0 \qquad (1.10)$$

Therefore, within the LCAO approximation, (1.6) transforms into a system of N_{at} equations with N_{at} unknown parameters c_{jk}. The resolution of (1.10) implies that the determinant of the $\{h_{jk} - \epsilon s_{jk}\}$ matrix has to be zero:

$$
\begin{vmatrix}
h_{11} - \epsilon s_{11} & h_{12} - \epsilon s_{12} & \cdots & h_{1N_{at}} - \epsilon s_{1N_{at}} \\
h_{21} - \epsilon s_{21} & h_{22} - \epsilon s_{22} & \cdots & h_{2N_{at}} - \epsilon s_{2N_{at}} \\
\vdots & \vdots & \vdots & \vdots \\
h_{N_{at}1} - \epsilon s_{N_{at}1} & h_{N_{at}2} - \epsilon s_{N_{at}2} & \cdots & h_{N_{at}N_{at}} - \epsilon s_{N_{at}N_{at}}
\end{vmatrix} = 0 \qquad (1.11)
$$

In conclusion, the energies that satisfy (1.11) are associated with molecular electronic states. Because (1.11) is an equation of N_{at} order, we obtain N_{at} energy values ϵ_j ($j = 1, \ldots, N_{at}$), that is, as many molecular levels as atomic orbitals.

In the case of the H_2O molecule, the hydrogen atoms contribute with two $1s$ atomic orbitals (one each) and the oxygen atom with $1s$, $2s$, and $2p$ orbitals. The participation of $O1s$ in the bonding is negligible (we later show the large energy difference between $O1s$ and $O2s$), hence we concentrate on two atomic orbitals from hydrogen and four from oxygen, namely $2s$, $2p_x$, $2p_y$, and $2p_z$. Thus $N_{at} = 6$ and our task consists in determining the corresponding 6 MOs. But before trying to solve the full 6 MOs problem, it is illustrative to start with the H–H subsystem, that is, ignoring the oxygen atom, as shown in Figure 1.4. In this case $N_{at} = 2$ and we have:

$$|\psi_1\rangle = c_{11}|\chi_1^{at}\rangle + c_{12}|\chi_2^{at}\rangle \qquad (1.12a)$$

$$|\psi_2\rangle = c_{21}|\chi_1^{at}\rangle + c_{22}|\chi_2^{at}\rangle \qquad (1.12b)$$

where $|\chi_j^{at}\rangle \equiv |1s_j\rangle$ with $j = 1, 2$. We thus have to solve the following expression:

$$(h_{11} - \epsilon s_{11})c_{11} + (h_{12} - \epsilon s_{12})c_{12} = 0 \qquad (1.13a)$$

$$(h_{21} - \epsilon s_{21})c_{21} + (h_{22} - \epsilon s_{22})c_{22} = 0 \qquad (1.13b)$$

Because the atomic orbitals are normalized, then $s_{11} = s_{22} = 1$ and the interatomic overlap $s_{12} = s_{21} = 0.2261$, computed at a H–H distance of 1.514 Å (McGlynn et al. 1972). On the other hand $h_{11} = h_{22} = -13.6$ eV, the ionization potential of hydrogen,

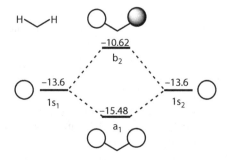

FIGURE 1.4 Orbital interaction diagram for H–H. Energies in eV.

and $h_{12} = h_{21} = K_{WH}s_{12}h_{11}$ by virtue of the Wolfsberg–Helmholtz approximation, where $h_{jk} = K_{WH}s_{jk}(h_{jj} + h_{kk})/2$ and $K_{WH} = 1.75$. We are simply calculating orbital energies following the extended Hückel (EH) approximation, as developed by R. Hoffmann (1963).

From the determinant:

$$\begin{vmatrix} -13.6 - \epsilon & 0.2261(-23.8 - \epsilon) \\ 0.2261(-23.8 - \epsilon) & -13.6 - \epsilon \end{vmatrix} = 0 \tag{1.14}$$

we obtain the two solutions for the MO energies, $\epsilon_1 = -15.48$ eV and $\epsilon_2 = -10.62$ eV, and from (1.13) and the normalization of the MOs, $\langle \psi_1 | \psi_1 \rangle = \langle \psi_2 | \psi_2 \rangle = 1$, we obtain the coefficients $c_{11} = c_{12} = (\sqrt{2(1 + s_{12})})^{-1} = 0.6385$ and $c_{21} = -c_{22} = (\sqrt{2(1 - s_{12})})^{-1} = 0.8038$. Thus,

$$|\psi_1\rangle = 0.6385\{|1s_1\rangle + |1s_2\rangle\} \tag{1.15a}$$

$$|\psi_2\rangle = 0.8038\{|1s_1\rangle - |1s_2\rangle\} \tag{1.15b}$$

Applying the symmetry operations of the C_{2v} point group to both $|\psi_1\rangle$ and $|\psi_2\rangle$ we find out that $|\psi_1\rangle$ and $|\psi_2\rangle$ transform according to the A_1 and B_2 irreducible representations, respectively (see Table 1.1). Accordingly, the symmetric/antisymmetric combination of atomic orbitals is labeled a_1 and b_2, respectively, as shown in Figure 1.4.

Let us now proceed with the calculation of the energy diagram for the water molecule by allowing the interaction between the two MOs from the H–H molecule and the oxygen atomic orbitals. If we apply the symmetry operations to the oxygen s and p atomic orbitals, we find that s and p_z transform according to the A_1 irreducible representation, and p_x and p_y according to the B_1 and B_2 irreducible representations, respectively, hence the labels a_1 (s, p_z), b_1 (p_x), and b_2 (p_y). Before trying to generate the corresponding secular 6×6 determinant we should have in mind that symmetry is of great help because according to group theory, two orbitals will not interact unless they are of the same symmetry. This strongly reduces the number of nonzero s_{jk} terms. With the convention $a_1(\text{H}1s) \equiv 1$, $b_2(\text{H}1s) \equiv 2$, $a_1(\text{O}1s) \equiv 3$, $b_1(\text{O}2p_x) \equiv 4$, $b_2(\text{O}2p_y) \equiv 5$, and $a_1(\text{O}2p_z) \equiv 6$ the nonzero s_{jk} terms are $s_{13} = 0.5885$, $s_{16} = 0.3074$, and $s_{25} = 0.4993$, again taken from tabulated values according to the water molecular structure, and $s_{jj} = 1$ for $j = 1, \ldots, 6$. In addition, $h_{11} = -15.48$ eV, $h_{22} = -10.62$ eV, as previously obtained for the H–H case (Figure 1.4), and $h_{33} = -32.30$ eV and $h_{44} = h_{55} = h_{66} = -14.80$ eV, where -32.30 and -14.80 eV correspond to the O1s and degenerate O2p ionization energies, respectively ($h_{13} = h_{31}$, $h_{16} = h_{61}$ and $h_{25} = h_{52}$ are obtained using the Wolfsberg–Helmholtz approximation mentioned earlier). From the corresponding determinant:

$$\begin{vmatrix} -15.48-\epsilon & 0 & 0.59(-41.81-\epsilon) & 0 & 0 & 0.31(-26.49-\epsilon) \\ 0 & -10.62-\epsilon & 0 & 0 & 0.50(-22.24-\epsilon) & 0 \\ 0.59(-41.81-\epsilon) & 0 & -32.30-\epsilon & 0 & 0 & 0 \\ 0 & 0 & 0 & -14.80-\epsilon & 0 & 0 \\ 0 & 0.50(-22.24-\epsilon) & 0 & 0 & -14.80-\epsilon & 0 \\ 0.31(-26.49-\epsilon) & 0 & 0 & 0 & 0 & -14.80-\epsilon \end{vmatrix} = 0 \tag{1.16}$$

the following energies (in eV) are obtained:

$$\epsilon_1 = -34.02,\ \epsilon_2 = -17.11,\ \epsilon_3 = -15.33,\ \epsilon_4 = -14.80,\ \epsilon_5 = -0.22,\ \epsilon_6 = 14.35$$

Figure 1.5 summarizes the orbital interaction diagram with the corresponding energies as calculated with the EH method. The resulting symmetry adapted water MOs are labeled according to the irreducible representation they belong to and to an integer number that just indicates a sequence. Note that the lowest MO is labeled $2a_1$ because $1a_1$ would correspond to the O1s core level, which participates negligibly in the molecular bonding.

Each MO can allocate two electrons, so that because oxygen participates with 6 electrons to the bonding (atomic number 8 minus the 2 from the 1s level) and

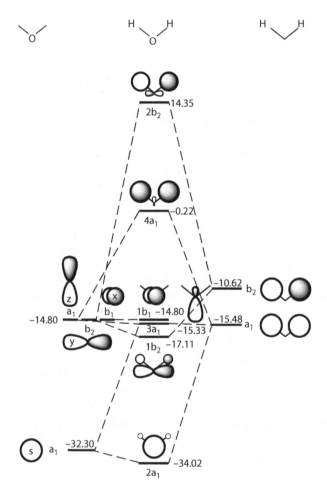

FIGURE 1.5 Orbital interaction diagram for H_2O. Energies (eV) obtained from an EH calculation are indicated and scaled. The orbital mixing is: $2a_1 \rightarrow 81\%$ O2s, 19% H1s, $1b_2 \rightarrow 66\%$ O2p_y, 34% H1s, $3a_1 \rightarrow 93\%$ O2p_z, 1% O2s, 6% H1s, $1b_1 \rightarrow 100\%$ O2p_x, $4a_1 \rightarrow 17\%$ O2s, 7% O2p_z, 76% H1s. $2b_2 \rightarrow 34\%$ O2p_y, 66% H1s. (Courtesy of E. Canadell.)

each hydrogen gives one, we have 8 electrons to distribute. These 8 electrons fill the $2a_1$, $1b_2$, $3a_1$, and $1b_1$ MOs and leave the rest, $4a_1$ and $2b_2$, empty. $1b_1$ and $4a_1$ are termed the highest occupied molecular orbital (HOMO) and lowest occupied molecular orbital (LUMO), respectively. $2a_1$ and $1b_2$ have bonding and $4a_1$ and $2b_2$ antibonding character, respectively, and $3a_1$ and $1b_1$ are nonbonding, also known as lone-pairs. The fact that the water molecule has two such lone-pairs makes it rather unique, allowing versatile interaction with other molecules, as we show in Section 1.4, through hydrogen bonding (H-bonding). This enables the water molecule to adapt to many configurations and here lies precisely the core of the complexity.

We now compare the relatively simple EH calculation with a state of the art (at the time this text is being written) density functional theory (DFT) calculation using the Gaussian code with the Becke three-parameter Lee–Yang–Parr (B3LYP) hybrid functional (Stephens et al. 1994) and the 6-311++G(2d,2p) basis. The obtained energies (in eV) are:

$$\epsilon_1 = -27.72, \quad \epsilon_2 = -14.97, \quad \epsilon_3 = -10.64, \quad \epsilon_4 = -8.77,$$
$$\epsilon_5 = -0.59, \quad \epsilon_6 = 0.79$$

Many simulations of the MO structure of water can be found in the literature but it is beyond the scope of the present book to summarize them. The point here is to be able to compare a simplified vis à vis a complex calculation. The energies obtained from both simulations are compared in Figure 1.6 to available photoemission data from Winter et al. (2004) on water vapor using synchrotron radiation (SR) in the vacuum ultraviolet (VUV) region. According to Koopmans' theorem, the molecular ionization energies correspond to orbital energies, hence the photoelectron spectrum of a molecule should be a direct representation of the MO energy diagram. This is true if the vibrational structure is ignored as it is in the present case. Figure 1.6 proves such an assessment illustrating the correctness of the MO approach. The calculated energies have been rigidly shifted (-3.83 and $+2.2$ eV for the DFT and EH calculations, respectively) so as to bring the $1b_1$ MO to the well-known experimental -12.6 eV value. The figure is completed with available X-ray emission spectroscopy (XES) and X-ray absorption spectroscopy (XAS) from Guo and Luo (2010) taken from liquid water. Table 1.2 compares both calculated and measured MO energies and includes a self-consistent Hartree–Fock (HF) calculation from Ellison and Shull (1955) as well as X-ray photoemission spectroscopy (XPS) results of water vapor taken with 1486.6 eV photons (Siegbahn 1974). We observe an overall agreement for the occupied MOs, and the accuracy of the powerful DFT method, which correctly predicts the experimental values. In the case of the $1a_1$ MO, the corrected DFT value is -524.2 eV, quite far from the experimental -540.0 eV (Lundholm et al. 1986). A major prediction difference involves the unoccupied, also termed virtual, states. DFT provides the best approximation but both the EH and HF methods fail to reproduce the experimental data.

Why EH fails can be inferred from the simple H–H subsystem discussed earlier. It can be shown, by solving (1.14) in the general case, that the bonding state lowers the initial $h_{11} = h_{22} = -13.6$ eV energy by $-h_{11}(1 - K_{WH})s_{12}/(1 + s_{12})$ and the

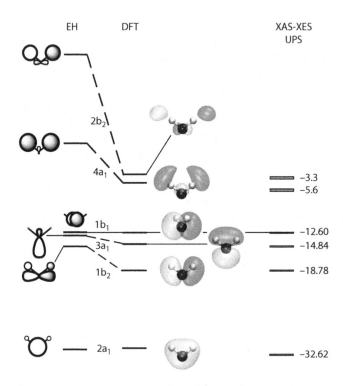

FIGURE 1.6 Comparison between the EH and DFT calculations (continuous black lines) of an isolated water molecule with ultraviolet photoemission spectroscopy (UPS) of water vapor from (Winter et al. 2004) (continuous thin gray lines) and X-ray absorption and emission spectroscopies (XAS and XES) of liquid water from Guo and Luo (2010; continuous thick gray lines). The calculated energies have been rigidly shifted by -3.83 eV (DFT) and $+2.2$ eV (EH) in order to make the $1b_1$ MO energy coincide with the experimental -12.6 eV value. The DFT calculations (courtesy of J. Iglesias and C. Rovira) have been performed with the Gaussian code using the B3LYP functional and the $6-311++G(2d,2p)$ basis. Density isosurfaces of the MOs are plotted at 0.05 e Å^3.

antibonding state increases it by $h_{11}(1 - K_{WH})s_{12}/(1 - s_{12})$. Thus, the binding energy difference of the antibonding orbital increases more strongly than for the bonding counterpart for increasing overlap, diverging in the $s_{12} = 1$ case. The HF method includes an extra electron in the virtual states, thus becoming occupied and introducing electronic repulsion. Although the EH method fails to reproduce the experimental data it helps to understand the underlying chemical basics due to its conceptual simplicity. By looking at the representation of the DFT MOs it may become difficult to recognize the origin of the orbitals involved, a matter that is relatively straightforward from the EH approximation. When using available codes, although they have been successfully used and thoroughly tested, it is easy to lose the essential details that a simple method such as EH offers.

TABLE 1.2
Comparison of Calculated and Experimental MO Energies of Water

MO	EH	HF[a]	DFT	UPS[b]	XPS[c]	XES-XAS[d]
$2b_2$	16.55	15.09	−3.04			≃ −3.3
$4a_1$	1.98	12.89	−4.42			≃ −5.6
$1b_1$	−12.60	−12.60	−12.60	−12.60	−12.62	
$3a_1$	−13.13	−14.01	−14.87	−14.84	−14.78	
$1b_2$	−14.91	−19.36	−18.80	−18.78	−18.55	
$2a_1$	−31.82	−37.00	−31.55	−32.62	−32.12	
$1a_1$		−558.08	−524.21		−539.9	

[a] (Ellison and Shull 1955), [b] (Winter et al. 2004), [c] (Siegbahn 1974), [d] (Guo and Luo 2010).
The calculated energies have been rigidly shifted by −3.83 eV (DFT) and +2.2 eV (EH) in order to make the $1b_1$ MO energy coincide with the experimental −12.6 eV value.

1.2.3 VIBRATIONAL STRUCTURE

In 3D space a molecule formed by N atoms has $3N$ degrees of freedom (3 per atom). Three of them correspond to rigid translations of the molecule, three to rigid rotations, and the rest, $3N - 6$, to internal vibrations. Thus, water has there vibrational modes. Let us try to derive in a simplified way the energies of such modes from fundamental classical mechanics. We further simplify the estimation by assuming a linear H–O–H molecule in 1D space, that would belong to the $D_{\infty h}$ point group instead of to C_{2v}. Such an imaginary molecule, shown in Figure 1.7(a), has only three (translational) degrees of freedom due to the 1D limitation, in our case along the y-axis. The MOs from water can also be derived from a hypothetical linear molecule through the Walsh diagrams (Albright, Burdett, and Whangbo 1985). This is achieved by bending the $1\sigma_g$, $1\sigma_u$, $1\pi_{uz}$, $1\pi_{ux}$, $2\sigma_g$, and $2\sigma_u$ MOs from the linear molecule, where the labeling corresponds to the $D_{\infty h}$ point group, which render the $1a_1$, $1b_2$, $2a_1$, $1b_1$, $3a_1$, and $2b_2$ MOs, respectively, taking into account the corresponding orbital mixing.

The total vibrational energy will be the sum of the kinetic (K) and potential (V) energies. The kinetic energy is given by the expression:

$$K = \frac{1}{2}\sum_{i=1}^{3} m_i \left\{\frac{dy_i}{dt}\right\}^2 = \frac{1}{2}\sum_{i=1}^{3}\left\{\frac{dq_i}{dt}\right\}^2 \quad (1.17)$$

where the mass-weighted Cartesian coordinates $q_i = \sqrt{m_i}\,y_i$ have been used with m_i representing the mass of the ith atom.

Within the harmonic approximation, where only small displacements around the equilibrium (eq) positions are considered, the interaction potential can be approximated by the Taylor series:

$$V \simeq V_0 + \frac{1}{2}\sum_{i=1}^{3}\sum_{j=1}^{3}\left.\frac{\partial^2 V}{\partial q_i \partial q_j}\right|_{eq} q_i q_j \quad (1.18)$$

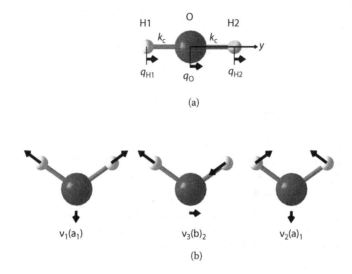

(a)

$v_1(a_1)$ $v_3(b)_2$ $v_2(a)_1$

(b)

FIGURE 1.7 (a) Idealized linear water molecule. The mass-weighted Cartesian coordinates q_i for the oxygen (O) and hydrogen (H1 and H2) atoms are shown. The O–H bonds are modeled by the k_c force constant. (b) Vibrational normal modes of the real V-shaped water molecule.

because $|\partial V/\partial q_i|_{eq} = 0$. According to Newton's law and making use of (1.18):

$$\frac{d^2 q_i}{dt^2} = -\frac{\partial V}{\partial q_i} = -\sum_{j=1}^{3} \left.\frac{\partial^2 V}{\partial q_i \partial q_j}\right|_{eq} q_j \qquad (1.19)$$

which can be written as:

$$\frac{d^2 q_i}{dt^2} + \sum_{j=1}^{3} v_{ij} q_j = 0 \qquad (1.20)$$

where $v_{ij} = \left.\left|\partial^2 V/\partial q_i \partial q_j\right|\right._{eq}$ stands for the (i, j) element of the potential matrix $\tilde{V}$. (1.20) represents three coupled differential equations that would provide the time-dependent mass trajectories $q_i(t)$. Because the potential matrix $\tilde{V}$ is real and symmetric it can be diagonalized. The determinant to be solved is:

$$\begin{vmatrix} v_{11} - \lambda & v_{12} & v_{13} \\ v_{21} & v_{22} - \lambda & v_{23} \\ v_{31} & v_{32} & v_{33} - \lambda \end{vmatrix} = 0 \qquad (1.21)$$

with λ representing the eigenvalues of $\tilde{V}$ and building the corresponding diagonal matrix Λ formed by λ_k for $k = 1, 2, 3$. The transformation matrix L, which fulfills the condition $L^{-1} \tilde{V} L = \Lambda$, is composed of the eigenvectors of $\tilde{V}$ and transforms the coordinate system q into a special coordinate system of normal coordinates Q, in which the $\tilde{V}$ matrix takes the canonical form. Thus, (1.20) becomes uncoupled because of the diagonalization and transformed into:

$$\frac{d^2 Q_k}{dt^2} + \lambda_k Q_k = 0 \qquad (1.22)$$

Solutions of (1.22) are of the form $Q_k = A_k \sin \sqrt{\lambda_k} t + \varphi_k$, where A_k and φ_k stand for the amplitude and phase of the wave, respectively, and $\sqrt{\lambda_k} = 2\pi \nu_k$, representing ν_k, the associated frequency. For our linear water molecule from Figure 1.7a the potential energy is given by:

$$V \simeq \frac{1}{2} k_c \left\{ \left(\frac{q_O}{\sqrt{m_O}} - \frac{q_{H1}}{\sqrt{m_{H1}}} \right)^2 + \left(\frac{q_{H2}}{\sqrt{m_{H2}}} - \frac{q_O}{\sqrt{m_O}} \right)^2 \right\} \qquad (1.23)$$

where k_c stands for the equivalent O–H force constant (imagine the bond as a spring) and q_{H1}, q_{H2}, q_O and m_{H1}, m_{H2}, m_O represent the mass-weighted coordinates and masses of both hydrogen and oxygen atoms, respectively, so that the determinant to be solved assuming $m_{H1} = m_{H2} = m_H$ is:

$$\begin{vmatrix} \frac{k_c}{m_H} - \lambda & -\frac{k_c}{\sqrt{m_H m_O}} & 0 \\ -\frac{k_c}{\sqrt{m_H m_O}} & 2\frac{k_c}{m_O} - \lambda & -\frac{k_c}{\sqrt{m_H m_O}} \\ 0 & -\frac{k_c}{\sqrt{m_H m_O}} & \frac{k_c}{m_H} - \lambda \end{vmatrix} = 0 \qquad (1.24)$$

whose solutions, once transformed into frequencies, are:

$$\nu_1 = \frac{1}{2\pi} \sqrt{\frac{k_c}{m_H}} \qquad (1.25a)$$

$$\nu_2 = 0 \qquad (1.25b)$$

$$\nu_3 = \frac{1}{2\pi} \sqrt{k_c \left\{ \frac{2}{m_O} + \frac{1}{m_H} \right\}} \qquad (1.25c)$$

where ν_1 and ν_3 represent the symmetric and asymmetric stretching of the molecule, respectively, and $\nu_2 = 0$ a rigid translation. Taking $k_c \sim 850$ N m^{-1} as derived for the actual V-shape molecule (Fifer and Schiffer 1970) we obtain $\nu_1 \sim 3,800$ and $\nu_3 \sim 4,030$ cm^{-1}, that is in the IR region. Note that $\nu_1 \sim \nu_3$ because $m_O \gg m_H$, according to (1.25c). This IR part of the spectrum is very important for us humans, because water vapor contributes significantly to the greenhouse effect, keeping a pleasant habitable environment and avoiding excessive global warming by forming micron-sized droplets in the clouds acting as an efficient IR reflector due to the high albedo.

The $\nu_1 \sim 3,800$ and $\nu_3 \sim 4,030$ cm^{-1} values compare surprisingly well, given the crudeness of the approximation, to the $\nu_1(a_1) = 3,832$ and $\nu_3(b_2) = 3,942$ cm^{-1} values from Benedict, Gailer, and Plyler (1956) and to the $\nu_1(a_1) = 3,823$ and $\nu_3(b_2) = 3,925$ cm^{-1} values obtained with the DFT calculation using the B3LYP functional and the 6–311++G(2d,2p) basis with the same geometry as for the calculation of the MOs energies, corresponding to the real molecule within the harmonic approximation. Both symmetric, $\nu_1(a_1)$, and asymmetric, $\nu_3(b_2)$, stretching modes of the real V-shaped water molecule are schematized in Figure 1.7b. Indeed, the bending mode $\nu_2(a_1)$ (also represented by δ) shown in the figure cannot be simulated by the linear molecule because of the 1D boundary condition. Its calculated frequency for the real molecule is $\nu_2(a_1) = 1,648$ (Benedict, Gailer, and Plyler 1956) and $\nu_2(a_1) = 1,638.88$ cm^{-1} according to the DFT calculations, which renders the equivalent force constants (k_c) 899.54, 982.75, and 171.47 N m^{-1} for the symmetric and asymmetric stretching and for the bending modes, respectively. We could have tried to work directly with the V-shaped molecule, but that would have implied solving a 9×9 determinant, which is a quite tedious task, and in view of the excellent results provided by the linear molecule this appears unnecessary. However, the accepted experimental values for the three frequencies are 3,657, 3,756, and 1,595 cm^{-1} (or 0.453, 0.466, and 0.198 eV, respectively) (Benedict, Gailer, and Plyler 1956; Fraley and Rao 1969) once anharmonicity is taken into account.

Does water vapor only exhibit three lines in the absorption spectra? The answer is no, that would be too easy (Benedict, Gailer, and Plyler 1956; Toth 1998; Maréchal 2007). In fact it shows a huge number of lines making it a very complex problem (with only three atoms!), but most of the lines can be rationalized in terms of the normal modes. A fraction of the lines corresponds to the isotopic composition (D–O–H, D–O–D, H–^{17}O–H, H–^{18}O–H, etc.) and a considerable part to coupling with other vibrations (rotations). Hence, with only the three normal modes we can build most of the vibrational spectra. In addition, we can obtain more lines if we are able to excite the water molecules externally by, for example, increasing the temperature (Bernath 2002). This has important consequences for research in astronomy and atmospheric and combustion science. Using the triple-resonance overtone excitation technique, which consists of the selective excitation by three successive laser pulses of 5–7 ns (1 ns = 10^{-9} s), it has been shown that the water molecules survive up to 60 ps (1 ps = 10^{-12} s) in states above the dissociation threshold of 41,145.94 cm^{-1} (Grechko et al. 2010). Such states are known as Feshbach resonances.

When the photoemission of water in the gas phase is performed with excitation lines exhibiting very small linewidths (about 3 meV or less) the resulting spectra become rather complex, raising reasonable doubts about Koopmans' theorem mentioned in the previous section. This has been shown by Karlsson et al. (1975) using a HeI line (21.22 eV photon energy). Instead of the three expected $1b_2$, $3a_1$, and $1b_1$ MOs ($2a_1$ cannot be excited with 21.22 eV photons) a large number of features are observed. In fact, each MO gives a group of peaks in the spectrum because of the possibility to couple with vibrational modes. Each resolved peak stands for a single vibrational line and represents a definite number of quanta of vibrational energy of the molecular ion. Ionization is a fast process, of the order of fs (1 fs = 10^{-15} s), the time required for the ejected electron to leave the immediate neighborhood of the molecular ion. The time is so short that motions of the atomic nuclei that make up vibrations and proceed on a much longer timescale of 10^{-13} s can be considered as frozen during ionization. This results in many accessible vibrational states, an effect that is known under the Franck–Condon principle. The fact that the number of peaks largely exceeds the number of MOs shows the limitation of Koopmans' theorem. Bonding orbitals exhibit a distinct vibrational structure as compared to nonbonding orbitals because removal of one electron strongly perturbs the orbitals involved. $1b_1$ shows the less rich spectra, due to its marked nonbonding character, as compared to $3a_1$. In the case of H$_2^{16}$O, the observed vibrational energies obtained from $1b_1$ are 402 and 177 meV (3,242 and 1,427 cm^{-1}), which are considerably lower than the values discussed above, due to the fact that they correspond to ionized water molecules. Similar values are also found with VUV absorption using SR (Mota et al. 2005).

1.3 HYDROGEN BONDING OFF: VAN DER WAALS INTERACTIONS

So far we have considered the ideal case of isolated water molecules, as in the gas phase, and next we allow an individual molecule to interact with a surface intentionally ignoring H-bonding, which is discussed in detail in the next section. But before, let us start with the general case of two interacting polar and polarizable molecules, with permanent dipoles μ_1 and μ_2 and polarizabilities α_1 and α_2, respectively. The total interaction free energy, angle-averaged for freely rotating molecules and known as the total van der Waals (vdW) interaction, is given by the expression (Israelachvili 1991):

$$E_{vdW} = -\frac{C_{vdW}}{r^6} = -\frac{C_K + C_D + C_L}{r^6}$$

$$= -\frac{1}{(4\pi\varepsilon_0)^2 r^6} \left\{ \frac{\mu_1^2 \mu_2^2}{3k_B T} + (\mu_1^2 \alpha_2 + \mu_2^2 \alpha_1) + \frac{3}{2}\hbar\omega_{red}\alpha_1\alpha_2 \right\} \quad (1.26)$$

where the vdW coefficient C_{vdW} is expressed as the sum of the C_K, C_D, and C_L coefficients, which represent the Keesom, Debye, and London contributions, respectively, and the reduced normalized frequency ω_{red} is defined as $\omega_{red}^{-1} = \omega_1^{-1} + \omega_2^{-1}$. The Boltzmann factor ($1.381 \times 10^{-23}$ J K^{-1}) and the ionization energy of the ith molecule are represented by k_B and $\hbar\omega_i$, respectively. C_{vdW} is defined as positive and the minus sign from (1.26) accounts for the attractive character of the interaction.

The Keesom term $C_K \propto \mu_1^2 \mu_2^2/3k_B T$ describes the permanent-dipole–permanent-dipole interaction and has an electrostatic origin. Its r^{-6} dependence arises from the spatial average when dipoles (μ_1, μ_2) are allowed to rotate freely. The interaction energy E_K of two fixed dipoles has a r^{-3} dependence, as can be simply derived from the Coulomb interaction between the individual charges of the dipoles:

$$E_K(r) = -\frac{\mu_1 \mu_2}{4\pi\varepsilon_0 r^3} \{2\cos\theta_1\cos\theta_2 - \sin\theta_1\sin\theta_2\} \quad (1.27)$$

where θ_1 and θ_2 represent the angles formed by dipoles μ_1 and μ_2 with a line connecting their centers, respectively. With this term, the interaction of water with surfaces of amino acids has been satisfactorily described, as discussed in Section 4.4. The Debye term, $C_D \propto \mu_1^2 \alpha_2 + \mu_2^2 \alpha_1$, describes the interaction between a permanent dipole and the dipole induced in the polarizable molecule by the permanent dipole having thus an electrostatic origin as well. However, the London term, $C_L \propto \frac{3}{2}\hbar\alpha_1\alpha_2\omega_{red}$, has a quantum-mechanical origin and accounts for the so-called dispersion forces. The underlying idea is that all molecules and atoms, even apolar such as methane and argon, have charge fluctuations that generate instantaneous dipoles ($\partial\mu/\partial t \neq 0$) although the time-averaged value is zero, $\langle\partial\mu/\partial t\rangle = 0$. Intuitively, such dipoles polarize the

TABLE 1.3
Keesom, Debye, London, and Total Van Der Waals Coefficients

Interacting Molecules	Dipole Moment [D]	Electronic Polarizability	Ionization Potential [eV]	C_K	C_D	C_L	C_{vdW}
H_2O–H_2O	1.85	1.48	12.6	9.6	1.0	3.3	13.9
NH_3–NH_3	1.47	2.26	10.2	3.8	1.0	6.3	11.1
CH_4–CH_4	0	2.60	12.6	0	0	10.2	10.2
H_2O–CH_4				0	0.9	5.8	6.7

Source: Data taken from Israelachvili, *Intermolecular and Surface Forces.* 1991. Sa Diego, CA: Acade Press. With permission.
Calculated from (1.26) in vacuum at $T = 293$ K. The electronic polarizabilities are given in units of $(4\pi\varepsilon_0) \times 10^{-30}$ m$^3 = 1.11 \times 10^{-40}$ C^2 m^2 J^{-1} and the vdW energy coefficients in units of 10^{-78} J m^6.

neighboring molecule, hence the mutual interaction. The involved timescale is much shorter as compared to the steady state imposed by the presence of at least one permanent dipole, as in the Keesom and Debye terms, where it is assumed that $\partial\mu/\partial t = 0$ (no charge fluctuations). For two identical molecules, $\mu = \mu_1 = \mu_2, \alpha = \alpha_1 = \alpha_2$ and $\omega = \omega_1 = \omega_2$, (1.26) transforms into the expression:

$$E_{vdW} = -\frac{1}{(4\pi\varepsilon_0)^2 r^6}\left\{\frac{\mu^4}{3k_BT} + 2\mu^2\alpha + \frac{3\hbar\omega\alpha^2}{4}\right\} \qquad (1.28)$$

In Table 1.3 the calculated values for all coefficients for H_2O derived from (1.28) are given and compared to the closely related NH_3 and CH_4 molecules. Note that for NH_3 and CH_4 the dispersion coefficient C_L dominates over the electrostatic C_K and C_D terms, whereas for H_2O the dipole–dipole contribution dominates. In general, dispersion forces exceed the permanent dipole-based forces except for small highly polar molecules, such as water. On the other hand, $C_{vdW}(H_2O–CH_4)$ is notably smaller than $C_{vdW}(H_2O–H_2O)$ and $C_{vdW}(CH_4–CH_4)$, indicating a poor affinity between H_2O and CH_4. This is an important point that is discussed in Chapter 4 in the context of the hydrophobic effect.

Once we know how two polar molecules mutually interact, we can calculate the interaction between a single molecule and a flat surface. Figure 1.8 shows a schematic representation that helps in obtaining the expression for the interaction energy. Our single molecule is arbitrarily set in the coordinate origin at a distance D from the surface of a semi-infinite solid, which is represented in gray in the figure. The net interaction energy is the sum of the interactions between the single molecule and all the molecules of the body. Instead of summing over the infinite pair interactions, it is simpler to integrate the pair interaction between the molecule and differential volumes of the solid. If we consider a circular ring with the x-axis as the rotation axis of cross-sectional area $dxdy$ and radius y, the ring volume is $2\pi ydxdy$ and the number of molecules in the ring will be $\rho2\pi ydxdy$, where ρ is the molecular density of the solid.

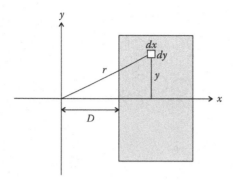

FIGURE 1.8 Scheme for the calculation of the interaction energy between an individual water molecule, set at the coordinate origin, and a flat surface of a semi-infinite solid at a distance D from the molecule.

The net interaction energy for a molecule at a distance D away from the surface is calculated as (Israelachvili 1991):

$$E_{\text{vdW}}(D) = -\rho 2\pi C_{\text{vdW}} \int_D^\infty dx \int_0^\infty \frac{y\,dy}{(x^2 + y^2)^3} = -\frac{\rho \pi C_{\text{vdW}}}{6D^3} \qquad (1.29)$$

using (1.26) as the pair interaction energy, where $r = (x^2 + y^2)^{1/2}$. Although (1.29) has been derived for a solid composed of polar and polarizable molecules, it can be applied quite generally to any solid, molecular or not.

Note that the distance exponent has changed from the original -6 to -3 upon integration, indicating a longer interaction range. The implication is clear: in the absence of repulsive interactions, a surface will attract molecules so that the average density of gas molecules near the surface will always be, in equilibrium, larger than it is in the gas phase. However, molecules close enough to the surface will be repelled due to the interacting electronic clouds. This effect can be included in (1.26) through a repulsive short-range term with a r^{-12} dependence, as formulated in the Lennard–Jones expression:

$$E_{\text{LJ}} = \frac{A}{r^{12}} - \frac{C}{r^6} \qquad (1.30)$$

where both A and C coefficients are positive. Some authors prefer to express E_{LJ} in terms of the distance of minimum energy, r_0, and the corresponding energy value at r_0, E_{LJ}^0. Both parameters are related to A and C through:

$$r_0 = \left\{ \frac{2A}{C} \right\}^{1/6} \qquad (1.31a)$$

$$E_{\text{LJ}}^0 = -\frac{C^2}{4A} \qquad (1.31b)$$

The simplest model that accounts for the repulsion is called the hard wall model, where the repulsion becomes infinite below a threshold distance. If the molecular density increases up to the point where the interaction among the water molecules

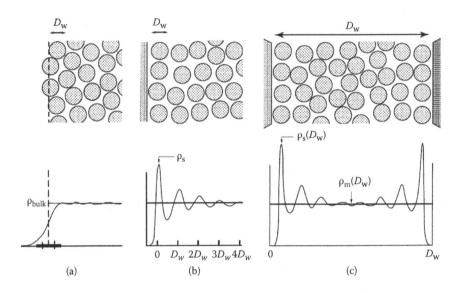

FIGURE 1.9 Liquid density profile: (a) at a vapor/liquid interface, (b) at an isolated solid/liquid interface and (c) between two hard walls separated a distance D_W. ρ_{bulk} is the bulk liquid density and ρ_s stands for the contact density at the surface. $\rho_s(D_W)$ and $\rho_m(D_W)$ are the contact and midplane densities, respectively. (Reprinted from Intermolecular & Surface Forces, 3rd edition, J. Israelachvili, 2011. With permission).

becomes effective (e.g., in the liquid state), then the surface will tend to structure the molecules forming quasi-discrete layers (Abraham 1978). This is illustrated in Figure 1.9, where water molecules are represented by circles of radius D_w. In absence of the constraining surface (Figure 1.9a), the vapor/liquid interface exhibits a characteristic quasi-uniform density profile. We come back to this interface in Section 2.1.1. Figure 1.9b shows the effect of a flat surface when the molecule–surface interaction is ruled by a hard wall model. The surface-induced layering is reflected in an oscillatory density profile extending several D_w along the direction perpendicular to the surface. This phenomenon, also termed structuration, is dealt with in Section 3.4. Structuration is termed solvation or hydration as well, in the case of water, and is not exclusive of flat extended surfaces. We show how CH_4 molecules are able to structure water around them forming clathrate hydrates in Section 3.4. as well as how the surfaces of proteins induce a layering effect in water, which in turn act on the conformation of the proteins (Chapter 6). In addition, we may think of hydrophobicity as a kind of water structuration through H-bonding. When water is constrained inside two identical flat and parallel surfaces separated by a distance D_W comparable to D_w, as shown in Figure 1.9c, then the molecules become confined between both surfaces and ordered. This is the case of water structuration upon confinement that is also discussed in Section 3.4.

When the interaction between water molecules and the surface involves orbital mixing then water molecules are said to be chemisorbed. Because of the mixing, new electronic levels are generated and the bonding character is stronger as compared to the gentle physisorption case. The water MOs that will play a more relevant role in the bonding are those less bound to the molecule, that is, those with lower binding energies. These are the frontier orbitals $1b_1$ and $3a_1$ shown in Figures 1.5 and 1.6. R. Hoffmann (1988) modeled the chemisorptive molecule–surface interactions following a rather simple approach where HOMO and LUMO mix with electronic bands of the solid surface around the Fermi level (E_F). In Section 3.1 we discuss in some detail the chemisorption of water monomers with close-packed surfaces of transition metals (Pt, Pd, Ag, Ru, etc.) and see that the bonding can essentially be described in terms of orbital mixing between $1b_1$ and $3a_1$ MOs and d metal states. You can have a quick look at Figures 3.2, 3.4, and 3.5 in order to have a first impression. A careful reading of references (Hoffmann 1987 and 1988) is strongly recommended.

1.4 HYDROGEN BONDING ON: CONDENSED WATER

1.4.1 WATER CLUSTERS

When the water molecules are brought close enough to each other (below ~ 3 Å or so) they can no longer be considered as isolated and they build clusters, $(H_2O)_n$, where n stands for the number of water molecules, and the familiar liquid and solid water (ice) condensed phases ($n \to \infty$). In this case the kinetic energy of the molecules is of the order (liquid) or well below (solid) of the interaction potential energy and the cohesion is achieved dominantly through H-bonding. This concept was introduced by Latimer and Rodebush (1920) based on the celebrated Lewis theory of valence electrons. As mentioned in their original work: "A free pair of electrons on one water molecule might be able to exert sufficient force on a hydrogen held by a pair of electrons on another water molecule to bind the two molecules together." The free pair corresponds to the lone pairs discussed in Section 1.2.2 and the resulting bond, of electrostatic origin, is strongly directional. H-bonding can be considered as rather strong when compared to other intermolecular interactions, with interaction energies in the range 0.1–0.3 eV per molecule, but is clearly weak when compared to the covalent O–H bond in water, with energies of about 5 eV per molecule. H-bonding is responsible for the surprising physical properties of water and ice, as we show later, and is abundant in biological systems, which has motivated their exploitation in supramolecular chemistry. We just mention here that the deoxyribonucleic acid (DNA) double-helix structure is held together by N–H $\cdots$ O and N–H $\cdots$ N H-bonds involving cytosine-guanine and adenine-thymine pairs (see Figures 6.6 and 6.7). A detailed description of H-bonding involving water molecules can be found in Maréchal (2007). The term H-bonding is also used when deuterium substitutes hydrogen atoms (D_2O, H–O–D).

Figure 1.10 shows an example of H-bonding (O–H $\cdots$ O), represented by a dotted line, on a water dimer, the most simple nontrivial cluster. The molecule with the O–H covalent bonding along the H-bond is called the proton donor (D), and the other molecule is termed the proton acceptor (A). Hence, they are inequivalent. One can imagine building clusters for any value of n but we are interested here in the small ones ($n \leq 10$) because they provide valuable and illustrative information on the cooperative effect. Such an effect was first proposed by Frank and Wen (1957) and can be thought of as a kind of positive feedback where the increase of the monomer dipole moment in a cluster increases the ability to make further H-bonds.

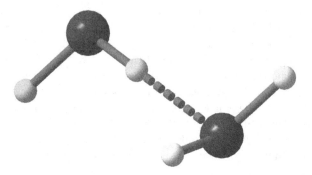

FIGURE 1.10 Water dimer. Oxygen and hydrogen atoms are represented by dark gray and white balls, respectively. Covalent and hydrogen bonds are represented by continuous and dashed thick lines, respectively.

Figures 1.11a and b show the mean oxygen–oxygen and covalent oxygen–hydrogen distances in water clusters as a function of n. From the figures we observe that the oxygen–oxygen distance decreases whereas the oxygen–hydrogen distance increases for an increasing number of water molecules. Figure 1.11a compares experimental (open circles) and theoretical values. For $n = 2$ the experimental oxygen–oxygen distance is 2.976 Å, as determined by electric resonance spectroscopy (Dyke, Mack, and Muenter 1977), and for $n \geq 3$, the distances have been obtained by means of far-infrared vibration-rotation–tunneling (FIR-VRT) spectroscopy (Lin et al. 1996). The FIR-VRT spectroscopy makes use of tunable far-IR lasers and is indicated to measure low-frequency intermolecular vibrations (~ 80 cm^{-1}; Cohen and Saykally 1992). The clusters are generated in the form of supersonic molecular beams by expansion of a gas mixture (e.g., argon–water) through nozzles.

The calculations shown in Figures 1.11a and b have been performed using different methods: HF [HF1 = (Ludwig 2001), HF2 = (Xantheas and Dunning 1993)] and DFT and Moller–Plesset second-order (MP2) perturbation theory (Xantheas 1995). Note that the oxygen–oxygen and oxygen–hydrogen values reach a plateau, corresponding to different values depending on the chosen method, already for the pentamer. The DFT and MP2 calculations give values close to distances in hexagonal ice (at 223 K) and HF calculations reproduce oxygen–oxygen values corresponding to liquid water but oxygen–hydrogen distances of the isolated molecule. Finally, theoretical calculations of the monomer dipole moment (Gregory et al. 1997) show that it increases from the known value of the isolated molecule, 1.8546 D (Clough et al. 1973), up to the values of condensed states, 2.6 D for hexagonal ice (Coulson and Eisenberg 1966), in the same n range (see Figure 1.11c). This is exactly the cooperative effect: the electric field of the surrounding molecules increases the electric dipole moment of the molecule, enhancing the ability to interact through H-bonding. Because the effect is completed for low values of n, there is no need to work with larger clusters in order to identify it.

As shown in Figure 1.11b the formation of H-bonding involves an elongation and thus a weakening of the covalent O–H bond (the hydrogen atom is now shared by two oxygens), which should have important consequences for the vibrational modes. Weakening implies a reduction of k_c, and as a consequence, the vibration frequencies of the stretching modes should decrease (red-shift), as indicated by (1.25). In the case of dimers, the observed stretching bands show absorption lines at 3,745, 3,735, 3,660 and 3,601 cm^{-1} (Buck and Huisken 2000). The proton donor molecule shows the most significant variations ($\nu_3 = 3,735$ and $\nu_1 = 3,601$ cm^{-1}), and the acceptor contributes with the $\nu_3 = 3,745$ and $\nu_1 = 3,660$ cm^{-1} frequencies, closer in energy to the free molecule values. For cyclic clusters it is more indicated to classify the

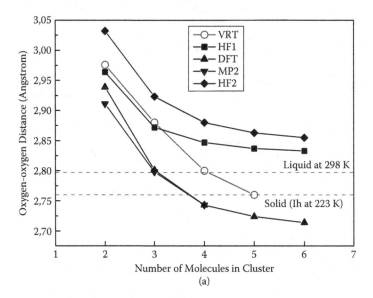

FIGURE 1.11 (a) Mean oxygen–oxygen and (b) covalent oxygen–hydrogen distances as a function of the number n of molecules in a water cluster. Experimental values, obtained by means of electric resonance ($n = 2$; Dyke, Mack, and Muenter 1977) and FR-VRT spectroscopies ($n \geq 3$; Lin et al. 1996) are represented by open circles. Full black squares, diamonds, triangles, and inverted triangles represent values obtained with HF (HF1, Ludwig 2001), (HF2, Xantheas and Dunning 1993), DFT and MP2 (Xantheas 1995) methods, respectively. Values corresponding to liquid at 298 K (Soper 2007), solid at 223 K (Röttger et al. 1994), to a disordered hexagonal structure (Kuhs and Lehmann 1986), and to isolated molecules (Benedict, Gailer and Plyler 1956) are indicated by dashed lines. (c) Theoretical calculations of the monomer dipole moment (Gregory et al. 1997) as a function of n. For $n = 1$ the experimental value of the isolated molecule, 1.8546 D, is taken from (Clough et al. 1973). For hexagonal ice, 2.6 D (Coulson and Eisenberg 1966) and liquid water, $\simeq 3.0$ D (Silvestrelli and Parrinello 1999; Badyal et al. 2000; Gubskaya and Kusalik 2002). (d) Experimental intramolecular stretching frequencies, in cm^{-1}, of small water clusters in the gas phase taken from Buck and Huisken (2000). Open squares and circles correspond to the free and bonded OH stretches, respectively (for $n = 2$ the values correspond to the proton donor). Also shown are the symmetric ν_1 and asymmetric ν_3 stretching modes for the isolated molecule ($n = 1$). The 3,220 cm^{-1} mode for Ih (Johari 1981) and that corresponding to free OH at the water surface at 3,690 cm^{-1} (Du et al. 1993) are given. Continuous lines connecting points are guides to the eye. *(Continued)*

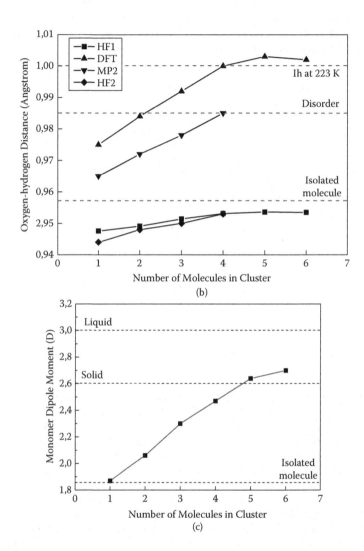

FIGURE 1.11 *(Continued)*

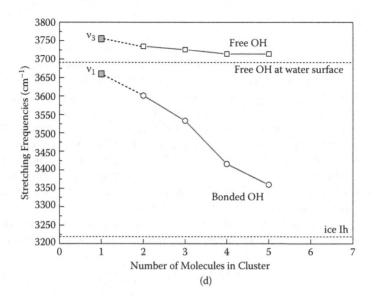

FIGURE 1.11 *(Continued)*

observed intramolecular stretching modes as either bonded or free OH stretches, corresponding to OH groups engaged or not in the H-bond, respectively, inasmuch as all molecules are both donors and acceptors but have free O–H groups (twofold configuration). Figure 1.11d shows the experimental values for free and bonded OH stretching frequencies of small water clusters in the gas phase taken from Buck and Huisken (2000). As expected, the larger decrease is observed for the bonded OH stretching mode, because it is directly involved in the H-bond. The predicted shifts of the bonded OH stretching frequencies vary according to the used theory levels. For $n = 5$ the calculations give shifts of ~ 200, ~ 400, and ~ 500 cm^{-1} using HF, MP2, and DFT, respectively (Xantheas 1995), so that in this case HF provides a better prediction.

With an increasing number of molecules in the cluster, water molecules coordinate both in twofold and threefold configurations leading to the formation of 3D structures. Molecules can act either as double donor and single acceptor or as single donor and double acceptor and as a consequence the IR absorption spectra becomes more complex. The limiting cases are liquid and solid water. As discussed in the next section, the IR spectrum of the OH stretch mode of hexagonal ice Ih exhibits a peak at 3,220 cm^{-1} (Johari 1981), so that a red-shift of about 440 cm^{-1} is observed when $n \to \infty$. Such a large shift is indicative of the stronger H-bonds in the tetrahedral arrangement of the solid and provides evidence that inelastic scattering techniques, in particular IR spectroscopy, are well suited to characterize H-bonding. The free OH stretch exhibits a smaller shift, of about 60 cm^{-1}, with a feature at 3,690 cm^{-1} at the free surface of liquid water (Du et al. 1993), as discussed in Chapter 2.

Once we have the gluing tool to assemble water molecules, let us explore the solid and liquid states, in this order, inasmuch as it is easier to rationalize matter with crystalline order.

1.4.2 SOLID WATER

Here, we discuss the most relevant issues regarding solid water, following the introductory style of this chapter. Those willing to have a deeper insight are suggested to read the book *Physics of Ice* by V. F. Petrenko and R. W. Whitworth (2006). As can any other solid, water can condense with or without long-range order depending on thermodynamic and kinematic conditions. The crystalline (long-range order) or amorphous (no long-range order) character of water ice is mainly associated with the oxygen sublattice. All reliably known crystalline forms of ice (sixteen) can be built from water molecules exhibiting pseudo-tetrahedral configurations bound through H-bonding. The existence of a large number of polymorphs is a common property of molecular solids due to the involved weak interactions and to the conformational degrees of freedom (Bernstein 2002; Fraxedas 2006).

The long-range order in ice can be quantified by an orientational order parameter ζ_i for each molecule, that can be defined as (Errington and Debenedetti 2001):

$$\zeta_i = 1 - \frac{3}{8} \sum_{j=1}^{3} \sum_{k=j+1}^{4} \left(\cos \theta_{jik} + \frac{1}{3} \right)^2 \tag{1.32}$$

where θ_{jik} is the angle formed by the oxygen of molecules j, i, and k with molecule i at the vertex of the angle. If a molecule is located at the center of a regular tetrahedron whose vertices are occupied by its four nearest neighbors, then $\cos \theta_{jik} = -1/3$. Thus, in a perfect tetrahedral network, $\zeta_i = 1$. Otherwise, $\zeta_i < 1$.

Ice Ih

Let us now discuss in some detail the form of water ice with which we are all familiar. At ambient pressure and below 273 K, water crystallizes in the hexagonal form known as Ih. Ice Ih belongs to the hexagonal space group $P6_3/mmc$, using Hermann–Mauguin symbols, and Figure 1.12 shows its crystal structure at the particular temperature of 223 K along the (a) c- and (b) a-axis, respectively. Ice Ih exhibits the wurtzite (ZnS) structure, which is characterized by two interpenetrating hcp lattices, and is isostructural to the hexagonal diamond modification known as lonsdaleite. The average structure of the oxygen sublattice is given by the fourfold fractional coordinates:

$$\left(\frac{1}{3}, \frac{2}{3}, z_O \right), \quad \left(\frac{1}{3}, \frac{2}{3}, \frac{1}{2} - z_O \right), \quad \left(\frac{2}{3}, \frac{1}{3}, \frac{1}{2} + z_O \right), \quad \left(\frac{2}{3}, \frac{1}{3}, 1 - z_O \right)$$

where the z_O parameter represents the puckering of the hexagonal rings lying in the plane perpendicular to the c-axis (basal plane). Flat hexagonal rings would be represented by $z_O = 0$. Perfect tetrahedral symmetry around each oxygen site would be parametrized by $z_O = 1/16 = 0.0625$, but in ice Ih this value is slightly smaller (0.0623 at 223 K) representing a slight flattening of the rings and an increase of the ideal tetrahedral angles (109.47°) as determined by Kuhs and Lehmann (1986).

Note both the hexagonal and open structure of the framework in the figure, this last property accounting for the surprising fact that ice Ih is less dense than liquid water, the underlying reason of why water ice floats on liquid water (icebergs). Concerning

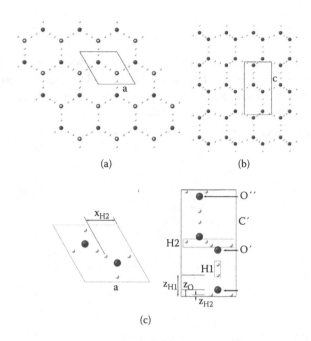

FIGURE 1.12 Crystal structure of ice Ih at 223 K along the (a) c- and (b) a-axis, respectively, including disorder in the hydrogen sublattice. Space group $P6_3/mmc$, $a = 4.5117$ Å and $c = 7.3447$ Å, as determined by Röttger et al. (1994). Atomic coordinates taken from Kuhs and Lehmann (1986). (c) Unit cell along the c- (left) and a-axis (right), respectively, showing the relevant parameters $z_O = 0.0623$, $z_{H1} = 0.1989$, $z_{H2} = 0.0167$, and $x_{H2} = 0.4540$. Oxygen and hydrogen atoms are represented by dark gray and white balls, respectively.

hydrogens, two different types, H1 and H2, are found in the unit cell (see Figure 1.12c). There are 4 H1-hydrogens on the O–O bonds along the c-axis with fractional coordinates $(\frac{1}{3}, \frac{2}{3}, z_{H1})$, where $z_{H1} - z_O$ stands for the O–H distance (O–H_1), and 12 H2-hydrogens represented by $x_{H2}, 2x_{H2}, z_{H2}$, where x_{H2} and z_{H2} are the x and z fractional coordinates on the O–O bonds along tetrahedral directions other than the c-direction. Note that in the figure each oxygen atom has four hydrogen atoms, instead of the expected two. This is related to disorder of the hydrogen lattice and is discussed below. Using the lattice constants $a = 4.5117$ Å and $c = 7.3447$ Å, as determined by Röttger et al. (1994) by X-ray diffraction using SR, we obtain the following distances: $d(O - O') = 2.759$ Å, $d(O - O'') = 2.761$ Å, $d(O - H1) = 1.004$ Å, and $d(O - H2) = 1.000$ Å and angles $\theta(O' - O - O'') = 109.36°$ and $\theta(O–O–O)=109.58°$. Thus, θ is well above the 104.5° value from the isolated molecule and close to the pure tetrahedral angle 109.5°. A list of relevant crystallographic structures of ice Ih is given for both H_2O and D_2O at the top of Table 1.4.

Figure 1.13 shows the temperature dependence of the lattice parameters a and c. Note that both a and c lattice constants decrease as temperature decreases (positive thermal expansion) but below 73 K the thermal expansion becomes negative. This effect is also observed in silicon, a material with the zinc blende structure characterized

TABLE 1.4

Known crystallographic phases of ice. Z represents the number of molecules per unit cell and T and P stand for temperature and pressure, respectively, at which the structures were determined. Ambient pressure is denoted by 0 GPa (1 atm = 1013 mbar = 1.01×10^{-4} GPa).

Ice	Space Group	Z	Cell Parameters Å	Density g cm^{-3}	T K	P GPa
Ih(H$_2$O)^a	$P6_3cm$	12	$a = 7.82, c = 7.36$			
Ih(H$_2$O)^b	$P6_3/mmc$	4	$a = 4.53, c = 7.41$	0.920	253	0
Ih(H$_2$O)^c			$a = 4.511, c = 7.346$		243	0
Ih(H$_2$O)^d			$a = 4.518, c = 7.355$		250	0
Ih(D$_2$O)^e			$a = 4.489, c = 7.327$	1.040	123	0
Ih(D$_2$O)^d			$a = 4.522, c = 7.363$		250	0
Ih(D$_2$O)^f			$a = 4.497, c = 7.324$		5	0
Ic(D$_2$O)^g	$Fd3m$	8	$a = 6.353$		80	0
Ic(D$_2$O)^h			$a = 6.358$	1.035	78	0
II(H$_2$O)^i	$R\bar{3}$	12	$a = 7.78$	1.170	123	0
			$\alpha = 113.1°$			
III(D$_2$O)^j	$P4_12_12$	12	$a = 6.666, c = 6.936$	1.294	250	0.28
III(D$_2$O)^k			$a = 6.671, c = 6.933$	1.292	250	0.3
IV(D$_2$O)^l	$R\bar{3}c$	16	$a = 7.60$	1.272	110	0
			$\alpha = 70.1°$			
V(H$_2$O)^m	$A/2a$	28	$a = 9.22, b = 7.54, c = 10.35$	1.231	98	0
			$\beta = 109.2°$			
V(D$_2$O)^k			$a = 9.074, b = 7.543, c = 10.237$	1.404	254	0.5
			$\beta = 109.07°$			
VI(H$_2$O)^n	$P4_2/nmc$	10	$a = 6.27, c = 5.79$	1.31	98	0
VI(D$_2$O)^o			$a = 6.181, c = 5.698$	1.526	225	1.1
VII(D$_2$O)^o	$Pn\bar{3}m$	2	$a = 3.344$	1.778	295	2.4
VII(D$_2$O)^p	$Pn3m$		$a = 3.350$		295	2.6
VIII(D$_2$O)^o	$I4_1/amd$	8	$a = 4.656, c = 6.775$	1.810	10	2.4
IX(D$_2$O)^j	$P4_12_12$	12	$a = 6.692, c = 6.715$	1.326	165	0.28
X(H$_2$O)^q	$Pn\bar{3}m$	2	$a = 2.76$	2.83	300	65.8
XI(D$_2$O)^r	$Cmc2_1$	8	$a = 4.502, b = 7.798, c = 7.328$		5	0
XI(D$_2$O)^f			$a = 4.465, b = 7.859, c = 7.292$	1.039	5	0
XII(D$_2$O)^s	$I\bar{4}2d$	12	$a = 8.304, c = 4.024$	1.437	260	0.5
XIII(D$_2$O)^t	$P2_1/a$	28	$a = 9.242, b = 7.472, c = 10.297$	1.391	80	0
			$\beta = 109.69°$			
XIV(D$_2$O)$^{[t]}$	$P2_12_12_1$	12	$a = 8.350, b = 8.139, c = 4.082$	1.535	80	0
XV(D$_2$O)$^{[u]}$	$P\bar{1}$		$a = 6.232, b = 6.244, c = 5.790$	1.476	80	0
			$\alpha = 90.06°, \beta = 89.99°, \gamma = 89.92°$			

^{a} (Bernal and Fowler 1933), ^{b} (Barnes 1929), ^{c} (Goto, Hondoh, and Mae 1990), ^{d} (Röttger et al. 1994), ^{e}(Peterson and Levy 1957), ^{f} (Line and Whitworth 1996), ^{g} (Arnold et al. 1968) ^{h} (Kuhs, Bliss, and Finney 1987), ^{i} (Kamb 1964), ^{j} (Londono, Kuhs, and Finney 1993), ^{k} (Lobban, Finney, and Kuhs 2000), ^{l} (Engelhardt and Kamb 1981), ^{m} (Kamb, Prakash, and Knobler 1967), ^{n} (Kamb 1965), ^{o} (Kuhs et al. 1984), ^{p} (Jorgensen and Worlton 1985), ^{q} (Hemley et al. 1987), ^{r} (Leadbetter et al. 1985), ^{s} (Lobban, Finney, and Kuhs 1998), ^{t} (Salzmann et al. 2006), ^{u} (Salzmann et al. 2009).

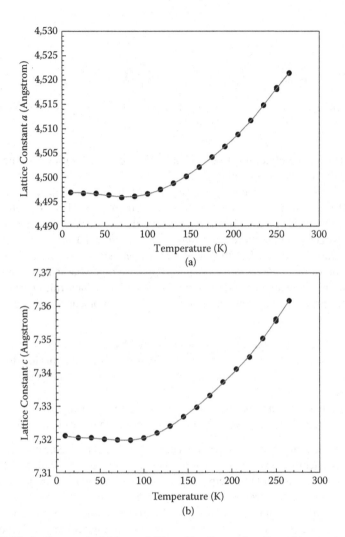

FIGURE 1.13 Lattice constant (a) a and (b) c of ice Ih as a function of temperature. Experimental data taken from Röttger et al. (1994) represented by full circles. Least-square fits to polynomials of order 7 are represented by continuous gray lines: $a(T) = 4.49682 + 9.29142 \times 10^{-6}T + 4.50629 \times 10^{-8}T^2 - 2.36196 \times 10^{-8}T^3 + 4.34126 \times 10^{-10}T^4 - 2.92556 \times 10^{-12}T^5 + 8.87984 \times 10^{-15}T^6 - 1.0181 \times 10^{-17}T^7$ and $c(T) = 7.32227 - 1.81508 \times 10^{-4}T + 7.77782 \times 10^{-6}T^2 - 1.6973 \times 10^{-7}T^3 + 1.83771 \times 10^{-9}T^4 - 9.88188 \times 10^{-12}T^5 + 2.60147 \times 10^{-14}T^6 - 2.67828 \times 10^{-17}T^7$.

by two interpenetrating fcc lattices, with a transition temperature of about 80 K. It is important to add here that ice Ih expands when hydrogen is substituted by deuterium (Röttger et al. 2012), although the opposite is expected: the isotope effect involves volume contraction with increased isotope mass. The volume of H_2O at $T = 0$ K is ~0.1% smaller than that of D_2O. The zero-point expansion of crystal lattices is a rather well-understood phenomenon and is almost always larger for the lightest

FIGURE 1.14 The six possible orientations of a water molecule at a given site in the Ih lattice. Lone-pairs are represented by broken bonds.

isotopes. This subtle effect has been satisfactorily described with DFT calculations including vdW interactions (Pamuk et al. 2012).

Ice Ih is oxygen-ordered but hydrogen (proton)-disordered. This is a further peculiarity of ice. How the idea of proton disorder emerged is a beautiful and elegant example of scientific thinking. It was experimentally known that the residual entropy of ice at cryogenic temperatures was rather high, about 3.4 J K^{-1} mol^{-1} (Giauque and Stout 1936, Flubacher, Leadbetter, and Morrison 1960). L. Pauling (1935) made an estimation of the theoretical residual entropy based on what is nowadays known as ice or Bernal–Fowler rules (Bernal and Fowler 1933). Such rules describe in an architectural practical way the configuration of water molecules in ice and can be summarized as:

(i) Each oxygen atom has two covalently bonded hydrogen atoms at about 1 Å.
(ii) Each water molecule is oriented so that its two hydrogen atoms are directed approximately toward two of the four oxygen atoms surrounding it tetrahedrally.
(iii) Only one hydrogen atom lies along each oxygen–oxygen axis.
(iv) All configurations satisfying the preceding conditions are equally probable.

According to such rules, there are six ways in which a water molecule can orient itself within the tetrahedral arrangement, which is shown in Figure 1.14. Of the four H-bonds in which one water molecule participates, two are occupied by its hydrogen atoms and two are unoccupied. The probability that a given direction is available to a hydrogen atom is therefore 1/2, and as there are two hydrogen atoms to be placed, the total probability reduces to 1/4. Thus, the total amount of configurations for N molecules (per mole) of water ice is $(6/4)^N$. The entropy S is given by $S = k_B \ln(6/4)^N = R \ln(3/2) = 3.37$ J K^{-1}mol^{-1}, where R stands for the molar gas constant (8.3145 J K^{-1}mol^{-1}), which is in excellent agreement with the experimentally obtained value, given the simplicity of the calculation. Thus, within the oxygen–oxygen axis the hydrogen atoms can be located close to any of the two oxygens with equal probability, hence the term half-hydrogen traditionally used and the dual representation from Figure 1.12. However, only two out of four can be closer in a given instant. Thus, the hydrogen atoms can adopt a random distribution over the ice lattice as long as they follow the Bernal–Fowler rules, sometimes also referred to as the Bernal–Fowler–Pauling rules. This disorder is also known as orientational disorder, because it is equivalent to the rotation of the water molecules.

The pseudo-tetrahedral configuration of the water molecules within the solid justifies the use of the hybrid sp^3 orbitals, built as the linear combination of the s, p_x, p_y, and p_z atomic orbitals, to describe the solid electronic structure. Four such orbitals are commonly used for molecules exhibiting a tetrahedral geometry such as CH_4 as well as for solids with fcc symmetry such as diamond, silicon, germanium, and the zinc blende series. In some texts there is some confusion when dealing with the electronic structure of water molecules, irrespective of their aggregation state. In the case of isolated molecules the sp^3 orbitals do not transform according to the C_{2v} point group and are not eigenfunctions of the effective h_i Hamiltonian. However, closely related sp^3 orbitals can be built from linear combinations of the eigenfunctions. If the transformation matrix is unitary then the resulting wavefunction should correctly describe the electronic density.

Let us now have a look to the electronic structure of ice Ih. Upon the formation of a molecular solid with long-range order, the MOs transform into energy bands in reciprocal ($\mathbf{k}$) space as a result of the intermolecular interactions, where $\mathbf{k}$ stands for the wavevector. For those interested in learning more about this subject, reference textbooks such as Ashcroft and Mermin (1976), Madelung (1978), and Martin (2004) are strongly recommended. Figure 1.15 shows a DFT–GGA (GGA stands for generalized gradient approximation) calculation of the electronic structure of ice Ih using a supercell of 16 water molecules and following the Bernal–Fowler rules, where each molecule participates in 4 H-bonds, two as a donor and two as an acceptor. The figure shows the band structure along the principal symmetry directions of the hexagonal lattice as well as the projected density of states (PDOS). The calculations reproduce earlier reported DFT calculations (Hahn et al. 2005; Prendergast, Grossman, and Galli 2005). The PDOS from Figure 1.15 shows two large peaks that can be associated

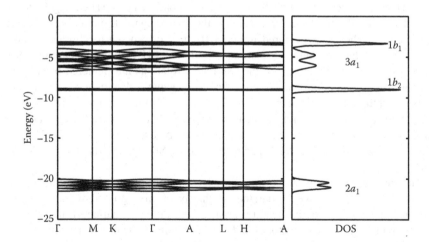

FIGURE 1.15 DFT–GGA calculation of the electronic structure of ice Ih using a supercell of 16 water molecules and following the Bernal–Fowler rules. (Left) Band structure along the principal symmetry directions of the hexagonal lattice and (right) PDOS. The zero of energy is arbitrary set to the highest computed value. The crystal structure used: space group $P6_3/mmc$, $a = 4.5117$ Å, and $c = 7.3447$ Å (Röttger et al. 1994), corresponding to 223 K, and atomic coordinates taken from Kuhs and Lehmann (1986). The associated MOs are indicated. (Courtesy of G. Tobías and E. Canadell. With permission.)

with both $1b_1$ and $1b_2$ MOs and two doublets associated with $2a_1$ and $3a_1$ MOs, respectively. *Stricto senso*, one should not use the terminology reserved for MOs. However, when the intermolecular interactions are weak, as is the case for molecular materials, this abuse of language is commonly accepted because it reveals the origin of the band. You can have a look now at Figure 1.21 to observe that the calculations correctly reproduce the experimentally observed spectra. A first impression suggests that the b-like features are not involved in the bonding whereas a-like are. This leads to the apparent contradiction that both lone-pairs, $1b_1$- and $3a_1$-like, have contrasting participations; $1b_1$-like have a limited contribution to the bonding. But this is only apparent.

Let us finish this part devoted to ice Ih by briefly discussing its vibrational spectrum. We show in Section 1.4.1, when dealing with the formation of small water clusters by means of hydrogen bonding, that the bonded OH stretching mode frequency decreased for an increasing number of water molecules in the cluster (see Figure 1.11d). In the limiting case of the solid, the corresponding mode builds an intense and broad band centered around $3{,}220 \, \text{cm}^{-1}$ (Johari 1981), clearly shifted from the $3{,}675 \, \text{cm}^{-1}$ value corresponding to the symmetric stretching mode of the free molecule. The vibrational spectra of ice can be divided, as a first approximation, in four clearly defined regions centered about 3,220, 1,650, 820, and below $400 \, \text{cm}^{-1}$ (Johari 1981; Petrenko and Whitworth 2006). As already mentioned, the first one corresponds to the bonded OH stretching mode. In Figure 1.11d the free OH stretching mode frequency tends to an asymptotic value of about $3{,}700 \, \text{cm}^{-1}$, a value that is not observed in the experimental IR spectra of ice, meaning that the concentration of free bonds has to be very low, as expected for the solid: all bonds are used as stated in the Bernal–Fowler model. The second region, centered around $1{,}650 \, \text{cm}^{-1}$, corresponds to H–O–H bending. Note the small increase ($55 \, \text{cm}^{-1}$) when compared to the scissoring mode of the free molecule ($1{,}595 \, \text{cm}^{-1}$). The region centered around $820 \, \text{cm}^{-1}$ corresponds to a libration mode. Librations are the three relative rotations of individual H_2O molecules around their three main axes (x, y, and z). In the solid such rotations are hindered because of intermolecular H-bonding and the large frequency value is due to the small moment of inertia of the water molecule: 2.9376×10^{-40}, 1.0220×10^{-40}, and 1.9187×10^{-40} g cm^2 referred to the x, y, and z directions, respectively (see Figure 1.2). Finally, the far IR region contains hindered translations with characteristic features at 215 and $155 \, \text{cm}^{-1}$ as determined with IR spectroscopy using SR (Miura, Yamada, and Moon 2010).

The case of ice is quite interesting because the translational, librational, and intramolecular bands are well separated, so that band mixing is negligible. The vibrational spectra also contain features arising from the combination of the previous modes and the presence of isotopes and defects results in the broadening of the features. The measured Debye temperature (Θ_D) for bulk ice Ih determined in the low temperature range is 218 K (Leadbetter 1965), which leads to a frequency of 152 cm^{-1}, according to the expression $k_B \Theta_D = \hbar \omega$. This means that the major contribution arises from low-frequency vibrations (translational–librational), something to be expected inasmuch as intramolecular vibrations are hardly excited at low temperatures. We just mention that the surface Debye temperature is markedly smaller, 132 K, as determined by means of helium atom scattering (HAS; Glebov et al. 2000), due to the lower coordination of the molecules at the surface.

Ice Polymorphs

Water has the ability to crystallize in at least 16 different phases (Ih, Ic, II–XV) depending on the experimental external conditions (temperature, cooling/heating rate, pressure, doping, etc.). Figure 1.16a shows the phase diagram of solid and liquid water, Table 1.4 summarizes the relevant crystallographic information of such phases concerning the oxygen sublattice for both H_2O and D_2O, and Figures 1.17 and 1.18 show projections of selected crystallographic phases along specific directions.

Sixteen is a considerable number and reveals the versatility of the water molecule to build several fully hydrogen-bonded structures, but in general molecules have the tendency to order in different configurations, even in the absence of hydrogen-bonding. The solids of the parent molecular hydrogen and oxygen show at least three (I, II, and III; Mao and Hemley 1994) and five (α, β, γ, δ, and ϵ; Gorelli et al. 2002) phases, respectively. Molecular hydrogen is formally the only quantum solid and has interest in astrophysics because of the elevated pressures and low temperatures encountered in planets such as Jupiter. However, the phase diagram of hydrogen is poorly understood because the experimental determination of the stable structures is challenging due to the weak scattering of X-rays and because of the small energy differences between the referred structures. A further example is solid N_2 with seven known molecular phases (α, β, γ, δ, δ_{loc}, ϵ, ζ) plus a nonmolecular η phase, which is semiconducting up to at least 240 GPa (Bini et al. 2000; Gregoryanz et al. 2002). A large number of examples of polymorphism on molecular organic materials can be found in Bernstein (2002) and Fraxedas (2006). Silica, SiO_2, also exhibits a large number of polymorphs: α- and β-quartz, α- and β-cristobalite, α-and β-tridymite, and coesite, among others. It is structurally related to water because of its tetrahedral coordination. Thus, it seems clear that Mother Nature has found in the versatile tetrahedral coordination a chameleonic way to adapt to a broad variety of physical conditions, so that it is no surprise that two relevant representatives, H_2O and SiO_2, are so abundant on Earth.

Figure 1.16b introduces a classification of the polymorphs according to the order state of the hydrogen sublattice (ordered or disordered) and to the pressure range (ambient and high) required to obtain them. The classification in ordered or disordered lattices is rather crude and binary (or Boolean if you wish), without specifying the degree of order, and refers to the main trend of the structure. Few phases exhibit a high degree of order and disordered phases may show partial order. The figure aims to provide a simplified scheme of the solid-state domain of the phase diagram of water, shown in Figure 1.16a, in order to readily identify them, indicating the most relevant phase transitions with arrows. It is important to point out that the tetrahedral configuration is maintained in all phases. However, upon increase of pressure (increase of density) vdW interactions strengthen while H-bonding weakens, as revealed by DFT calculations (Santra et al. 2011). Next we briefly discuss the main characteristic features of such phases. Detailed descriptions on such polymorphs can be found in Petrenko and Whitworth (1999) and Zheligovskaya and Malenkov (2006).

In the ambient pressure column of Figure 1.16b we find three phases, namely Ih, Ic, and XI (see Figure 1.17). Ice XI is the proton-ordered low-temperature analogue of Ih, whereas the metastable Ic has the zinc blende structure, which is characterized by two interpenetrating fcc lattices, and is thus isostructural to diamond, silicon, and

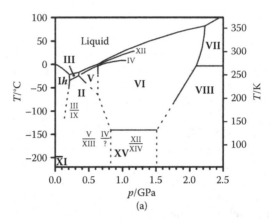

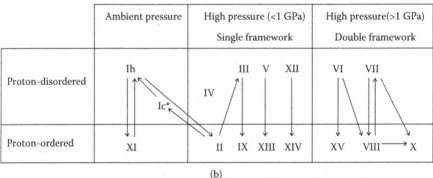

FIGURE 1.16 (a) Phase diagram of water and ice up to 2.5 GPa including phase boundaries (solid lines) and extrapolated phase boundaries (dashed lines). Metastable and stable phases outside their regions of stabilities are indicated by a smaller font size. (Reprinted from C. G. Salzmann et al. *Phys. Chem. Chem. Phys.* 13:18468–18480, 2011. With permission of the Royal Society of Chemistry.) (b) Ice polymorphs classified according to the ordered–disordered character of the proton sublattice and the pressure range needed to generate them. The most relevant transitions between phases are indicated by arrows.

germanium, to mention a few. Ice Ic can be prepared in different ways, for example, by condensation of water vapor on cold metallic substrates between 130 and 150 K, but upon annealing it transforms irreversibly into Ih. A remarkable property of pure ice Ih is that proton disorder is maintained upon cooling down to temperatures close to 0 K. However, when doping with minute quantities of hydroxide anions (KOH or KOD) the hydrogen sublattice orders: the hydroxide anions are able to catalyze the ordering of protons. Above about 72 K ice XI transforms back into ice Ih. In the case of D_2O the temperature rises to 76 K. This phase is particularly interesting because it is ferroelectrically ordered (Jackson and Whitworth 1997), with the water dipole moments adding up to yield a net moment, and is believed to be present in the solar system, above 20 AU, where the temperature lies below the 72 K transition temperature, involving Uranus, Neptune, and Pluto.

In the single-framework high-pressure column of Figure 1.16b (the use of the term single framework becomes evident below) we find the proton-disordered III, V, IV, and XII polymorphs as well as the proton-ordered ices II, IX, XIII, and XIV. The high-pressure range spans up to about 1 GPa. The most remarkable structural property of ice II, which is proton-ordered, is its columnar structure, with cavity diameters inside the columns of about 3 Å, which can allocate guest atoms such as helium or neon or hydrogen molecules, forming host–guest hydrates. Ices IX, XIII, and XIV are the low-temperature proton-ordered analogues of ices III, V, and XII, respectively. Ice III exhibits partial proton-ordered (Lobban, Finney, and Kuhs 2000) and the ordering of ice IX is antiferroelectric according to neutron diffraction experiments (Londono, Kuhs, and Finney 1993), thus the water dipole moments cancel each other. Ice V has perhaps the most complex structure of all ices, with a monoclinic space group and 28 molecules per unit cell. It is classified as proton-disordered but it exhibits some ordering that increases with decreasing temperature and increasing external pressure (Lobban, Finney, and Kuhs 2000). The crystal structure of ice V is rather distorted, with O–O–O angles in the 84–128° range, which results in the bending of the H-bonds (Kamb, Prakash, and Knobler 1967). When doped with hydrochloric acid, ices V and XII transform into ices XIII and XIV, respectively (Salzmann et al. 2006). Ices XIII and XIV, together with ice XV, are the three last discovered polymorphs (see Figure 1.18).

At higher hydrostatic pressures, well above approximately 1.2 GPa, water accommodates in two independent interpenetrating frameworks with no common hydrogen bonds in order to maintain its tetrahedral structure. In this way the induced phases (VI, VII, VIII, and XV) notably increase their densities, as can be observed from Table 1.4. This appealing distribution has also been termed self-clathrate configuration. In the case of ice VI the structure is built up of H-bonded chains of water molecules, which are linked laterally to one another to form an open framework. The cavities of the framework are filled with a second identical framework (Kamb 1965). Ices VII and VIII are built from two sublattices each with the structure of ice Ic. Ices VI and VII are proton-disordered whereas ice VIII is proton-ordered (antiferroelectric). Projections of such phases are shown in Figures 1.17 and 1.18. Note from the phase diagram in Figure 1.16a that ice VII is stable above 0°C. This is an example of room temperature (RT) ice but at elevated pressures (above ~2 GPa). We discuss RT ice in Chapter 3 but in this case associated with nanometer-sized confinement. Ice XV is obtained upon cooling of ice VI doped with hydrochloric acid (Salzmann et al. 2009). Further increasing the applied hydrostatic pressure (e.g., in a diamond anvil cell) leads to the interesting situation where protons become allocated in the center of the H-bonds, generating a new phase, ice X, which is characterized by such symmetric H-bonds. Proton centering occurs from about 60 GPa to 150 GPa, according to X-ray diffraction of single crystals with SR (Loubeyre et al. 1999).

Let us finish this part dedicated to the polymorphs of ice to the solid phases of the closely related molecules ammonia and methane. Ammonia crystallizes in at least five different forms. The low temperature (about 210 K)–low pressure (below 2 GPa) cubic I phase transforms into the rotationally disordered hexagonal (II) and fcc (III) phases with increasing pressure and temperature. Above 4 GPa at RT, ammonia III transforms into an ordered orthorhombic solid, ammonia IV. Phase V, which is isostructural to phase IV, is achieved at elevated pressures (14 GPa at RT; Ninet and Datchi 2008). On the other hand, methane crystallizes at RT and at 1.6 GPa in phase

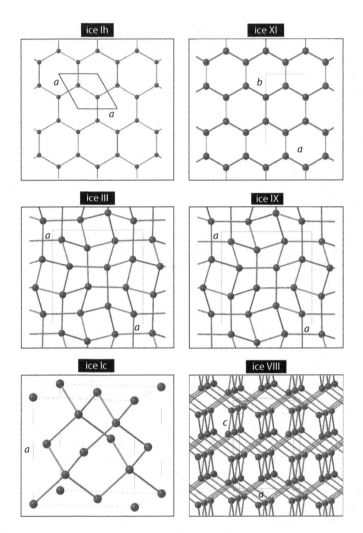

FIGURE 1.17 Crystal structures of ices: Ih (Kuhs and Lehmann 1986) and XI (Leadbetter et al. 1985) viewed along the *c*-axis, III (Lobban, Finney, and Kuhs 2000) and IX (Londono, Kuhs, and Finney 1993) viewed along the *c*-axis and Ic (Kuhs, Bliss, and Finney 1987) and VIII (Kuhs et al. 1984). Hydrogen/deuterium atoms are omitted for clarity. The proton-disordered phases Ih and III are shown on the left side and the related proton-ordered XI and IX phases are shown on the right side (Ih→ XI and III→ IX). Ice Ic is compared to the VIII phase, which is built by two Ic sublattices.

I, with carbon atoms occupying fcc lattice sites and hydrogen atoms free to rotate. By isothermal compression at RT, phase I transforms into phase A (rhombohedral) at approximately 5 GPa, and above 12 GPa phase A transforms into phase B (hexagonal hcp), and at 25 GPa phase B transforms into the so-called high-pressure phase (HP), which is monoclinic (Spanu et al. 2009). Both materials have not been as extensively studied as ice, so that the number of their polymorphs could easily be higher.

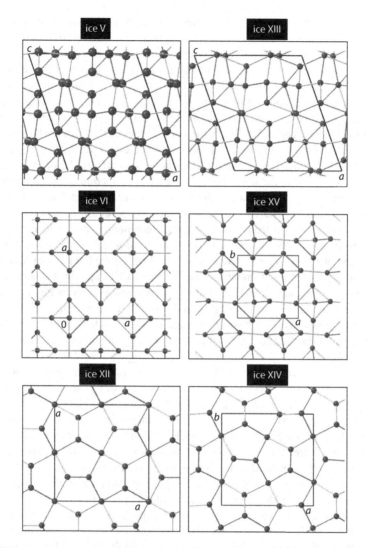

FIGURE 1.18 Crystal structures of ices: V (Kamb, Prakash, and Knobler 1967) and XIII (Salzmann et al. 2006) viewed across the (010) plane, VI (Kamb 1965) viewed along the c-axis and XV (Salzmann et al. 2009) across the (001) plane and XII (Lobban, Finney, and Kuhs 1998) and XIV (Salzmann et al. 2006) along the c-axis. Hydrogen/deuterium atoms are omitted for clarity. Proton-disordered phases are shown on the left side and the related proton-ordered phases are shown on the right side (V→ XIII, VI→ XV, and XII→ XIV).

Ice Polyamorphs

When solid ice exhibits no long-range order in the oxygen sublattice it is termed amorphous ice. The major difference between liquid and amorphous states is the dynamic range: in an amorphous solid atoms and molecules remain static whereas in the liquid they are in permanent motion exhibiting very short residence times. There are several known forms of amorphous ice, hence the term polyamorphism, that can be classified in three main categories according to their densities referred to that of liquid water (Loerting et al. 2011). Those with densities below 1 g cm^{-3} are generically referred to as low-density amorphous (LDA) ices and those with densities above such value are termed high-density amorphous (HDA) ices. There are four known LDA ices, with densities of about 0.94 g cm^{-3} at ambient pressure, namely amorphous solid water (ASW), hyperquenched glassy water (HGW), and two with the generic acronym LDA (LDA-I and LDA-II). In the same line, two HDA ices are known: HDA and very-high-density amorphous (VHDA). In this case the distinction is clear, because the densities are about 1.15 and 1.26 g cm^{-3}, respectively. Amorphous ices are considered metastable in the sense that an increase of temperature leads to molecular rearrangement and to stable phases with crystalline order. Three distinct HDA ices are known: unannealed HDA (uHDA), expanded HDA (eHDA), and relaxed HDA (rHDA). According to neutron scattering measurements the atomic level structures of ASW, HGW, and LDA are essentially identical on length scales up to about 10 Å (Bowron et al. 2006). The low-density ices build a tetrahedral random network of H-bonded water molecules. Concerning the HDA and VHDA phases the networks are (on average) tetrahedrally coordinated up to 3 Å but beyond this region, the oxygen coordination number rises to 5 for HDA and 6 for VHDA, in an interstitial configuration. Such interstitials are also on average fully H-bonded water molecules, so there must be a considerable degree of H-bond bending in these materials compared to LDA.

Figure 1.19 summarizes the most salient transformations involving amorphous ices. Note that the generic LDA and HDA acronyms are used. The diagram is divided in two regions: the light gray background highlights the metastable zone and the rest contains the stable forms of water. A first visual inspection stresses the pivotal role of both ice Ic (with long-range order) and HDA (with no long-range order) in the metastable zone. Note that ice Ic is the intermediary between all amorphous ices and ice Ih, so that in the crystallization process ice Ic is always present. As mentioned above, one way to prepare cubic ice is by condensing water vapor on a metallic substrate held in the 130–150 K range. If the substrate temperature is decreased typically below 110 K, then ASW ice is obtained, which exhibits a microporous character. If instead of using water vapor we send micrometer-sized droplets at supersonic speeds on substrates cooled down at 77 K, HGW is formed (Hallbrucker, Mayer, and Johari 1989). Such a strategy is used because one cannot supercool bulk water into a glassy state because crystallization occurs first. Both ASW and HGW have the same two alternatives: transform either into ice Ic after overcoming a glass transition at 136 K or into HDA under pressure (about 0.6 GPa) at 77 K. A glass transition defines the transformation of a glass into a state with liquidlike properties (Debenedetti and Stillinger 2001).

Ice Ic can transform directly into HDA (uHDA) under pressure at 77 K but HDA needs some intermediate states in order to become ice Ic. Once HDA is formed, the isothermal release of pressure leads to recovered HDA that transforms into LDA (LDA-I) by annealing at 125 K. LDA can choose either to transform back into HDA by compression at about 135 K (eHDA), closing a loop, or to ice Ic after overcoming a new glass transition at 129 K and annealing at 150 K. In this case a second loop is detected, because ice Ic could transform back to HDA. A further loop involves

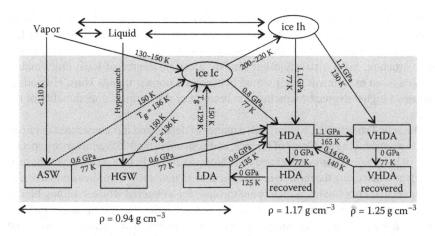

FIGURE 1.19 Diagram of the ASW, HGW, LDA, HDA and VHDA amorphous forms of ice. The light gray background indicates the metastability region. Outside it, the stable forms of water are shown. The transformation paths between phases are indicated together with indicative temperature and hydrostatic pressure values. Glass transitions are represented by discontinuous lines and the corresponding transition temperature by T_g. (Adapted from V. F Petrenko and K. W. Whitworth *Physics of Ice*, Oxford UH: Oxford University Press 2006. With permission.)

HDA and VHDA. Isobaric annealing of HDA (uHDA) at 1.1 GPa leads to VHDA, that can be recovered by releasing pressure at 77 K. VHDA can be transformed back into HDA (eHDA) by increasing both temperature and pressure up to 140 K and 0.14 GPa, respectively. The larger loop involves ice Ih, because annealing ice Ic leads to ice Ih and it can become HDA (uHDA) by applying 1.1 GPa at 77 K (Whalley et al. 1987).

Amorphous ices can also be obtained from high-pressure polymorphs by decompression and heating as well as from irradiation, of great interest for astronomy. By decompressing ice VIII to 1 bar at 80 K and then heating to 125 K LDA is obtained (Yoshimura et al. 2006). Ice Ic can be transformed into amorphous ice either under keV ion-bombardment (Baratta et al. 1991) or ultraviolet irradiation below 70 K (Kouchi and Kuroda 1990). In addition, Mastrapa and Brown (2006) found that amorphization of crystalline ice upon irradiation with 0.8 MeV protons is temperature dependent. Below 50 K irradiation produces amorphous ice but above such temperature the transformation is incomplete and above 70 K the crystalline phase remains almost intact, according to near-IR experiments.

1.4.3 LIQUID WATER

Liquid water is an eccentric representative of the liquid state, as put by F. Franks (2000), because of its salient and surprising physical properties such as the density increase (volume contraction) upon melting, a maximum density at 4°C, large heat capacity, a minimum of isothermal compressibility around 46.5°C, high surface tension, and many more. Liquid water is the dynamical disordered condensed phase of water with a given degree of short-range (local) order imposed by H-bonding. If we could somehow switch the H-bonding interaction off, as hypothesized in Section 1.3, then water molecules would densely pack through vdW interactions resulting in a shell structure. Such a structure has already a certain degree of local order and in this *gedanken* experiment water would behave as a normal liquid. Thus, H-bonding imposes a higher degree of structuration because of the intrinsic directionality of the bonds.

The actual structure of water is still a matter of debate and it is considered as one of the most outstanding problems of science, as discussed in the 125th anniversary special issue of the *Science Journal* (Kennedy and Norman 2005). In fact, we make an abuse of language when using the term *structure*, because *stricto senso* it implies knowledge of the molecular coordinates, which is not the case for a material whose building blocks are in permanent motion, thus with no long-range order. Such long-range order is lost upon the melting of ice but one expects the local tetrahedral configuration to be preserved as suggested by the small latent heat of fusion as compared to the large latent heat of evaporation and to the low density of free O–H bonds as evidenced by IR measurements, which show a characteristic absence of the 3,700 cm^{-1} feature otherwise encountered in water clusters (see Figure 1.11d). Computer simulations provide a scenario where water molecules build a continuous, disordered, and dynamic network of H-bonds in which each molecule is linked with up to four (rarely five) others. The distortion of the network enables a higher occupancy with a relevant contribution from vdW interactions rendering the liquid denser than the solid. This may be called the standard structure of liquid water. Conde, Vega, and Patrykiejew (2008) estimate the order parameter ζ to be 0.85, implying a large degree of tetrahedral order (see Section 1.4.2 for the definition of ζ). However, a more complex picture has emerged when using spectroscopic techniques that probe matter within time-scales well below the characteristic H-bond lifetime of $\sim$1 ps (10^{-12} s). Note that the picture of the idealized tetrahedral configuration has to be considered only as model, based on the ice rules, because MOs are distributed in space (see Figure 1.6), so that the probability to bind in nontetrahedral sites (e.g., trigonal) is nonnegligible. The simplest and perhaps clearest example of trigonal geometry is given by hydronium, which forms a truly covalent bond through the trigonal site (Agmon 2012).

Let us first start with photoemission results, having in mind that the photon-induced ionization process is in the femtosecond range (1 fs = 10^{-15} s). In Figure 1.6 and in Table 1.2 we compared gas-phase photoemission data to EH and DFT calculations and Figure 1.20 shows the measured photoemission spectra for both the gas and liquid phases performed in vacuum using 60 eV photons from a SR source on free-standing liquid water microdroplets with diameters of about 6 μm (Winter et al. 2004). The advantage of using microdroplets is that measurements are performed continuously on fresh samples, so that beam-induced damage, a known problem when using intense beams on molecular species, is strongly reduced. Top, center, and bottom panels display the measured liquid, gas, and difference spectra, respectively.

The direct comparison shows that the spectral features of the liquid are red-shifted (to lower binding energies) by about 1.5 eV and broadened with respect to the gas-phase features. The binding energy of the HOMO band ($1b_1$-like) is 11.16 eV with a photoionization threshold of about 10 eV as derived from extrapolating the slope of the $1b_1$ signal shown in the inset of the figure. The energy shifts can be attributed to the

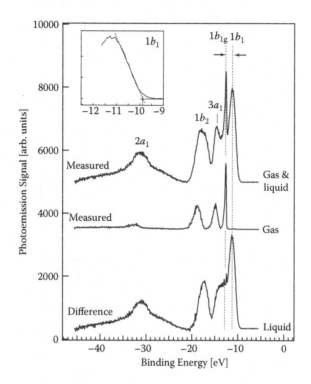

FIGURE 1.20 Photoemission spectra taken with 60 eV photons from gas-phase water sampled for the maximum liquid signal (top), from the pure gas-phase 0.5 mm aside from the liquid microjet (center), and the difference spectrum (bottom). Binding energies are referred to the vacuum level. The gas-to-liquid binding energy shift between $1b_1$ features is indicated. The inset shows the onset of the photoemission signal. Reprinted from B. Winter, et al. *J. Phys. Chem. A* 108:2625–2632, 2004, American Chemical Society. With permission.)

electronic polarization by the surrounding water molecules during the photoemission process, to changes in the surface dipoles of the microdroplets, which modify the work function and to H-bonding (Winter et al. 2004). The peak broadening largely reflects different local environments of water molecules in the liquid. The $1b_1$-like feature, which is the most weakly bound, is particularly sensitive to such changes (see figure) although its absolute width is the smallest due to the nonbonding character of this orbital. The $3a_1$-like feature is also considerably affected, a fact that was already discussed when describing the band structure of water ice, the broadening arising from the H-bonding contribution, whereas broadening for $1b_2$ and $2a_1$ is smaller.

What happens when the photoemission spectrum of the liquid is compared to that of ice? This is shown in Figure 1.21, where valence band photoemission taken with 530 eV photons is shown for liquid water, and amorphous and crystalline ice, all spectra being arbitrarily normalized to the $1b_1$ peak position of the gas phase (top of the figure; Nordlund et al. 2008). Instead of using water microdroplets as discussed above, water droplets were deposited on gold substrates at 1°C and held in thermodynamic

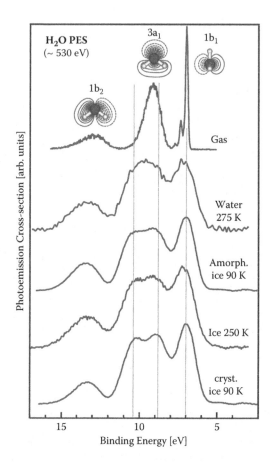

FIGURE 1.21 Valence band photoemission spectra of water in aggregation states of increasing structural order, measured with 530 eV photons. At such photon energy the O $2s$ photoionization cross-section is about 2.75 times that of O $2p$ levels. Vertical gray lines are shown at the energy position of the split $3a_1$ peak positions for crystalline ice. The energy is normalized to the lone pair peak in crystalline ice. (Reprinted from D. Nordlund et al. *Chem. Phys. Lett.* 460:86–92, 2008. With permission from Elsevier.)

equilibrium by means of a background vapor pressure in the analysis chamber of about 5 torr. The amorphous ice films were grown at ∼100 K on clean Pt(111) surfaces and crystalline ice was obtained by heating the amorphous ice up to 150 K. The valence band spectrum of crystalline ice at 90 K (bottom of the figure) shows the expected broadening of the $1b_1$-like band and a splitting of the $3a_1$-like feature. This splitting reflects the relevant role of the $3a_1$ MOs in the intermolecular bonding and is due to symmetry effects. Approaching the melting temperature ($T_m = 273$ K) such splitting is smeared out, as a consequence of the increasing degree of disorder, but it is still observable in liquid water. It is important to notice that close to T_m, a liquid layer on top of the ice crystal builds up (this point is discussed in the next chapter). However, at 250 K, such a liquid layer is absent. We can conclude that the liquid and solid are rather similar from the electronic point of view. The electronic structure is thus dominated by the local structure involving few coordination shells.

The complementary local probes X-ray Raman spectroscopy (XRS) and XES exhibit attosecond (1 attos = 10^{-18} s) and fs timescales, respectively, and have provided a new scenario. XRS corresponds to transitions from a core level to unoccupied states, whereas XES measures the decay from an occupied valence state to an emptied (excited) core level with the emission of an X-ray photon. Figure 1.22a shows the temperature dependence in the lone pair $1b_1$ region of the XES spectra for D_2O taken with 550 eV photons (Huang et al. 2009). Two features are observed, denoted by $1b_1'$ and $1b_1''$ in the figure. The energy positions of $1b_1'$ and $1b_1''$ are close to those of $1b_1$ in crystalline ice and water vapor, respectively. Hence, they can be assigned to tetrahedral ($1b_1'$) and H-bond distorted ($1b_1''$) local structures, respectively. The temperature-dependent XRS spectra are shown in Figure 1.22b. Crystalline ice exhibits a strong post-edge at about 541 eV whereas gas-phase water concentrates nearly all of the intensity in the pre-edge (535 eV) and main-edge (about 537 eV) regions. Note that the spectra shown in Figure 1.22a were taken nonresonantly, that is, with a photon energy different from those corresponding to the features of Figure 1.22b. If XES spectra are taken in resonance valuable information is obtained, as shown in Figure 1.22c, where resonant XES spectra are compared with nonresonant (550 eV) XES.

Pre-edge excitation essentially eliminates the $1b_1'$ contribution, whereas excitation on the main edge gives a slight enhancement of the $1b_1''$ (distorted), and excitation on the post-edge enhances the $1b_1'$ peak. Because the absorption post-edge feature in ice is stronger than in the liquid, the resonant XES is consistent with the $1b_1'$ peak being related to tetrahedrallike species. The pre-edge peak in XRS has, in turn, been assigned to distorted H-bonding configurations. This assignment is consistent with the observed absence of $1b_1'$ and the strong enhancement of the $1b_1''$ peak when resonantly exciting on the pre-edge feature. Such assignment is consistent with the experimentally observed temperature dependence of both XRS and XES: in XRS, the post-edge decreases and the preedge increases with increasing temperature (see Figure 1.22b).

The picture that has emerged from such experiments is that liquid water seems to be organized in two structurally distinct motifs involving two- and four-neighbor H-bonded structures (Wernet et al. 2004; Nilsson and Pettersson 2011). At RT the dense strongly distorted two H-bonded configuration involving one strong donor and one strong acceptor dominates building chains or rings in a hyperdense flexible H-bond network (Maréchal 2007). The tetrahedrally coordinated molecules build small clusters (nanoicebergs) and become more relevant when temperature decreases. From the historical perspective it is interesting to refer to the work published by Röntgen (1892) where he already proposed that liquid water is a mixture of two phases: *Eismolecüle* (aggregates with tetrahedral order) and liquid.

1.4.4 COMPUTER WATER

One strategy to better understand water is to define theoretical counterparts of water molecules in a parallel computer space interacting with each other through force fields with the aim to reproduce the experimentally obtained physical properties of water as

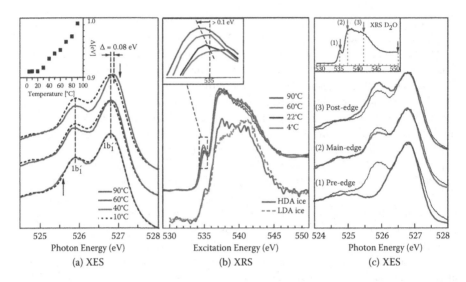

FIGURE 1.22 (a) The lone-pair $1b_1$ region of the O 1s soft XES spectra of liquid D_2O at 10, 40, 60, and 90°C using a nonresonant excitation energy of 550 eV. Spectra are normalized to the $1b_1''$ peak height. The positions of the corresponding $1b_1$ state of crystalline ice (525.6 eV) and gas-phase water (527 eV) are indicated with arrows. The two lone-pair peaks in liquid water are denoted, respectively, $1b_1'$, close to the corresponding position in crystalline ice, and $1b_1''$, close to gas-phase water. (Inset) Energy difference between the $1b_1''$ and $1b_1'$ peaks as function of temperature. (b) (Upper) XRS spectra of liquid H_2O at 4, 22, 60, and 90°C. (Inset) Magnification of the pre-edge (535 eV) spectral feature indicating a shift toward lower energy with increasing temperature. (Lower) XRS spectra of LDA and HDA ice. (c) XES spectra at various excitation energies (full lines) compared with nonresonant excitation (dashed lines) at 550 eV of D_2O water at 25°C. (Inset) XRS spectrum of D_2O with arrows marking the corresponding excitation energies. (Reprinted from C. Huang, et al. *Proc. Natl. Acad. Sci. USA* 106:15214–15218, 2009. With permission.)

well as to predict its behavior in complex or still nonexplored conditions. The simplest approximation consists of defining rigid and nonpolarizable water molecules with specific structural (lengths and angles) and interaction (charges and LJ) parameters. The electrostatic interaction is Coulombic and the dispersion and repulsion forces are represented by LJ potentials. The two-body or pairwise interaction potentials are empirical, fitted to reproduce bulk-phase experimental data using classical molecular dynamics (MD). The aim is to obtain reliable and cheap models, cheap in terms of computational time and cost, over a wide range of experimental conditions. Figure 1.23 schematizes water molecules including from three up to six interaction sites and Table 1.5 summarizes the characteristic parameters of the most popular rigid nonpolarizable water models.

The simplest three-site models (e.g., TIPS, SPC, TIP3P, and SPC/E) include the O–H distance, d_{O-H}, the H–O–H angle, $\widehat{HOH}$, and the charges at the hydrogen and oxygen atoms, q_H and q_O, respectively. The four-site models (e.g., BF, TIPS2, TIP4P,

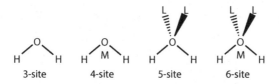

FIGURE 1.23 Simple n-site rigid nonpolarizable water models ($n = 3, 4, 5, 6$).

TIP4P-Ew, TIP4P/Ice, and TIP4P/2005) place the negative charge q_M in a point M at a distance d_{O-M} from the oxygen along the H–O–H bisector and the 5-site models (e.g., BNS, ST2, TIP5P, and TIP5P-E) place the negative charge q_L on dummy atoms (L) representing the lone-pairs of the oxygen atom, with a tetrahedral-like geometry, at distances d_{O-L}. The LJ site is usually located at the oxygen atom with parameters r_0 and E_{LJ}^0, as defined in Section 1.3. The relevance of research on water models is reflected or weighted through the fact that an article devoted to the comparison of different models such as BF, SPC, ST2, TIPS2, TIP3P, and TIP4P (Jorgensen et al. 1983) has been cited more than 11,000 times. This article is not only highly cited but is in fact the most often cited (by far) when using the term "water" as the topic.

Given the simplicity of such water models and the fact that they are fitted to particular physical parameters it is not expected that they can reproduce all known parameters such as melting temperatures, maximum in the density of water at ambient pressure, heat of vaporization, dielectric constant, self-diffusion coefficient, structure of water and ice Ih, phase diagrams, critical parameters, and the like. Most models are indeed essentially indicated for the physical properties that were used to define the empirical potentials but have a varying degree of success with the rest of the properties. However, some models have a remarkable success. This is the case of TIP4P/2005 which predicts quite nicely the orthobaric densities, critical temperature, surface tension, densities of the different solid phases of water, phase diagram, melting properties, isothermal compressibility, coefficient of thermal expansion, and the structure of water and ice, covering a temperature range from 120 to 640 K and pressures up to 30,000 bar (Vega et al. 2009). However, it underestimates the dielectric constant. As shown in Table 1.5, TIP4P/2005 is at the end a minor modification of the model proposed by Bernal and Fowler (1933), something to be underlined in order to recognize the robustness of their original model. Such water models can be improved by including bond flexibility (polarizability) although no dramatic improvements have been observed. With such inclusion the list of available models increases substantially, with about 50 registered members, some of them summarized in (Guillot 2002). The large number of available models indicates that the best model is yet to be found, or perhaps that other more involved strategies have to be sampled. We can conclude here that although water is apparently a simple molecule, it dislikes being modeled in a simple way.

An alternative more fundamental way to proceed is to perform *ab initio* (first principles) calculations, where in principle no experimental information is used (Szalewicz, Leforestier, and van der Avoird 2009). The extensively used DFT method calculates the electronic structure by including electronic correlation effects. Exchange

TABLE 1.5
Parameters of Selected Rigid Nonpolarizable Water Models

Model	d_{O-H} Å	$\widehat{HOH}$ Degrees	d_{O-M} Å	d_{O-L} Å	q_H e	q_O e	q_M e	q_L e	r_0 Å	E_{LJ}^0 K
$n = 3$										
SPC	1.0	109.47			+0.41	−0.82			3.166	78.20
SPC/E	1.0	109.47			+0.424	−0.848			3.166	78.20
TIPS	0.957	104.52			+0.40	−0.80				
TIP3P	0.957	104.52			+0.417	−0.834			3.151	76.54
$n = 4$										
BF	0.96	105.7	0.15		+0.49		−0.98			
TIPS2	0.957	104.52	0.15		+0.535		−1.07			
TIP4P	0.957	104.52	0.15		+0.52		−1.04		3.154	78.02
TIP4P-Ew	0.957	104.52	0.125		+0.524		−1.048			
TIP4P/Ice	0.957	104.52	0.158		+0.590		−1.179			
TIP4P/2005	0.957	104.52	0.155		+0.556		−1.113		3.159	93.20
$n = 5$										
BNS	1.0	109.47		1.0	+0.196			−0.196		
ST2	1.0	109.47		0.8	+0.236			−0.236		
TIP5P	0.957	104.52		0.70	+0.241			−0.241	3.120	80.51
TIP5P-E	0.957	104.52		0.70	+0.241			−0.241		
$n = 6$										
TIP6P	0.980	108.00	0.230	0.89	+0.477		−0.866	−0.044	3.115	85.98

$\widehat{HOH}$=109.47° for all 5-site models and 111.00° for TIP6P. [SPC]=(Berendsen et al. 1981), [SPC/E]=(Berendsen, Grigera, and Straatsma 1987), [TIPS]=(Jorgensen 1981), [TIP3P]=(Jorgensen et al. 1983), [BF]=(Bernal and Fowler 1933), [TIPS2]=(Jorgensen 1982), [TIP4P]=(Jorgensen et al. 1983), [TIP4P-Ew]=(Horn et al. 2004), [TIP4P/Ice]=(Abascal et al. 2005), [TIP4P/2005]=(Abascal and Vega 2005), [BNS]=[ST2]=(Stillinger and Rahman 1974), [TIP5P]=(Mahoney and Jorgensen 2000), [TIP5P-E]=(Rick 2004), [TIP6P]=(Nada and van der Eerden 2003).

correlations can be implemented through GGA functionals such as B3LYP (Lee, Yang, and Parr 1988), the most popular approximation in chemistry and introduced in Section 1.2.2, and Perdew–Burke–Ernzerhof (PBE; Perdew, Burke, and Ernzerhof 1996), indicated for extended systems. B3LYP and PBE, together with the earlier used local density approximation are usually known as the standard approximations to exchange correlations. Car and Parrinello combined DFT with MD leading to the widely used Car–Parrinello MD (CPMD) method (Car and Parrinello 1985).

Although the well-documented success of such approximations in a host of examples in physics, chemistry, and material science, when the simulated species is water, in particular in the liquid state, then some discrepancies, sometimes severe, between theory and experiment appear. Radial distribution functions and self-diffusivity of liquid water at a given temperature compare well with the experimental results but at a temperature about 20% lower. Using the B3LYP and PBE standard functionals equilibrium densities of 0.75 and 0.88 g cm^3 are obtained, respectively, thus 25% and 12% below the experimental value, respectively. Concerning ice GGAs describe the ambient pressure ice Ih phase reasonably well and predict the proton order–disorder phase transition temperatures between ice Ih and XI and ice VII and VIII (Singer et al. 2005). In this case the Ih/XI transition is predicted at 98 K, higher than the observed transition at 72 K. One reason for the discrepancies is the known problem of the standard exchange correlation functionals to correctly describe vdW forces, hence the intense activity for the search of their efficient incorporation in DFT methods (Klimeš and Michaelides 2012). As discussed in Section 1.3, the vdW dispersion or London term has a quantum-mechanical origin and can be viewed as an attractive interaction in response to instantaneous charge density fluctuations. Standard exchange correlation functionals fail to describe dispersion because instantaneous density fluctuations are not considered and, in addition, they are local. We face this situation here by exploring the case of the water dimer. Figure 1.24 compares the computed total energy of a water dimer as a function of the intermolecular distance for both (a) non H-bonded and (b) H-bonded geometries. In order to characterize the vdW interaction energy the molecules have been oriented in such a way that H-bonding is avoided (see Figure 1.24a). The DFT simulations shown here use PBE and revPBE functionals (Zhang and Yang 1998) and the vdW interactions are taken into account in the functional labeled DRSLL-revPBE (Dion et al. 2004). DRSLL-PBE uses PBE instead of revPBE (Wang et al. 2011).

Figure 1.24(a) clearly shows that vdW-based functionals exhibit minima in the total energy, which are not shown for PBE and revPBE. DRSLL-revPBE and DRSLL-PBE exhibit minima at about 3.7 and 3.4 Å, respectively, with a binding energy of 10 and 25 meV, respectively. Using (1.31) the derived vdW coefficients accounting for the attractive contribution (C) are 8.2×10^{-78} and 12.3×10^{-78} J m^6 corresponding to the 10 and 25 meV energy minima, respectively. The 12.3×10^{-78} J m^6 value compares rather well with the 13.9×10^{-78} J m^6 coefficient obtained from (1.28) and shown in Table 1.3. Although the energies are considerably weaker as compared to those obtained for the H-bond configuration (see Figure 1.24b), we have here an indication that vdW interactions can significantly contribute to the cohesion energy for example, by increasing the occupation of interstitial (non-H-bonded) sites. In the case of ice polymorphs, the contribution to the lattice energy arising from vdW forces increases

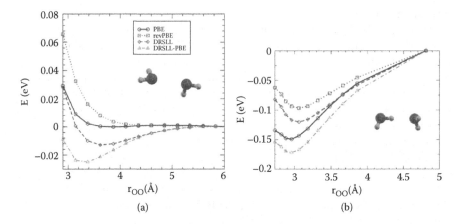

FIGURE 1.24 Total energy of the water dimer as a function of the intermolecular separation for two different molecular orientations calculated for PBE (circles), revPBE (squares), DRSLL-revPBE (diamonds), and DRSLL-PBE (triangles). (a) Non-H-bonded and (b) H-bonded configurations as shown in the insets. Energies have been shifted to have the zero at the largest separation. (Reprinted from J. Wang et al., *J. Chem. Phys.* 134:024516, 2011. American Institute of Physics. With permission.)

monotonically with applied pressure. By accounting for vdW forces with DFT, the phase transition pressures closely agree with experiments (Santra et al. 2011).

1.5 ELIXIR OF LIFE

LIFE MATRIX

Life has been defined as a self-sustained chemical system capable of undergoing Darwinian evolution (Joyce 1994) and, as we know it, it relies on liquid water. In its role as solvent it maintains and allows many crucial chemical reactions. It permits proteins to develop their specific functionalities, it induces the formation of membranes due to hydrophobic interactions (as we show in the next chapter), it regulates temperature, and so on (Ball 2008). But is life possible without water? Why not? The fact that we do not understand life without water does not prevent the existence of other forms of life based on other molecules. I recommend the book *Life as We Do Not Know It* by P. Ward (2005) as a fascinating introduction to this subject. Let us entertain here a few considerations concerning this point and see how clever Mother Nature is having made the choice of water (perhaps She had no other choice!).

In this chapter we have explored the molecular structure of the water molecule, in particular the well-known pseudotetrahedral orbital distribution. Such a distribution plays a fundamental role because the two-donor two-acceptor H-bonding configuration confers a high bonding flexibility (but no flexible bonds) that allows water to adapt to many different configurations. Adaptability is an essential property for life and water largely exhibits such a property. Proteins must be adaptable to fulfill their functions and they cannot make it without water, as discussed in Section 6.2.

The large number of polymorphs and polyamorphs, together with our difficulty in understanding the structure of liquid water, is the clearest example. If we consider ammonia, closely-related to water due to pseudotetrahedral bonding configuration, then only one lone-pair is available (it is a three-donor, one-acceptor molecule). This, together with an associated weaker H-bonding (less than 10 kJ mol^{-1}), reduces the required flexibility, which can be stated by the existence of only five known polymorphs for solid ammonia, as discussed in Section 1.4.2. However, ammonia is an interesting candidate as a solvent inasmuch as it is liquid over a wide range of temperatures (195–240 K at 1,013 hPa), a range that becomes even wider at higher pressures, but its increased ability to dissolve hydrophobic organic molecules (as compared to water) prevents its participation in the structuration of such molecules in the form of membranes. Methane-based life has been also hypothesized, in particular after the discovery of abundant liquid methane on Titan, the largest moon of Saturn (Lorenz and Mitton 2002). Titan is related to the Earth because it has a dense atmosphere (first pointed out by the Catalan astronomer Comas i Solà in 1903) essentially formed by nitrogen gas (about 1,500 hPa) that permits the stable presence of liquids on its surface, although too cold for liquid water to exist (94 K at the surface). Titan's atmosphere is hazy, formed by organic compounds originated by the photochemical action of ultraviolet sunlight on methane. Titan is considered as a benchmark to study the origin of life on Earth and may host forms of life that would consume hydrogen and acetylene exhaling methane (McKay and Smith 2005). If methanogenic forms of life are confirmed, H-bonding would lose its uniqueness label.

Life as we know it is also carbon-based. Again, the ability that enables carbon atoms to build either sp^2 or sp^3 orbitals allows it to participate in myriad configurations so necessary for life not only from the structural point of view but also from energy transfer (CO, CO_2, amino acids, hydrocarbons, fatty acids, saccharides, proteins, etc.). Thus, it seems clear that life requires chameleonic atoms/molecules able to adapt to a host of physical and chemical conditions and carbon and water indeed belong to such a select party. Concerning silicon-based life, that has been proposed given the proximity of Si and C in the periodic table, one has to admit that silicon is less flexible because it only builds sp^3 bonding, so it may not be indicated. However, silicon builds a large number of stable solids, in particular when combined with oxygen, so that it can be considered a kind of physical support for life. Other forms of life are thus not excluded, but if they exist they would hardly be rivals to the successful combination of water and carbon-based molecules.

WATER FOR LIFE

Due to the ever-increasing human population we should care for the limited water resources on Earth. By the end October 2011, we were 7 billion people, with disputes from several countries about the boy or girl deserving such a magic number. Unless serious political actions at the global level are undertaken water will become a matter of dispute and its lack will continue to be the cause of death. Nearly four of every ten people in the world have no source of safe drinking water and two in ten lack adequate sanitation. Inadequate water supply, sanitation, and hygiene are the cause of death of millions of people every year, most of them children. In December 2003, the United Nations (UN) General Assembly, in resolution A/RES/58/217, proclaimed the period

2005–2015 the International Decade for Action *Water for Life* and on July 28, 2010 recognized, in resolution A/RES/64/292, the right to safe and clean drinking water and sanitation as a human right that is essential for the full enjoyment of life and all human rights and called upon States and international organizations to provide financial resources, capacity-building and technology transfer, through international assistance and cooperation, in particular to developing countries, in order to scale up efforts to provide safe, clean, accessible and affordable drinking water and sanitation for all.

For our future we have to make rational use of water, because it is one of our most precious treasures, exploring efficient strategies to purify, filter, store, and reuse water (see Section 5.1). In developing countries the urban population will grow dramatically, generating demand well beyond the capacity of already inadequate water supply and sanitation infrastructure and services. According to the UN World Water Development Report (http://www.unesco.org), by 2050, at least one in four people is likely to live in a country affected by chronic or recurring shortages of fresh water. In developed countries wealth seems to go hand in hand with an abusive use of water. In addition, massification in naturally dry areas, irrational use of the soil (water demanding plantations), and so on are breaking the natural equilibrium. It is an extremely serious problem and we are all concerned.

1.6 SUMMARY

- The molecular orbital picture correctly describes the electronic structure of the isolated water molecule. Density–functional theory calculations of energy levels accurately reproduce the experimental photoemission results performed on water vapor.
- The two lone-pairs make water a rather unique molecule, allowing versatile interaction with other molecules through hydrogen bonding. Water molecules can adapt to many configurations leading to a large degree of complexity. Mother Nature has found in the tetrahedral coordination a chameleonic way to adapt to a broad variety of physical conditions; that's why H_2O and SiO_2 are so abundant on Earth.
- Water clusters are ideal objects that help to understand in a discrete way (molecule by molecule) the traveling from isolated water molecules to their condensed phases. One example is the cooperative effect, where the increase in monomer dipole moment increases the ability to make further hydrogen bonds.
- Solid water orders at least in 16 crystallographic phases, depending on temperature, pressure, and doping conditions. The tetrahedral configuration is preserved in all the phases and nothing prevents the discovery of new phases in the future. Three families of metastable amorphous ices have been described, with densities below, above, and well above the density of liquid water, respectively.
- The tetrahedral configuration of liquid water has been challenged as being the sole configuration based on spectroscopic observations involving timescales below 1 ps. It has been argued that in ambient conditions water molecules

prefer to arrange in chains and rings, building hyperdense flexible hydrogen-bonding networks. The linear–tetrahedral duality is still a matter of intense debate.

- Water molecules will be physisorbed on any nonrepulsive surface: the average density of gas molecules near the surface will always be larger than in the gas phase.
- Complex simplicity: the knowledge of bulk properties of water is far from complete. Many questions remain unanswered and theoretical models reproduce only part of the properties. With this situation, it becomes clear that the surface of water has to be a complex system, a point that becomes evident throughout the rest of the book.

REFERENCES

1. Abascal, J.L.F. and Vega, C. 2005. A general purpose model for the condensed phases of water: TIP4P/2005. *J. Chem. Phys.* 123:234505.
2. Abascal, J.L.F., Sanz, E., García Fernández, R., and Vega, C. 2005. A potential model for the study of ices and amorphous water: TIP4P/Ice. *J. Chem. Phys.* 122: 234511.
3. Abraham, F. F. 1978. The interfacial density profile of a Lennard-Jones fluid in contact with a (100) Lennard-Jones wall and its relationship to idealized fluid/wall systems: A Monte Carlo simulation. *J. Chem. Phys.* 68:3713–3716.
4. Agmon, N. 2012. Liquid water: From symmetry distortions to diffusive motion. *Acc. Chem. Res.* 45:63–73.
5. Albright, T.A., Burdett, J.K., and Whangbo, M.-H. 1985. *Orbital Interactions in Chemistry.* New York: John Wiley & Sons.
6. Andreas, B., Azuma, Y., Bartl, G., Becker, P., Bettin, H., Borys, M. et al. 2011. Determination of the Avogadro constant by counting the atoms in a ^{28}Si crystal. *Phys. Rev. Lett.* 106:030801.
7. Arnold, G.P., Finch, E.D., Rabideau, S.W., and Wenzel, R.G. 1968. Neutron–diffraction study of ice polymorphs. III. Ice Ic. *J. Chem. Phys.* 49:4365–4369.
8. Ashcroft, N.W. and Mermin, N.D. 1976. *Solid State Physics.* New York: Holt-Saunders International Editions.
9. Badyal, Y.S., Saboungi, M.-L., Price, D.L., Shastri, S.D., Haeffner, D.R., and Soper, A.K. 2000. Electron distribution in water. *J. Chem. Phys.* 112:9206–9208.
10. Ball, Ph. 2001. *Life's Matrix: A Biography of Water.* Berkeley: University of California Press.
11. Ball, Ph. 2008. Water as an active constituent in cell biology. *Chem. Rev.* 108:4–108.
12. Baratta, G.A., Leto, G., Spinella, F., Strazzulla, G., and Foti, G. 1991. The 3.1 microns feature in ion-irradiated water ice. *Astron. Astrophys.* 252:421–424.
13. Barnes, W.H. 1929. The crystal structure of ice between 0 degrees C and −183 degrees C. *Proc. Roy. Soc. London A* 125:670–693.
14. Benedict, W.S., Gailer, N., and Plyler, E.K. 1956. Rotation-vibration spectra of deuterated water vapor. *J. Chem. Phys.* 24:1139–1165.
15. Berendsen, H.J.C., Grigera, J.R., and Straatsma, T.P. 1987. The missing term in effective pair potentials. *J. Phys. Chem.* 91:6269–6271.
16. Berendsen, H.J.C., Postma, J.P.M., van Gunsteren, W.F., and Hermans, J. 1981. *Intermolecular Forces*, B. Pullmann (Ed.). Dordrecht: Reidel, 331.

17. Bernal, J.D. and Fowler, R.H. 1933. A theory of water and ionic solution, with particular reference to hydrogen and hydroxyl ions. *J. Chem. Phys.* 1:515–548.
18. Bernath, P. F. 2002. The spectroscopy of water vapour: Experiment, theory and applications. *Phys. Chem. Chem. Phys.* 4:1501–1509.
19. Bernstein, J. 2002. *Polymorphism in Molecular Crystals.* Oxford: Oxford Univ. Press.
20. Bini, R., Ulivi, L., Kreutz, J., and Jodl, H.J. 2000. High-pressure phases of solid nitrogen by Raman and infrared spectroscopy. *J. Chem. Phys.* 112:8522–8529.
21. Borucki, W.J., Koch, D.G., Batalha, N., Bryson, S.T., Rowe, J., Fressin, F. et al.. 2012. Kepler-22b: A 2.4 Earth-radius planet in the habitable zone of a Sun-like star. *Astrophys. J.* 745:120.
22. Bowron, D.T., Finney, J.L., Hallbrucker, A., Kohl, I., Loerting, T., Mayer, E., and Soper, A.K. 2006. The local and intermediate range structures of the five amorphous ices at 80 K and ambient pressure: A Faber-Ziman and Bathia-Thornton analysis. *J. Chem. Phys.* 125:194502.
23. Brovchenko, I. and Oleinikova, A. 2008. *Interfacial and Confined Water.* Amsterdam: Elsevier.
24. Buck, U. and Huisken, F. 2000. Infrared spectroscopy of size-selected water and methanol clusters. *Chem. Rev.* 100:3863–3890.
25. Burbridge, E.M., Burbridge, G.R., Fowler, W.A., and Hoyle, F. 1957. Synthesis of the elements in the Stars. *Rev. Mod. Phys.* 29:547–650.
26. Butt, H.S., Graf, K. and Kappel, M. 2013. *Physics and Chemistry of Interfaces.* Weinheim: Wiley-VCH.
27. Campins, H., Hargrove, K., Pinilla-Alonso, N., Howell, E.S., Kelley, M.S., and Licandro, J. et al. 2010. Water ice and organics on the surface of the asteroid 24 Themis. *Nature* 464:1320–1321.
28. Car, R. and Parrinello, M. 1985. Unified approach for molecular dynamics and density–functional theory. *Phys. Rev. Lett.* 55:2471–2474.
29. Carr, M.H., Belton, M.J.S., Chapman, C.R., Davies, M.E., Geissler, P., Greenberg, R. et al. 1998. Evidence for a subsurface ocean on Europa. *Nature* 391:363–365.
30. Charbonneau, D., Berta, Z.K., Irwin, J., Burke, C.J., Nutzman, P., Buchhave, L.A. et al. 2009. A super–Earth transiting a nearby low–mass star. *Nature* 462:891–894.
31. Clough, S.A., Beers, Y., Klein, G.P., and Rothman, R.S. 1973. Dipole moment of water from Stark measurements of H_2O, HDO and D_2O. *J. Chem. Phys.* 59:2254–2259.
32. Cohen, R.C. and Saykally, R.J. 1992. Vibration–rotation–tunneling spectroscopy of the van der Waals bond: A new look at intermolecular forces. *J. Phys. Chem.* 96:1024–1040.
33. Colaprete, A., Schultz, P., Heldmann, J., Wooden, D., Shirley, M., Ennico, K., et al.. 2010. Detection of water in the LCROSS ejecta plume. *Science* 330:463–468.
34. Conde, M.M., Vega, C., and Patrykiejew, A. 2008. The thickness of a liquid layer on the free surface of ice as obtained from computer simulation. *J. Chem. Phys.* 129:014702.
35. Cotton, F.A. 1990. *Chemical Applications of Group Theory.* New York: John Wiley and Sons.
36. Coulson, C.A. and Eisenberg, D. 1966. Interactions of H_2O molecules in ice. I. The dipole moment of an H_2O molecule in ice. *Proc. R. Soc. London Ser. A* 291:445–453.
37. Debenedetti, P.G. and Stillinger, F.H. 2001. Supercooled liquids and the glass transition. *Nature* 401:259–267.
38. Dion, M., Rydberg, H., Schröder, E., Langreth, D.C., and Lundqvist, B.I. 2004. Van der Waals density functional for general geometries. *Phys. Rev. Lett.* 92:246401.
39. Du, Q., Superfine, R., Freysz, E., and Shen, Y.R., 1993. Vibrational spectroscopy of water at the vapor/water interface. *Phys. Rev. Lett.* 70:2313–2316.
40. Dyke, T.R., Mack, K.M., and Muenter, J.S. 1977. The structure of water dimer from molecular beam electric resonance spectroscopy. *J. Chem. Phys.* 66:498–510.

41. Eisenberg, D. and Kauzmann, W. 1969. *The Structure and Properties of Water*. London: Oxford University Press.

42. Ellison, F.O. and Shull, H. 1955. Molecular calculations. I. LCAO MO self-consistent field treatment of the ground state of H_2O. *J. Chem. Phys.* 23:2348–2357.

43. Engelhardt, H. and Kamb, B. 1981. Structure of ice IV, a metastable high-pressure phase. *J. Chem. Phys.* 75:5887–5899.

44. Errington, J.R. and Debenedetti, P.G. 2001. Relationship between structural order and the anomalies of liquid water. *Nature* 409:318–321.

45. Fifer, R.A. and Schiffer, J. 1970. Intramolecular interactions in the water molecule: The strength-stretch interaction force constant of water molecules in hydrogen-bonded systems. *J. Chem. Phys.* 52:2664–2670.

46. Flubacher, P., Leadbetter, A.J., and Morrison, J.A. 1960. Heat capacity of ice at low temperatures. *J. Chem. Phys.* 33:1751–1755.

47. Fraley, P.E. and Rao, K.N. 1969. High resolution infrared spectra of water vapor ν_1 and ν_3 bands of $H_2^{16}O$. *J. Mol. Spectrosc.* 29:348–364.

48. Frank, H.S., and Wen, W.–Y. 1957. Structural aspects of ion–solvent interaction in aqueous solutions: A suggested picture of water structure. *Discuss. Faraday Soc.* 24:133–140.

49. Franks, F. 2000. *Water: A Matrix of Life*. Cambridge: The Royal Society of Chemistry.

50. Fraxedas, J. 2006. *Molecular Organic Materials*. Cambridge, UK: Cambridge University Press.

51. Giauque, W.F. and Stout, J.W. 1936. The entropy of water and the third law of thermodynamics. The heat capacity of ice from 15 to 273K. *J. Am. Chem. Soc.* 58:1144–1150.

52. Glebov, A., Graham, A.P., Menzel, A., Toennies, J.P., and Senet, P. 2000. A helium atom scattering study of the structure and phonon dynamics of the ice surface. *J. Chem. Phys.* 112:11011–11022.

53. Gorelli, F.A., Santoro, M., Ulivi, L., and Hanfland, M. 2002. Crystal structure of solid oxygen at high pressure and low temperature. *Phys. Rev. B* 65:172106.

54. Goto, A., Hondoh, T., and Mae, S. 1990. The electron density distribution in ice Ih determined by single–crystal X-ray diffractometry. *J. Chem. Phys.* 93:1412–1417.

55. Grechko, M., Maksyutenko, P., Rizzo, T.R., and Boyarkin, O.V. 2010. Feshbach resonances in the water molecule revealed by state-selective spectroscopy. *J. Chem. Phys.* 133:081103.

56. Gregory, J.K., Clary, D.C., Lin, K., Brown, M.G., and Saykally, R.J. 1997. The water dipole moment in water clusters. *Science* 275:814–817.

57. Gregoryanz, E., Goncharov, A.F., Hemley, R.J., Mao, H., Somayazulu, M., and Shen, G. 2002. Raman, infrared, and x-ray evidence for new phases of nitrogen at high pressures and temperatures. *Phys. Rev. B* 66:224108.

58. Gubskaya, A.V. and Kusalik, P.G. 2002. The total molecular dipole moment for liquid water. *J. Chem. Phys.* 117:5290–5302.

59. Guillot, B. 2002. A reappraisal of what we have learnt during three decades of computer simulations on water. *J. Mol. Liq.* 101:219–260.

60. Guo, J. and Luo, Y. 2010. Molecular structure in water and solutions studied by photon-in/photon-out soft X-ray spectroscopy. *J. Electron Spectrosc. Related Phenom.* 177:181–191.

61. Hahn, P.H., Schmidt, W.G., Seino, K., Preuss, M., Bechstedt, F., and Bernholc, J. 2005. Optical absorption of water: Coulomb effects versus hydrogen bonding. *Phys. Rev. Lett.* 94:037404.

62. Hallbrucker, A., Mayer, E., and Johari, G.P. 1989. The heat capacity and glass transition of hyperquenched glassy water. *Phil. Mag. B* 60:179–187.

63. Hartogh, P., Lis, D.C., Bockelée-Morvan, D., de Val-Borro, M., Biver, N., Küppers, M. et al. 2011. Ocean-like water in the Jupiter-family comet 103P/Hartley 2. *Nature* 478:218–220.

64. Hemley, R.J., Jephcoat, A.P., Mao, H.K., Zha, C.S., Finger, L.W., and Cox, D.E. 1987. Static compression of H_2O-ice to 128 GPa (1.28 Mbar). *Nature* 330:737–740.

65. Hoffmann, R. 1963. An extended Hückel theory. I. Hydrocarbons. *J. Chem. Phys.* 39:1397–1412.

66. Hoffmann, R. 1987. How chemistry and physics meet in the solid state. *Angew. Chem. Int. Ed. Engl.* 26:846–878.

67. Hoffmann, R. 1988. A chemical and theoretical way to look at bonding on surfaces. *Rev. Mod. Phys.* 60:601–628.

68. Horn, H.W., Swope, W.C., Pitera, J.W., Madura, J.D., Dick, T.J., Hura, G.L., and Head–Gordon, T. 2004. Development of an improved four-site water model for biomolecular simulations: TIP4P-Ew. *J. Chem. Phys.* 120:9665–9678.

69. Huang, C., Wikfeldt, K.T., Tokushima, T., Nordlund, D., Harada, Y., Bergmann, U., et al. 2009. The inhomogeneous structure of water at ambient conditions. *Proc. Natl. Acad. Sci. USA* 106:15214–15218.

70. Israelachvili, J. 1991. *Intermolecular & Surface Forces*. San Diego, CA: Academic Press.

71. Israelachvili, J. 2011. *Intermolecular & Surface Forces*. Amsterdam: Elsevier.

72. Jackson, S.M. and Whitworth, R.W. 1997. Thermally-stimulated depolarization studies of the ice XI-ice Ih phase transition. *J. Phys. Chem. B* 101:6177–6179.

73. Johari, G.P. 1981. The spectrum of ice. *Contemp. Phys.* 22:613–642.

74. Jorgensen, J.D. and Worlton, T.G. 1985. Disordered structure of D_2O ice VII from in situ neutron powder diffraction. *J. Chem. Phys.* 83:329–333.

75. Jorgensen, W.L. 1981. Quantum and statistical mechanical studies of liquids. 10. Transferable intermolecular potential functions for water, alcohols, and ethers. Application to liquid water. *J. Am. Chem. Soc.* 103:335–340.

76. Jorgensen, W.L. 1982. Revised TIPS for simulations of liquid water and aqueous solutions. *J. Chem. Phys.* 77:4156–4163.

77. Jorgensen, W.L., Chandrasekhar, J., Madura, J.D., Impey R.W., and Klein, M.L. 1983. Comparison of simple potential functions for simulating liquid water. *J. Chem. Phys.* 79:926–935.

78. Joyce, G.F. 1994. *Origins of Life: The Central Concepts*, D.W. Deamer and G.R. Fleischacker (Eds.). Boston: Jones and Bartlett.

79. Kamb, B. 1964. Ice II: A proton-ordered form of ice. *Acta Cryst.* 17:1437–1449.

80. Kamb, B. 1965. Structure of ice VI. *Science* 150:205–209.

81. Kamb, B., Prakash, A., and Knobler, C. 1967. Structure of ice V. *Acta Cryst.* 22:706–715.

82. Karlsson, L., Mattsson, L., Jadrny, R., Albridge, R.G., Pinchas, S., Bergmark, T., and Siegbahn, K. 1975. Isotopic and vibronic coupling effects in the valence electron spectra of $H_2^{16}O$, $H_2^{18}O$ and $D_2^{16}O$. *J. Chem. Phys.* 62:4745–4752.

83. Kennedy, D. and Norman, C. 2005. What don't we know? *Science* 309:75–102.

84. Klimeš, J. and Michaelides, A. 2012. Perspective: Advances and challenges in treating van der Waals dispersion forces in density functional theory. *J. Chem. Phys.* 137:120901.

85. Kouchi, A. and Kuroda, T. 1990. Amorphization of cubic ice by ultraviolet irradiation. *Nature* 344:134–135.

86. Kuhs, W.F. and Lehmann M.S. 1986. The structure of ice-Ih. *Water Sci. Rev.* 2:1–65.

87. Kuhs, W.F., Bliss, D.V., and Finney, J.L. 1987. High-resolution neutron powder diffraction study of ice Ic. *J. Phys. Colloque* 48:C1-631-C1-636.

88. Kuhs, W.F., Finney, J.L., Vettier, C., and Bliss, D.V. 1984. Structure and hydrogen ordering in ices VI, VII and VIII by neutron powder diffraction. *J. Chem. Phys.* 81:3612–3623.

89. Langmuir, I. 1950. Control of precipitation from cumulus clouds by various seeding techniques. *Science* 112:35–41.

90. Latimer, W.M. and Rodebush, W.H. 1920. Polarity and ionization from the standpoint of the Lewis theory of valence. *J. Am. Chem. Soc.* 42:1419–1433.

91. Leadbetter, A.J. 1965. The thermodynamic and vibrational properties of H_2O ice and D_2O ice. *Proc. R. Soc. London A* 287:403–425.

92. Leadbetter, A.J., Ward, R.C., Clark, J.W., Tucker, P.A., Matsuo, T., and Suga, H. 1985. The equilibrium low–temperature structure of ice. *J. Chem. Phys.* 82:424–428.

93. Lee, C., Yang, W., and Parr, R.G. 1988. Development of the Colle-Salvetti correlation-energy formula into a functional of the electron density. *Phys. Rev. B* 37:785–789.

94. Levine, I.N. 2008. *Quantum Chemistry*. Upper Saddle River, NJ: Prentice Hall.

95. Lin, K., Brown, M.G., Cruzan, J.D., and Saykally, R.J. 1996. Vibration–rotation–tunneling spectra of the water pentamer: Structure and dynamics. *Science* 271:62–64.

96. Line, C.M.B. and Whitworth, R.W. 1996. A high resolution neutron powder diffraction study of D_2O ice XI. *J. Chem. Phys.* 104:10008–10013.

97. Lobban, C., Finney, J.L., and Kuhs, W.F. 1998. The structure of a new phase of ice. *Nature* 391:268–270.

98. Lobban, C., Finney, J.L., and Kuhs, W.F. 2000. The structure and ordering of ices III and V. *J. Chem. Phys.* 112:7169–7180.

99. Loerting, T., Winkel, K., Seidl, M., Bauer, M., Mitterdorfer, C., Handle, P.H. et al. 2011. How many amorphous ices are there? *Phys. Chem. Chem. Phys.* 13:8783–8794.

100. Londono, J.D., Kuhs, W.F., and Finney, J.L. 1993. Neutron diffraction studies of ices III and IX on under-pressure and recovered samples. *J. Chem. Phys.* 98:4878–4888.

101. Lorenz, R. and Mitton, J. 2002. *Lifting Titan's Veil*. Cambridge, UK: Cambridge University Press.

102. Loubeyre, P., LeToullec, R., Wolanin, E., Hanfland, M., and Hausermann, D. 1999. Modulated phases and proton centring in ice observed by X-ray diffraction up to 170 GPa. *Nature* 397:503–506.

103. Ludwig, R. 2001. Wasser: von Clustern in die Flüssigkeit. *Angew. Chem.* 113:1856–1876.

104. Luisi, P.L. 2006. *The Emergence of Life, from Chemical Origins to Synthetic Biology*. Cambridge, UK: Cambridge University Press.

105. Lundholm, M., Siegbahn, H., Holmberg, S., and Arbman, M. 1986. Core electron spectroscopy of water solutions. *J. Electron Spectrosc. Related Phenom.* 40:163–180.

106. Lynden-Bell, R.M., Morris, S.C., Barrow, J.D., Finney, J.L., and Harper, C.L. 2010. *Water and Life*. Boca Raton, FL: CRC Press.

107. Madelung, O. 1978. *Introduction to Solid-State Theory*, M. Cardona, P. Fulde, and H.J. Queisser (Eds.). Berlin: Springer-Verlag.

108. Mahoney M.W. and Jorgensen, W.L. 2000. A five-site model for liquid water and the reproduction of the density anomaly by rigid, nonpolarizable potential functions. *J. Chem. Phys.* 112:8910–8922.

109. Mao, H. and Hemley, R. J. 1994. Ultra-high pressure transitions in solid hydrogen. *Rev. Mod. Phys.* 66:671–692.

110. Maréchal, Y. 2007. *The Hydrogen Bonding and the Water Molecule*. Amsterdam: Elsevier.

111. Martin, R.M. 2004. *Electronic Structure: Basic Theory and Practical Methods*. Cambridge, UK: Cambridge University Press.

112. Mastrapa, R.M.E. and Brown, R.H. 2006. Ion irradiation of crystalline H_2O-ice: Effect on the 1.65-μm band. *Icarus* 183:207–214.

113. McGlynn, S.P., Vanquickenborne, L.G., Kinoshita, M., and Carroll, D.G. 1972. *Introduction to Applied Quantum Chemistry*. New York: Holt, Rinehart and Winston.

114. McKay, C.P. and Smith, H.D. 2005. Possibilities for methanogenic life in liquid methane on the surface of Titan. *Icarus* 178:274–276.

115. Miura, N., Yamada, H., and Moon, A. 2010. Intermolecular vibrational study in liquid water and ice by using far infrared spectroscopy with synchrotron radiation of MIRRORCLE 20. *Spectrochim. Acta* A 77:1048–1053.

116. Mota, R., Parafita, R., Giuliani, A., Hubin–Franskin, M.J., Lourenco, J.M.C., Garcia, G. et al. 2005. Water VUV electronic state spectroscopy by synchrotron radiation. *Chem. Phys. Lett.* 416:152–159.

117. Mottl, M.J., Glazer, B.T., Kaiser, R.I., and Meech, K.J. 2007. Water and astrobiology. *Chemie der Erde* 67:253–282.

118. Murphy, W.F. 1977. The Rayleigh depolarization ratio and rotational Raman spectrum of water vapor and the polarizability components for the water molecule. *J. Chem. Phys.* 67:5877–5882.

119. Nada, H. and van der Eerden, J.P.J.M. 2003. An intermolecular potential model for the simulation of ice and water near the melting point: A six-site model of H_2O. *J. Chem. Phys.* 118:7401–7413.

120. Nilsson, A. and Pettersson, L.G.M. 2011. Perspective on the structure of liquid water. *Chem. Phys.* 389:1–34.

121. Ninet, S. and Datchi, F. 2008. High pressure-high temperature phase diagram of ammonia. *J. Chem. Phys.* 128:154508.

122. Nordlund, D., Odelius, M., Bluhm, H., Ogasawara, H., Pettersson, L.G.M., and Nilsson, A. 2008. Electronic structure effects in liquid water studied by photoelectron spectroscopy and density functional theory. *Chem. Phys. Lett.* 460:86–92.

123. Pamuk, B., Soler, J.M., Ramírez, R., Herrero, C.P., Stephens, P.W., Allen, P.B., and Fernández-Serra, M.V. 2012. Anomalous nuclear quantum effects in ice. *Phys. Rev. Lett.* 108:193003.

124. Pauling, L. 1935. The structure and entropy of ice and of other crystals with some randomness of atomic arrangement. *J. Am. Chem. Soc.* 57:2680–2684.

125. Perdew, J., Burke, K., and Ernzerhof, M. 1996. Generalized gradient approximation made simple. *Phys. Rev. Lett.* 77:3865–3868.

126. Peterson, S.W. and Levy, H.A. 1957. A single-crystal neutron diffraction of heavy ice. *Acta Cryst.* 10:70–76.

127. Petrenko, V.F. and Whitworth, R.W. 2006. *Physics of Ice*. Oxford, UK: Oxford University Press.

128. Prendergast, D., Grossman, J.C., and Galli, G. 2005. The electronic structure of liquid water within density–functional theory. *J. Chem. Phys.* 123:014501.

129. Rick, S.W. 2004. A reoptimization of the five-site water potential (TIP5P) for use with Ewald sums. *J. Chem. Phys.* 120:6085–6093.

130. Rivkin, A.S. and Emery, J.P. 2010. Detection of ice and organics on an asteroidal surface. *Nature* 464:1322–1323.

131. Röntgen, W.C. 1892. Ueber die Constitution des flüssigen Wassers. *Ann. Phys. Chem.* 45:91–97.

132. Röttger, K., Endriss, A., Ihringer, J., Doyle, S., and Kuhs, W.F. 1994. Lattice constants and thermal expansion of H_2O and D_2O ice Ih between 10 and 265 K. *Acta Cryst.* B50:644–648.

133. Röttger, K., Endriss, A., Ihringer, J., Doyle, S., and Kuhs, W.F. 2012. Lattice constants and thermal expansion of H_2O and D_2O ice Ih between 10 and 265 K. Addendum. *Acta Cryst.* B 68:91.

134. Salzmann, C.G., Radaelli, P.G., Hallbrucker, A., Mayer, E., and Finney, J.L. 2006. The preparation and structures of hydrogen ordered phases of ice. *Science* 311:1758–1761.

135. Salzmann, C.G., Radaelli, P.G., Mayer, E., and Finney, J.L. 2009. Ice XV: A new thermodynamically stable phase of ice. *Phys. Rev. Lett.* 103:105701.

136. Salzmann, C.G., Radaelli, P.G., Slater, B., and Finney, J.L. 2011. The polymorphism of ice: Five unresolved questions. *Phys. Chem. Chem. Phys.* 13:18468–18480.

137. Santra, B., Klimeš, J., Alfè, D., Tkatchenko, A., Slater, B., Michaelides, A., Car, R., and Scheffler, M. 2011. Hydrogen bonds and van der Waals forces in ice at ambient and high pressures. *Phys. Rev. Lett.* 107:185701.

138. Schaefer, V.I. 1946. The production of ice crystals in a cloud of supercooled water droplets. *Science* 104:457–459.

139. Seager, S. and Deming, D. 2010. Exoplanet atmospheres. *Ann. Rev. Astron. Astrophys.* 48:631–672.

140. Siegbahn, K. 1974. Electron spectroscopy, an outlook. *J. Electron. Spectrosc. Relat. Phenom.* 5:3–97.

141. Silvestrelli, P.L. and Parrinello, M. 1999. Structural, electronic and bonding properties of liquid water from first principles. *J. Chem. Phys.* 111:3572–3580.

142. Singer, S.J., Kuo, J.-L., Hirsch, T.K., Knight, C., Ojamäe, L., and Klein, M.L. 2005. Hydrogen–bond topology and the ice VII/VIII and ice Ih/XI proton-ordering phase transitions. *Phys. Rev. Lett.* 94:135701.

143. Smith, P.H., Tamppari, L.K., Arvidson, R.E., Bass, D., Blaney, D., Boynton, W.V., et al. 2009. H_2O at the Phoenix Landing Site. *Science* 325:58–61.

144. Soper, A.K. 2007. Joint structure refinement of X-ray and neutron diffraction data on disordered materials: Application to liquid water. *J. Phys. Condens. Matter* 19:335206.

145. Spanu, L., Donadio, D., Hohl, D., and Galli, G. 2009. Theoretical investigation of methane under pressure. *J. Chem. Phys.* 130:164520.

146. Stephens, P.J., Devlin, F.J., Chabalowski, C.F., and Frisch, M.J. 1994. Ab initio calculation of vibrational absorption and circular dichroism spectra using density functional force fields. *J. Phys. Chem.* 98:11623–11627.

147. Stillinger, F.H. and Rahman, A. 1974. Improved simulation of liquid water by molecular dynamics. *J. Chem. Phys.* 60:1545–1557.

148. Szalewicz, K., Leforestier, C., and van der Avoird, A. 2009. Towards the complete understanding of water by a first–principles computational approach. *Chem. Phys. Lett.* 482:1–14.

149. Tinetti, G., Vidal–Madjar, A., Liang, M.–C., Beaulieu, J.–P., Yung, Y., Carey, S. et al. 2007. Water vapour in the atmosphere of a transiting extrasolar planet. *Nature* 448:169–171.

150. Toth, R.A. 1998. Water vapor measurements between 590 and 2582 cm^{-1}: Line positions and strengths. *J. Mol. Spectrosc.* 190:379–396.

151. Tsiper, E.V. 2005. Polarization forces in water deduced from single molecule data. *Phys. Rev. Lett.* 94:013204.

152. Vega, C., Abascal, J.L.F., Conde M.M., and Aragonés, J.L. 2009. What ice can teach us about water interactions: A critical comparison of the performance of different water models. *Faraday Discuss.* 141:251–276.

153. Wang, J., Román-Pérez, G., Soler, J.M., Artacho, E. and Fernández–Serra, M.V. 2011. Density, structure, and dynamics of water: The effect of van der Waals interactions. *J. Chem. Phys.* 134:024516.

154. Ward, P. 2005. *Life as We Do Not Know It*. New York: Penguin Books.

155. Wernet, P., Nordlund, D., Bergmann, U., Cavalleri, M., Odelius, M., Ogasawara, H. et al. 2004. The structure of the first coordination shell in liquid water. *Science* 304:995–999.

156. Whalley, E., Klug, D.D., Floriano, M.A., Svensson, E.C., and Sears, V.F. 1987. Recent work on high–density amorphous ice. *J. Physique* 48:C1–429–C1–434.

157. Winter, B., Weber, R., Widdra, W., Dittmar, M., Faubel, M., and Hertel, I.V. 2004. Full valence band photoemission from liquid water using EUV synchrotron radiation. *J. Phys. Chem. A* 108:2625–2632.

158. Xantheas, S.S., 1995. *Ab initio* studies of cyclic water clusters $(H_2O)_n$, $n = 1 - 6$. III. Comparison of density functional with MP2 results. *J. Chem. Phys.* 102:4505–4517.

159. Xantheas, S.S. and Dunning Jr., T.H. 1993. *Ab initio* studies of cyclic water clusters $(H_2O)_n$, $n = 1 - 6$. I. Optimal structures and vibrational spectra. *J. Chem. Phys.* 99:8774–8792.

160. Yoshimura, Y., Stewart, S.T., Somayazulu, M., Mao, H., and Hemley, R.J. 2006. High-pressure x-ray diffraction and Raman spectroscopy of ice VIII. *J. Chem. Phys.* 124:024502.

161. Zhang, Y. and Yang, W. 1998. Comment on generalized gradient approximation made simple. *Phys. Rev. Lett.* 80:890–890.

162. Zheligovskaya, E.A. and Malenkov, G.G. 2006. Crystalline water ices. *Russ. Chem. Rev.* 75:57–76.

2 Interfaces of Condensed Pure Water

He had come up so fast and absolutely without caution
that he broke the surface of the blue water and was in the sun
E. Hemingway, The Old Man and the Sea

In this chapter the main interfaces involving pristine (pure and ion-free) condensed water (liquid and solid) are explored. Among the many interfaces that liquid water can build, two of them are of particular interest. One is the extremely important water vapor/liquid water interface. Needless to say how important it is inasmuch as it accounts for more than 70% of the Earth's surface and we experience this interface every single day of our lives. The other interesting interface involves organic molecules. In the case of amphiphilic molecules (with both hydrophilic and hydrophobic groups), water is able to organize such molecules in a well-defined structure, acting as a structuring media. Such ability enables the formation of micelles and biomembranes, so it is easily understandable how important they are for cellular life. On the other hand, they are the basis of cleaning agents such as detergents. We also show that liquid water can be undercooled (remain liquid below 0°C) on solid water (ice), and the presence of such a layer has important consequences for the low friction of ice as well as for the depletion of the ozone layer. We also discuss the amazing tendency of the surface of ice to become ordered at very low temperatures, overcoming proton disorder.

2.1 LIQUID WATER

In the absence of external perturbations, such as mechanical and acoustical vibrations, liquid water confined in a beaker or container with sufficiently large dimensions in ambient conditions exhibits nearly perfect flat surfaces, away from the walls of the container. The term nearly reflects the fact that water molecules are in permanent motion at the molecular level producing capillary waves at the surface with amplitudes of the order of the mean molecular diameter ($\sim$0.3 nm). The measured roughness is less than such a value, as determined by X-ray reflectivity experiments (Braslau et al. 1985). The intermolecular cohesive forces are of the vdW and H-bonding type, as previously discussed in Chapter 1, and we are ignoring now the liquid/walls interfaces but must consider the effects of gravity and of the atmospheric pressure because we are considering that the whole system is in equilibrium.

If an external vertical periodic oscillation is applied to the container then standing waves will be generated on its free surface. Such waves are known as Faraday waves in honor of M. Faraday (1831), who first studied them in a systematic way. They are also known under the more general term parametrically driven surface waves

(Cross and Hohenberg 1993). Patterns of various symmetries have been observed in the literature and can be relatively easily produced in the laboratory. Depending on the driving frequency of the oscillation and on the viscosity, surface tension, and density of the fluid, parallel stripes, square and hexagonal patterns, as well as quasiperiodic patterns are generated and the chaotic regime can be entered depending on the experimental conditions illustrating the intrinsic nonlinear character. Faraday observed that the frequency of the waves was half the excitation frequency. This subharmonic response to vertical oscillations is also found in pendulums when the pivots are vertically and periodically excited (Benjamin and Ursell 1954).

In zero gravity conditions, inside Space Shuttles or the International Space Station, for example, but liberated from any container, water molecules are subject only to their own intermolecular cohesive forces, forming spheres because such a geometrical body maximizes the volume-to-surface ratio (in outer space molecules would spread in all directions due to the absence of atmosphere). We have often seen videos of astronauts observing, playing with, and drinking such oscillating water spheres, which are nothing else but huge droplets, wandering within the spacecraft. Some examples can be found on the Internet (e.g., at http://www.nasa.gov or http://www.esa.int). Because the absence of gravity on Earth can be simulated by freefall, it becomes understandable why water forms drops when falling. In this state, water can be considered as weightless, adopting the shape of a ball distorted by the resistance of air. The mechanical stability of the droplets is due to the rather strong H-bonding at the surface, building a kind of structured shield, which makes a remarkable difference when compared to other liquids. We discuss this point below, in terms of surface tension. However, when the liquid approaches a surface still in the absence of gravity, additional interaction channels become available. If such forces are strong enough, the liquid will spread out over the surface, a phenomenon known as wetting, which is discussed in the following chapters. In the case of a bottle half-filled with water in a zero gravity environment water will spread out over the internal walls of the bottle leaving the center empty.

Extrapolating bulk properties to the surface of any material, liquid or solid, is not straightforward although reasonable as a first approximation. This is clearly observed for ordered solids, where the truncation of the 3D periodicity at the surface leads to surface reorganization in terms of surface reconstruction and relaxation with different physical properties from those of the bulk. However, for molecular organic materials, the molecular character usually implies weak intermolecular interactions so that such a restructuration is only rarely observed. Thus, for liquid water, with no long-range order, one should expect the surface to be essentially similar to the bulk but do not forget that the H-bond network is interrupted and that liquid water can be very eccentric.

2.1.1 VAPOR/LIQUID INTERFACES

Interface Profile

In general, when we think of an interface we have the tendency to visualize it in an idealized way as being sharp and abrupt, mathematically described by a Heaviside step function, which is 0 for $z < 0$ (vapor side) and 1 for $z \geq 0$ (liquid side), where z stands for the direction perpendicular to such interface. This is quite the case for a host of solid inorganic/inorganic and organic/inorganic interfaces but when considering liquids with their vapors, such an approximation is too simplistic, although essentially correct, due to their intrinsic dynamic character. Figure 2.1 shows the (sigmoidal-shaped) calculated density profile $\rho(z)$ of the vapor/liquid interface represented by the expression:

$$\rho(z) = \rho(z_G) \left\{ 1 + \tanh \frac{z - z_G}{\delta} \right\} \tag{2.1}$$

where z_G denotes the position of the Gibbs dividing surface, the location where the average density is one half of the bulk density, and δ stands for a thickness parameter. Here we arbitrarily assume that $z_G = 0$ for simplicity.

If $\delta \to 0$, $\rho(z)$ transforms into the Heaviside function. As already mentioned, the water molecules are in permanent motion wandering without fixed positions and this is enhanced at the interface because there is a dynamical exchange of molecules with those in the vapor phase. Thus, for $z < z_G$ the molecules are less bound resulting in a rough profile and for $z > z_G$ the molecules are prisoners of the *diktat* imposed by vdW and H-bonding interactions until they make their way to the vapor phase. This will have important consequences for the charge distribution at the interface. The δ parameter is related to the width of the interface, which is usually defined as the spatial region accounting for 10 to 90% of the bulk density, which results in 2.2δ. In the example given in Figure 2.1 such interfacial width is 3.83 Å, that is, of the order of the mean water molecular diameter, as pointed out earlier (Braslau et al. 1985), thus, not absolutely sharp but quite narrow.

This interface has been characterized by different experimental techniques but the surface-sensitive nonlinear spectroscopic technique called sum-frequency generation (SFG) deserves special attention. SFG is very powerful and versatile and is based on the different structural symmetries of the surface and the bulk. Two laser beams, one fixed in the visible range and a second one tunable in the infrared range, are

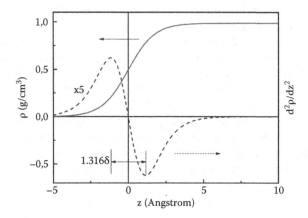

FIGURE 2.1 Density profile of the water vapor/liquid interface (continuous grey line) represented by the hyperbolic tangent function (2.1) with z_G arbitrarily set to zero, $\rho(z_G) = 0.493$ g cm^{-3}, and $\delta = 1.744$ Å from Fan et al. (2009) obtained using the SPC/E model. The second derivative of the density profile (discontinuous line) has been multiplied by a factor 5.

focused on an interface and the sum of the frequencies is detected from the output reflected beam. The sum frequency is resonance enhanced when a vibrational mode of the interfacial molecules matches the input frequency. The sum frequency is in the visible region thus the detection of the vibrational resonances becomes easier. The SFG vibrational spectrum in the OH stretch region shows a characteristic narrow feature at $\simeq 3,700$ cm^{-1}, which has been associated with the free OH dangling bonds pointing out of the liquid by comparison with the values obtained for small clusters (see Figure 1.11d), which amounts to about one quarter of the total molecules (Du et al. 1993). Using SFG spectroscopy with isotopic dilution by means of different D$_2$O:HOD:H$_2$O mixtures, it has been possible to discriminate between the free OD stretching modes of D$_2$O and HOD, with a shift of 17 cm^{-1} (2,745 and 2,728 cm^{-1} for D$_2$O and HOD, respectively; Stiopkin et al. 2011). Additional SFG experiments combined with MD simulations incorporating nuclear quantum effects show that the OH-bond orientation at the H$_2$O/vapor interface is similar to the OD-bond orientation at the D$_2$O/vapor interface. However, the OH-and-OD bonds have distinct orientations at the HDO/vapor interface: OD-bonds tend to orient toward the bulk phase and OH-bonds tend to orient toward the vapor phase (Nagata et al. 2012).

Other broader features in the 3,000–3,600 cm^{-1} region account for surface molecules with their saturated H-bonds facing the liquid (free lone-pairs dangling bonds), in roughly the same amount, and fully coordinated subsurface molecules (Ji et al. 2008). *Ab initio* CPMD calculations point toward the existence of such single-donor molecules (one dangling OH) as well as acceptor-only (two dangling OH) molecules at the interface (Kuo and Mundy 2004) although no signature for a significant amount of acceptor-only configurations could be found in a different study also based on CPMD simulations (Kühne et al. 2011). In addition, the calculations from Kuo and Mundy (2004) reproduce the substantial surface relaxation at the vapor/liquid interface, about 6% expansion of the intermolecular O–O distance, experimentally observed with extended X-ray absorption fine structure (EXAFS) spectroscopy using liquid microjets (Wilson et al. 2002).

The presence of dangling bonds, that is, unsaturated H-bonds, inevitably drives our attention toward the charge distribution near the surface of a system that is neutral inasmuch as we are considering the case of neat water, that is, with no room for dissociated molecules or ions. According to theoretical calculations using different water models, for $z < z_G$ (density below 50% of bulk density) the water dipoles point, on average, toward the vapor phase whereas for $z > z_G$ (density above 50% of bulk density) the opposite situation is found (Sedlmeier et al. 2008; Fan et al. 2009). This implies that, within 3–4 Å, the interface splits into two regions with opposite dipolar distribution along z building a double layer. The distribution of the mean cosine of the angle α between the water dipole moment and the normal to the vapor/liquid interface can be roughly approximated by the second derivative of (2.1):

$$\langle \cos \alpha \rangle \approx \frac{2\rho(z_G)}{\delta} \frac{\sinh z/\delta}{\cosh^3 z/\delta} \tag{2.2}$$

assuming $z_G = 0$ for simplicity where $\langle \cos \alpha \rangle$ stands for the average cosine of α. This equation is represented in Figure 2.1 with a discontinuous line. In spite of the crudeness of the approximation, (2.2) reveals the fact that the spatial dipolar separation is a consequence of the finite width of the interface. If we consider the ideal truncated liquid within the Bernal–Fowler model (maximizing the number of H-bonds), then one would expect an equal distribution of dangling OHs and lone-pairs at the surface, because three out of four bonds per molecule would be formed with the underlying molecules.

Surface Charge State

An immediate consequence of the interfacial dipolar distribution is the drop of the surface potential χ_s at the vapor/liquid interface. χ_s is generally defined as the difference between the liquid-phase inner potential and the vapor-phase outer potential. The outer potential can be determined from the work required to bring an unperturbing unit charge from infinity to a point just outside the vapor/liquid interface whereas the inner potential from the work required to bring an unperturbing unit charge from infinity through the vapor/liquid interface into the bulk liquid. MD simulations, both using different water models as well as of *ab initio* nature, consistently predict $\chi_s < 0$ with values ranging from few tens to several hundreds mV (Wilson, Pohorille, and Pratt 1988; Sedlmeier et al. 2008; Kathmann, Kuo, and Mundy 2008). However, simulations using *ab initio* and classical MD and DFT have been interpreted in terms of the surface being dominated by hydronium ions, thus exhibiting an acidic character (Buch et al. 2007; Jungwirth 2009) but we postpone the discussion on water dissociation until we get to Chapter 3. Unfortunately, experiments devised to obtain χ_s do not provide such a unitary response, with scattered both positive and negative χ_s values (Paluch 2000) although there is general agreement upon the negative sign of the surface. In distilled water air bubbles are known to migrate toward the positive electrode in an electrophoresis cell, so that they must be negatively charged. G. Quincke (1861) pioneered such studies and since then, many authors have performed more accurate measurements arriving at the very same conclusion: at near-neutral pH air bubbles are negatively charged. A closely related parameter, the ζ-potential, gives systematically negative values for air bubbles (see Section 3.5.2 for the definition of the ζ-potential).

Teschke and de Souza (2005) have measured the interaction force between an electrically neutral cantilever tip of an atomic force microscope (AFM) and the water/air interface of air bubbles attached to polytetrafluoroethylene surfaces observing a long-range attraction, within a range of about 250 nm, and discrete (stepwise) medium range repulsion upon approximation to the interface. These results have been interpreted in terms of the presence of water clusters in the interfacial region, that can be rather large (>25 nm) assuming an oscillatory spatial variation of the dielectric permittivity. In general, and this should not be interpreted as a criticism to the results from Teschke and de Souza (2005), when measuring force curves in the low nN regime in water (cantilever and substrate submerged), one has to be extremely cautious because of the potential presence of contaminants. The charge state of the tip is not absolutely controlled because of the complex chemistry at the tip surface and deionized water of the highest quality is always contaminated to some extent. In addition, the presence of ions is also important, leading to attraction/repulsion regions in the force curves. No matter how clean you work, contamination is always there. This is one of the real facts that surface scientists have to face: admit the presence of contamination and we show some examples in this book (even in ultrahigh vacuum (UHV) you find surface contamination!).

Surface Tension

The accepted description of the vapor/liquid interface given above, with a dynamic exchange of molecules between both phases, seems to be at odds with the known robustness of such an interface, which permits objects and beings to float on top of it (see Appendix A for a discussion of surface tension and Archimedes' principle). The surprisingly elevated elasticity of the water surface when compared to other liquids, quantified by the surface tension or free energy (γ_{lv}), where l and v stand for liquid and vapor, respectively, arises from the cohesion induced by H-bonding. γ_{lv} is defined as the work that must be performed isothermally to create the unit area of the liquid surface. At the surface, molecules are only partially surrounded by other molecules, as discussed above, so that the number of adjacent molecules is smaller than in the bulk, an energetically unfavorable situation. Let us first refresh some relevant related concepts and discuss next in some detail several characteristics of γ_{lv}.

The Earth's atmosphere can be considered as an ideal gas, where the intermolecular forces can be neglected. In such an approximation, the total pressure can be represented by the sum of the vapor partial pressures p_v of the gases present in the atmosphere (N_2, O_2, H_2O, CO_2, Ar, etc.). In equilibrium, a steady state is reached where the flux of molecules moving from one state (e.g., liquid) to the other (e.g., vapor) is balanced by an identical flux in the opposite direction. The corresponding vapor partial pressure is called the saturation vapor pressure p_v^{sat}. The relative humidity (RH) is defined in % as:

$$RH = 100 \times \frac{p_v}{p_v^{sat}} \tag{2.3}$$

and is an important parameter to be considered when performing experiments in ambient conditions as discussed in Chapter 3. The temperature dependence of p_v^{sat} can be obtained from the Clausius–Clapeyron equation, which is derived in the following expression:

$$\ln \frac{p_v^{sat}(T)}{p_0} = -\frac{m_v l_v}{k_B T} \tag{2.4}$$

where p_0 is a reference value and m_v and l_v stand for the molecular mass and the latent heat (or enthalpy) of vaporization, respectively, and m_v is related to the molar volume of the liquid V_m through the expression $V_m = (m_v/\rho)(R/k_B)$. In the case of water, l_v is the amount of heat required to generate steam from liquid water (40.7 kJ mol^{-1} at the boiling point, the highest heat of vaporization of any molecular liquid). A more realistic approximation should include the temperature dependence of l_v, but from the practical point of view we can calculate the numerical values of p_v^{sat}, in Pa, from the empirical expression (Wagner and Pruss 1993):

$$\ln \frac{p_v^{sat}}{p_c} = \frac{T_c}{T} \left\{ a_1 \tau + a_2 \tau^{1.5} + a_3 \tau^3 + a_4 \tau^{3.5} + a_5 \tau^4 + a_6 \tau^{7.5} \right\} \tag{2.5}$$

where $p_c = 22.064 \times 10^6$ Pa, $T_c = 647.096$ K, $\tau = 1 - T/T_c$, and $a_1 = -7.85951783$, $a_2 = 1.84408259$, $a_3 = -11.7866497$, $a_4 = 22.6807411$, $a_5 = -15.9618719$, and

$a_6 = 1.80122502$. For example, at $T = 273.16$ K (0°C), $p_v^{sat} = 611.657$ Pa and at $T = 373.12$ K (100°C), $p_v^{sat} = 1.01325 \times 10^5$ Pa. Note that according to (2.3) RH is temperature dependent.

However, p_v^{sat} also depends on the curvature κ of the liquid/gas interface; a dependence is given by the Kelvin equation (Thomson 1871):

$$\ln \frac{p_v^{sat}(T, \kappa)}{p_v^{sat}(T, \infty)} = \frac{m_v \gamma_{lv}}{\kappa \rho k_B T} \qquad (2.6)$$

κ is defined in terms of the so-called principal radii of curvature r_1 and r_2 through the expression $1/\kappa = 1/r_1 + 1/r_2$. For a droplet of radius r_d, $\kappa = r_d/2$. Hence, for an spherical droplet (2.6) transforms into:

$$\ln \frac{p_v^{sat}(T, \kappa)}{p_v^{sat}(T, \infty)} = \frac{2 m_v \gamma_{lv}}{r_d \rho k_B T} = \frac{2\lambda_K}{r_d} \qquad (2.7)$$

The Kelvin length $\lambda_K = m_v \gamma_{lv}/\rho k_B T$ characterizes the range of capillary forces. For water $\lambda_K \sim 0.5$ nm (Butt and Kappl 2009).

The value of γ_{lv} for a flat surface of pure water ($\kappa \to \infty$) at RT is 72.75 mN m^{-1} (Dorsey 1897). According to (2.6), the smaller κ is, the greater the p_v^{sat} required to keep a droplet in equilibrium. This has important consequences for atmospheric physics. Combining the exponents from both (2.4) and (2.7) we obtain the quantity:

$$\frac{1}{k_B T} \left\{ m_v l_v - \frac{2 m_v \gamma_{lv}}{r_d \rho_v} \right\}$$

which indicates that the energy required to escape a curved surface is lower when compared to a flat surface.

One can estimate the smallest r_d, called the critical radius r^*, above which an embryo droplet can grow and below which it would evaporate, by maximizing the change of the Gibbs free energy (ΔG; Butt, Graf, and Kappl 2003):

$$\Delta G = -\frac{4}{3}\pi r_d^3 n k_B T \ln \frac{p_v}{p_v^{sat}} + 4\pi r_d^2 \gamma_{lv} \qquad (2.8)$$

where n is the number density ($\simeq 33.4$ nm^{-3}). The first and second terms of (2.8) correspond to the bulk and surface contributions, respectively. The resulting critical radius has the expression:

$$r^* = \frac{2\gamma_{lv}}{n k_B T \ln(p_v/p_v^{sat})} \qquad (2.9)$$

Thus, the greater the p_v/p_v^{sat} ratio, termed supersaturation, the smaller r^* is, and thus the easier the formation of droplets. At $T = 273$ K and using $\gamma_{lv} = 72.75$ mN m^{-1} we obtain $r^* = 1.7$ and $r^* = 0.8$ nm for supersaturation values of 2 and 4, respectively, which involve about 690 and 70 water molecules, respectively.

Note that (2.7) assumes a constant surface tension; however, it has been shown both theoretically and experimentally that it depends on r_d: for a droplet with a finite radius r_d, $\gamma_{lv}(r_d) < \gamma_{lv}(r_d \to \infty)$ (Tolman 1949; Mecke and Dietrich 1999; Fradin et al. 2000). Mecke and Dietrich (1999) developed a DFT-based model that takes into account the long-range power-law decay of the dispersion forces between molecules in a fluid. According to this model the surface tension is a function of the length scale (λ) exhibiting two regimes: (i) γ_{lv} decreases due to the effect of attractive long-range forces for λ values above a characteristic value λ_0, and (ii) γ_{lv} increases for $\lambda < \lambda_0$ due to the distortion of the density profile caused by surface bending. Below λ_0, $\gamma_{lv}(\lambda)/\gamma_{lv}(\infty) \sim \lambda^{-2}$ enabling $\gamma_{lv}(\lambda) > \gamma_{lv}(\infty)$. The capillary-wave theory also predicts an increase, although moderate, of γ_{lv} for a fluid in a confined geometry (Kayser 1986). Grazing-incidence X-ray scattering measurements using SR have shown a reduction of up to 75% in γ_{lv} down to a length scale of $\simeq 2$ nm and a tendency to increase below such a value (Fradin et al. 2000). Note that $\lambda_0 \simeq 2$ nm is close to the r^* value obtained above from (2.9), thus giving an estimation of the critical length scale.

A different approach has been used by Fraxedas et al. (2005) to explore the variation of γ_{lv} with length scale. They have used an AFM in the dynamical amplitude modulation mode (AM-AFM) with ultrasharp tips (tip radius below 10 nm) to both image (topography) and apply perpendicular forces to (nanoindentation) confined water nanodroplets. The main outcome of this work is that confined droplets behave as Hookean springs with force constants significantly larger than the 0.073 N m^{-1} value. Figure 2.2 shows the topography (a) and phase (b) images of a water nanodroplet confined on a nanobeaker. The apparent height of the droplets is about 25 nm (see line profile). Note the well-defined phase contrast defining the location of water in the line profile in Figure 2.2b, which gives a measured diameter of approximately 150 nm.

Figure 2.3 shows a force plot performed on the confined water nanodroplet from Figure 2.2. Upon loading the cantilever tip is approached to the nanodroplet and a first $\sim$3 nN decrease is observed, which is due to the capillary force between the nanodroplet and the water layer adhered on the tip (see Appendix B for the expressions of capillary forces between spheres and flat surfaces). Increasing the applied force leads to two clearly differentiated slopes, $\sim$1.3 Nm^{-1} and $\sim$4.0 Nm^{-1}, the latter corresponding to the bottom of the confining walls. Upon unloading a long water meniscus is formed. The force curves differentiate between the water adhered to the tip surface and the nanodroplets. The Hookean response is clearly observed in the $\sim$1.3 Nm^{-1} range. Because the elastic response to nanoindentation with ultrasharp tips reveals the contribution of the surface, it can be assumed that, as a first approximation, $\sim$1.3 Nm^{-1} is a measure of the strength of the liquid/vapor interface. In such experiments the nanometric scale length is provided by the small radius ($<$10 nm) of the probing tip.

γ_{lv} also depends on temperature and on the presence of contaminants. Figure 2.4 shows the temperature dependence of γ_{lv} for $\kappa \to \infty$, which is given by the empirical expression (Kestin et al. 1984):

$$\gamma_{lv} = \gamma^* \left[1 - \frac{T}{T^*}\right]^\nu \left[1 + B\left(1 - \frac{T}{T^*}\right)\right] \qquad (2.10)$$

where $\gamma^* = 0.2358$ N m^{-1}, $T^* = 647.0667$ K, $\nu = 1.2564$, and $B = -0.625$. When

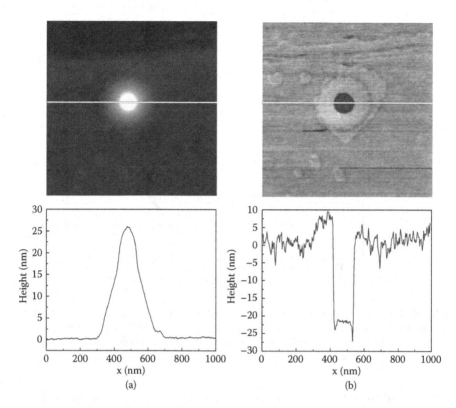

FIGURE 2.2 Tapping mode AFM image [1.0 μm × 1.0 μm] of a water nanodroplet confined in a nanobeaker measured at RH ~ 50%: (a) topography and (b) phase. Line profiles of the corresponding images are also shown. (Reprinted from J. Fraxedas, et al. *Surf. Sci.*, 588: 41–48, 2005. With permission from Elsevier.)

the free water/air interface undergoes thermal fluctuations, for example, by external heating, curious patterns (cells) may appear, which are known as Bénard–Marangoni instabilities, an effect that is briefly discussed in Appendix C.

Concerning the influence of contaminants on γ_{lv} a revealing story comes next. It was Agnes Pockels who made the first studies "with very homely appliances", as expressed by Lord Rayleigh (Pockels 1891). He was referring to the fact that this singular woman performed her research at home, given the restricted access of women to universities by the end of the nineteenth century. To measure the tension she developed the trough, precursor of the trough used in Langmuir–Blodgett instruments. Rayleigh repeated the experiments performed by Pockels and confirmed the importance of this phenomenon in the explanation of the mechanism of surface tension. We come back to this point later when describing oil/water interfaces. In combination with organic solvents, γ_{lv} closely follows the expression (Khossravi and Connors 1993):

$$\gamma_{lv} = \gamma_{lv}^w + \frac{\gamma_{lv}^s - \gamma_{lv}^w}{2} \frac{K_w x_s(x_w + 2K_s x_s)}{x_w^2 + K_w x_w x_s + K_w K_s x_s^2} \quad (2.11)$$

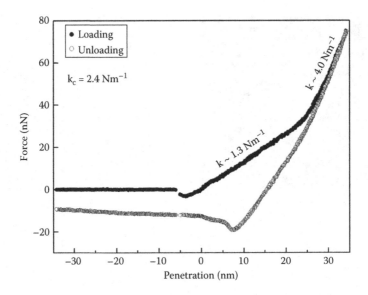

FIGURE 2.3 Force plot performed on the confined water nanodroplet from Figure 2.2 using a cantilever with a $k_c \simeq 2.4$ N m^{-1} force constant and tip radius of less than 10 nm. Applied forces F are given by $F = k_c \Delta$, where Δ is the cantilever deflection. (Reprinted from J. Fraxedas, et al. *Surf. Sci.*, 588: 41–48, 2005. With permission from Elsevier.)

where γ_{lv}^w and γ_{lv}^s stand for the surface tension of pure water and solvent, respectively, x_w and x_s are the corresponding bulk molar fractions, and K_w and K_s correspond to dimensionless exchange constants for water and solvent, respectively. As an example, $K_w = 33.4$ and $K_s = 14.5$ for acetonitrile and $K_w = 138.0$ and $K_s = 7.1$ for acetone.

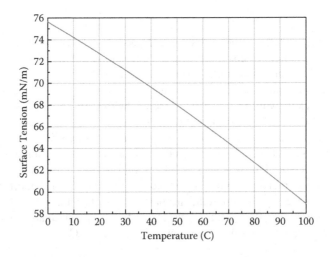

FIGURE 2.4 Temperature dependence of γ_{lv} calculated using (2.10). (From J. Kestin et al. *J. Phys. Chem. Ref. Data*, 13: 175–183, 1984. With permission.)

γ_{lv} can be intentionally increased by adding electrolytes to water. For most of them the increase is proportional to their concentration (Jarvis and Scheiman 1968), a point that is discussed in Section 3.5 in terms of the Hofmeister series.

2.1.2 WATER/OIL INTERFACES

Amphiphilic Molecules

"When a very small quantity of an oil, such as olive oil, is placed upon a large clean surface of water, the oil spreads rapidly upon the water surface until a definite area has been covered and then the oil shows little or no tendency to spread further." With this sentence I. Langmuir (1917) started the discussion on oil films on water before concluding that the spreading of an oil on water is caused by the presence of certain active groups in the molecule. The interaction of oils, fatty acids, and in general of amphiphilic molecules with water is of great relevance. Such molecules have both a nonpolar and a polar termination. Nonpolar groups are hydrophobic (lipophilic) whereas polar groups are hydrophilic. The hydrophobic part is usually composed of a $-CH_3$ termination (remember that H_2O and CH_4 exhibit a poor mutual affinity) and the hydrophilic parts contain chemical functions such as $-OH$, $-COOH$, $-NH_2$, and so on, which are the active groups. The hydrophilic tail is attracted toward the water surface, but the hydrophobic termination tries to avoid it.

We make use and we are made of amphiphilic molecules. We use soap, which is a salt of a fatty acid (e.g., sodium stearate) and detergents. We know that oil and water are immiscible and the action of the surfactants, another term describing amphiphilic molecules, is to dissolve the oil in water. The work principle is rather simple (that's why it is so successful): the lipophilic part of the amphiphilic molecule interacts with the oil molecules making a coating and building micelles around them. Such micelles are soluble in water because of the hydrophilic tails of the molecules, in the case of sodium stearate containing carboxylate groups. Detergents work in a similar way. Typical anionic detergents contain alkylbenzenesulfonates. The alkylbenzene portion of these anions is lipophilic and the sulfonate is hydrophilic.

When water is mixed with surfactants and nonpolar solvents micelles can be reversed, leading to surfactant-coated water droplets. This is a way to confine water nanodroplets. Within such reversed micelles, two kinds of water can be differentiated: interfacial and bulklike. This has been shown by means of polarization and frequency selective IR pump-probe experiments (Moilanen et al. 2009). In large reverse micelles (diameters of about 20 nm), the dynamics of water discriminate between slow interfacial water and fast core water. As the size decreases, the slowing effect of the interface and the collective nature of water reorientation begin to slow the dynamics of the core water molecules. In the smallest reverse micelles, these effects dominate and all water molecules end up exhibiting similar reorientational dynamics. The crossover between the slow–fast and the collective reorientation scenarios occurs at diameters of about 4 nm. Nonpolar amino acids are also examples of small amphiphilic molecules, inasmuch as they contain polar amino and carboxylic groups and nonpolar methyl-based terminations. Amino acids build peptides and finally proteins, which can be regarded as complex amphiphilic molecules with many hydrophilic and lipophilic groups. Phospholipids are a further example, which form bilayers, the base of membranes, a point discussed in Section 6.4.1.

The surface of water induces a preferential orientation of the amphiphilic molecules acting as a structuring agent. If the surface concentration increases, for example, by the mechanical action of movable barriers or pistons, the molecules tend to order and the interfacial tension becomes smaller than the surface tension of pure water. This is the 2D equivalent of the pressure versus volume isotherm for a gas–liquid–solid system. If the molecular density is sufficiently low, the floating film will behave as a 2D gas phase (disordered) where the molecules are far enough apart, resulting in negligible intermolecular interactions. As the monolayer (ML) is compressed the pressure rises, signaling a change in phase to a 2D liquid state. Upon further compression, the pressure begins to rise more steeply as the liquid expanded phase transforms to a condensed phase. Hence the molecules at the interface become anchored, strongly oriented, and with no tendency to form a layer more than one molecule thick. This is the principle of the widely used Langmuir–Blodgett (LB) technique, which is a way of preparing ultrathin organic films with a controlled layered structure with the aim to transfer them to solid surfaces (Langmuir 1917; Blodgett 1935).

Under certain experimental circumstances, the Langmuir ML can form a 2D crystalline network. This is particularly important in the case of aliphatic alcohols, $C_nH_{2n+1}OH$, if the unit cell projected onto a plane parallel to the water/alcohol interface closely matches the basal plane of ice Ih. In this case the alcohol sublattice acts as an heterogeneous 2D nucleus inducing the formation of ice. Thus, the water surface starts by structuring the alcohol layer but ends up by becoming structured. This phenomenon has been investigated by Gavish et al. (1990), where it was observed that the mean freezing temperature of decorated drops of supercooled water increased with the number of carbons in the alcohol molecules, with a dependence on the parity (even or odd) of the alcohol chain length. For $n = 31$ the freezing point approaches 273 K whereas for odd n values there is an asymptotic limit by 265 K. The parity dependence is attributed to the orientation of the alcohol head group with respect to the molecular tilt angle. This results in evidence that (pseudo)epitaxy is important in the nucleation mechanism of water ice.

Another amazing property of the air/water interface is its ability to induce chiral supramolecular assemblies from achiral amphiphilic molecules (Ariga et al. 2008). It is tempting to associate this property with the origin of chirality but most of the efforts have been focused toward the preparation of artificial materials using chirality to control enantiomer recognition or to produce devices such as light-emitting diodes, solar cells, and field-effect transistors (Guo et al. 2005). One simple way, associated with the LB technique, is making use of the precise mechanical control through the surface pressure–molecular area isotherms (Arnett, Harvey, and Rose 1989). In this way spontaneous chiral symmetry breaking of calcium arachidate monolayers (Viswanathan, Zasadzinski, and Schwartz 1994) or of a barbituric acid derivative (Huang et al. 2004) have been achieved. We show in Section 4.4 how water is able to identify the handedness of amino acid surfaces.

If the surfactants are nanoparticles (NPs) coated with ligands with specific terminal groups the water/oil interface can be used to induce self-assembly of such NPs (Wang, Duan, and Möhwald 2005). Figure 2.5 shows some examples. Gold and silver

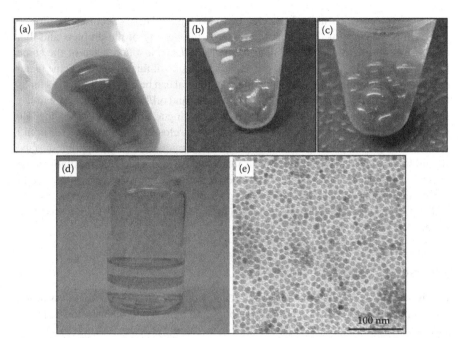

FIGURE 2.5 Photographs of self-assemblies at the water/toluene interface of: (a) 12 nm Au@DTBE NPs, (b) 40 nm Ag@DTBE NPs, and (c) their mixture with a molar ratio of 1:1. (d) Photograph and (e) TEM image of a 8-nm Fe₃O₄@BMPA NP ML, formed at the water/toluene interface. DTBE and BMPA stand for 2,2′-dithiobis[1-(2-bromo-2-methylpropionyloxy)ethane] and 2-bromo-2-methylpropionic acid, respectively. (Adapted from D. Wang, H. Duan, and H. Möhwald. *Soft Matter* 1:412–416, 2005. With permission of the Royal Society of Chemistry.)

NPs with diameters of 12 and 40 nm, respectively, capped with 2,2′-dithiobis[1-(2-bromo-2-methylpropionyloxy)ethane] (DTBE) at the water/toluene interface are shown in Figures 2.5a and b, respectively, and their mixture with a molar ratio of 1:1 is shown in Figure 2.5c. Note the metallic luster of the films. On the other hand, Figure 2.5d shows a ML of 8 nm Fe_3O_4 NPs capped with 2-bromo-2-methylpropionic acid (BMPA) which has formed at the water/toluene interface. The TEM image depicted: Figure 2.5e demonstrates the closed-packed structure of the ML.

Nonpolar Aliphatic Molecules

We finish the section devoted to liquid water by seeing what happens when aliphatic nonpolar chain molecules such as dodecane, hexadecane, and the like, interact with water. Electrophoresis experiments have shown that oil droplets, dispersed without any surfactant in the aqueous phase, are negatively charged leading to the conclusion that hydroxyl ions, released by the dissociation–association equilibrium of the water molecules, adsorb at the oil/water interface (Marinova et al. 1996). On the other hand, oil-in-water emulsions have been characterized by the electroacoustics technique with similar conclusions (Beattie and Djerdjev 2004). An electroacoustic effect is generated when an alternating MHz electric field is applied to a concentrated colloidal suspension. Because of the difference between the density of the droplets and that of the surrounding fluid, this oscillatory motion creates a sound wave of the same MHz frequency. By measuring the phase and amplitude of this sound wave, the mean diameter of the emulsion droplets and their ζ-potential can be obtained. Thus, as far as the interfacial charge is concerned, the vapor/liquid and oil/liquid interfaces are rather similar. But they bear still more similarities. According to Stillinger (1973), "the low pressure interface next to the flat repelling surface is closely related to the free liquid surface". In other words, the interfacial structure of water at a flat hydrophobic surface should be similar to that of the water/air interface. According to MD simulations the intrinsic gap between water and oil surfaces is a rather rigid structure, with a width being slightly larger than the water molecular diameter (about 1.4 times at 300 K) and with corrugations well below that size (Bresme et al. 2008). We discuss this point in detail in Chapter 4.

2.2   SOLID WATER

In this section we concentrate on the surface of bulk ice Ih, with a thickness along the c-direction well above one bilayer. Mono- and bilayers of water ice on different surfaces are considered in Chapter 3.

2.2.1   SOLID/VACUUM INTERFACES

Let us now imagine that we somehow cut the infinite periodic 3D ice Ih crystal from Figure 1.12b so as to expose the ideal (0001) hexagonal basal plane, as shown in Figure 2.6a.

The figure shows the layered structure of ice Ih along the crystallographic c-axis, and a detail of one layer, termed a bilayer because it contains two molecular planes, is given in Figure 2.6b. We observe the short distance between both molecular planes (about 0.9 Å) and the zig-zag structure along the a direction. The water molecules in the bottom half of the outermost ice bilayer retain their four hydrogen bonds, three bonds within the bilayer and one to the bilayer immediately below, but the oxygen-ordered molecules in the top half of the surface bilayer are only three-fold coordinated, lacking an H-bond to the missing bilayer above them. Each surface molecule contributes with either a donor dangling hydrogen (d-H) or with an acceptor dangling lone-pair (d-O). Within the proton-disorder scenario, the d-H and d-O atoms would form a disordered pattern anchored to a triangular lattice that corresponds to the upper half of the top bilayer.

Prior knowledge of tetrahedrally coordinated solids, thus built from directional sp^3 bonds, demonstrates that the generated (ideal) surfaces, usually termed 1×1, are not energetically favorable and that the surface atoms rearrange themselves in order to saturate the dangling bonds thus relaxing and reconstructing the surface. Several examples are available, but it suffices to recall both the (111) and (001) surfaces of (fcc) silicon, that, when prepared in UHV under certain experimental conditions (annealing temperature, temperature ramps, etc.), exhibit several reconstructions. The most important are the rather involved 7×7 (111) and 2×1 (001) surface reconstructions,

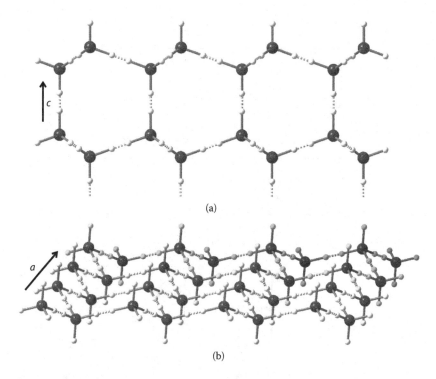

(a)

(b)

FIGURE 2.6 Two projections of the ideal truncated (0001) surface of ice Ih showing proton disorder.

involving not only the surface atoms but subsurface atoms as well (Srivastava 1997). However, in the case of molecular organic materials the ideal truncated surface becomes the actual one in the vast majority of cases, because the intermolecular interactions are weak (Fraxedas 2006). Then, what should we expect from the surface of water ice Ih, which is a pseudo-tetrahedrally coordinated molecular solid?

At very low temperatures one should expect proton-order, because according to the third law of thermodynamics the residual entropy should tend to zero, even though the system is proton-disordered, and it was Fletcher (1968; 1992) who proposed that the oxygen-ordered (0001) ice surface should undergo reconstruction to a striped proton-ordered phase in which d-H and d-O atoms form alternating rows along the a-direction. A scheme of this phase is illustrated in Figure 2.7. In this case the residual entropy is simply $S = k_B \ln 2$ because there are only two possible configurations. Based on simple estimates employing near-neighbor dipole–dipole interactions, Fletcher predicted the surface-ordering transition to take place near 30 K for the basal plane and near 70 K for the prismatic face. This model corresponds to a low-energy state in an antiferromagnetic classical Ising model on a 2D triangular lattice, in which d-H and d-O are viewed as dipoles in up and down configurations, respectively. Such dipole distribution is frustrated, because when two dipoles are aligned antiparallel (interaction energy is $-\mu^2/4\pi\varepsilon_0 a^3$), the third one must be forcedly parallel to one of them (interaction energy is $\mu^2/4\pi\varepsilon_0 a^3$). If instead of dipoles we were considering

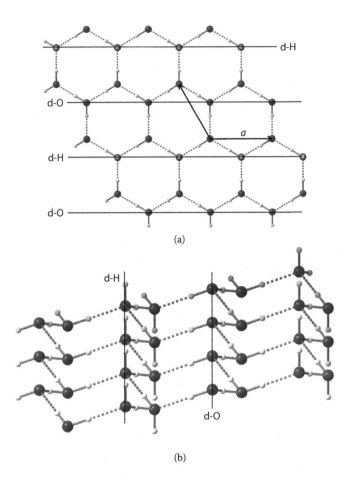

FIGURE 2.7 Striped proton-ordered phase of the oxygen-ordered (0001) ice surface with d-H and d-O atoms forming alternating rows along the *a* direction as proposed by Fletcher (1992). (a) Top view along the *c*-axis and (b) perspective view to visualize the upper and lower planes of the bilayer.

spins, we would be talking of Kagomé lattices. The interesting point is that Wannier (1950) analytically proved that such a lattice cannot undergo an order–disorder transition, implying that order should be preserved. We should thus expect proton-order to be present at least to some extent at higher temperatures, indeed below the melting temperature.

Theoretical calculations based on MD (Buch *et al.* 2008) and DFT simulations (Pan *et al.* 2010) confirm such a prediction, suggesting that such a ground state of the surface should remain proton-ordered well above the bulk order–disorder temperature corresponding to the bulk XI–Ih transition (72 K). High-resolution helium atom scattering (HAS) studies performed in UHV conditions in the 25–125 K range seem to point toward the existence of such a striped phase, although the spectral evidence is rather weak (Glebov et al. 2000). In this work, thick ice films (~100 nm) with predominant (1 × 1) termination were achieved by dosing ultrapure H_2O at RT from the vapor using a highly collimated, differentially pumped effusive source, with the substrate held at 150 K during exposure and at 125 K after exposure. It is needless to say how important it is to produce reliable surfaces and HAS permits us to clearly identify the ideal termination of the ice Ih basal plane, although no direct information on proton order–disorder can be obtained. In the same work, they observed a dispersionless surface phonon branch (phonon energy independent of wave vector) at ~50 cm^{-1}, which has been assigned to vibrations of individual molecules on the ice surface, and dispersive (energy proportional to the wave vector) surface Rayleigh phonons. On the other hand, electron energy loss spectroscopy (EELS) studies reveal the presence of vibrational features at 100, 470, 665, and 825 cm^{-1} (Yamada et al. 2003). The 100 cm^{-1} mode has been ascribed to a hindered-translational vibration of the outermost water molecules along the surface normal and 470, 665, and 825 cm^{-1} to surface hindered-rotational modes (see Figure 3.7).

2.2.2 SOLID/LIQUID/VAPOR INTERFACES

Contact Angle

At the triple point the three phases of water (vapor, liquid, and solid) coexist, building three distinct interfaces: solid/liquid, solid/vapor, and liquid/vapor. The latter has been discussed above in terms of surface tension. The liquid and a solid surfaces build a (contact) angle θ_c, as schematized in Figure 2.8, which is given by the Young's relation:

$$\gamma_{sv} = \gamma_{sl} + \gamma_{lv}\cos\theta \qquad (2.12)$$

where γ_{sv} and γ_{sl} stand for the solid/vapor and solid/liquid free energies per unit area. According to van Oss et al. (1992), $\gamma_{sv} = 69.2$ mJ m^{-2} with an estimated contact angle at 0°C of about 24° and $\gamma_{sl} \sim 0.04$ mJ m^{-2}, a rather small quantity. However, an experimental determination of γ_{sl} gives a value of 29 mJ m^{-2} (Hardy 1977), in agreement with recent predicted values of 28 mJ m^{-2} (Luo, Strachan, and Swift 2005) and calculations using the TIP4P model (Handel et al. 2008), where γ_{sl} equals 23.3, 23.6, and 24.7 mJ m^{-2} for the basal, prism, and (11$\bar{2}$0) faces, respectively.

Quasi-liquid Layer

When the temperature is increased close to but below T_m, the long-range order of the oxygen sublattice is lost and a liquidlike layer is formed at the surface. The term quasi-liquid layer (QLL) is also used. This process is also called premelting and what

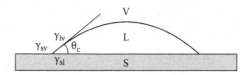

FIGURE 2.8 Scheme of the contact angle (θ_c) of a liquid (L) on a solid (S) surface. V stands for the vapor phase. The corresponding interfacial energies, γ_{lv}, γ_{sl}, and γ_{sv}, are shown.

is important is that the obtained liquid is in a supercooled state, because it exists at temperatures below T_m. Note that it is experimentally well established that, under conventional conditions, a solid cannot be superheated (remain solid above T_m) and this is due to the presence of a surface, which prevents superheating by building a liquid layer. However, Iglev et al. (2006) have shown that it is possible to superheat ice at least for short time intervals (about 250 ps) by exciting the OH stretching mode with ultrashort laser pulses. The existence of a liquid layer at the ice surface was hypothesized a long time ago by Faraday (1859) when trying to understand the phenomenon of regelation (the updated word is sintering) that happens when two pieces of ice are brought together and as a result the material between them becomes solid (that is why we can build snow balls). Such a liquidlike layer has been investigated using different experimental methods as well as simulations. Wei, Miranda, and Shen (2001) performed SFG measurements on the basal surface of ice. In the OH stretching region two peaks are clearly observed: a strong but relatively broad peak at ~3,150 cm^{-1}, which is ascribed to bonded OH and a sharp feature around 3,695 cm^{-1}, which has been found in the surface of liquid water as discussed in the previous section, associated with the stretch vibration of the d-OH protruding from the surfaces (Du et al. 1993). MD studies have assigned a feature at 510 cm^{-1} to the libration of d-OH at the surface (Ikeda-Fukazawa and Kawamura 2004). There is no general consensus concerning the thickness and onset temperature (T_s) of such a layer. The scatter of values is made evident in Table 2.1, where a comparison from different sources is shown. Note that the thickness values span from 1 to 70 nm.

But before comparing the values let us explore what state-of-the-art simulations are up to (again, at the time this text was written). Figure 2.9 shows the time evolution of the QLL of the basal plane of ice Ih (right and left of the simulation box) at 0, 1, 4, and 9 ns at 268 K as derived from molecular simulations using the TIP4P/Ice model (Vega, Martín-Conde, and Patrykiejew 2006; Conde, Vega, and Patrykiejew 2008).

TABLE 2.1
Onset Temperatures (T_s) and Thickness at Given Temperatures T of the QLL Layer on the Basal Plane of Ice Ih.

T_s [K]	Thickness [nm] (T [K])	Method
173	~0.75 (270)	MD (TIP4P/Ice)[a]
~200		SFG[b]
240	~70 (272)	AFM[c]
253	~2-3 (273)	NEXAFS[d]
<256	0.7-1.1 (272)	AFM[e]
260	~50 (273)	GAXS[f]
<263	~15 (273)	IR[g]
	1.5 (270)	MD (TIP4P)[h]

[a] (Conde, Vega, and Patrykiejew 2008), [b] (Wei and Shen 2002), [c] (Döppenschmidt, Kappl, and Butt 1998), [d] (Bluhm et al. 2002), [e] (Pittenger et al. 2001), [f] (Dosch, Lied, and Bilgram 1995), [g] (Sadtchenko and Ewing 2002), [h] (Furukawa and Nada 1997).

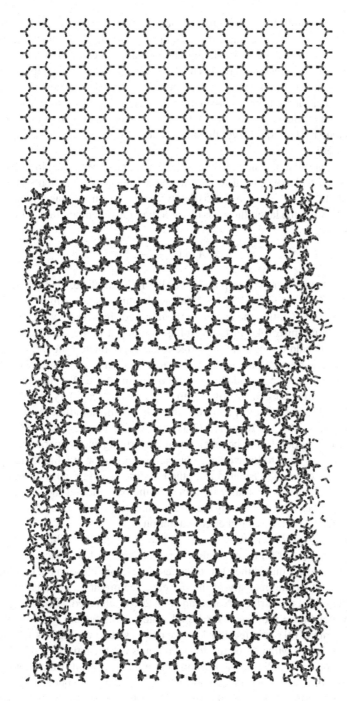

FIGURE 2.9 MD simulations of the free surface of Ih using the TIP4P/Ice model and at 268 K at 0, 1, 4, and 9 ns from top to bottom (Conde, Vega, and Patrykiejew 2008). The plane exposed to the vacuum is the basal plane. The simulation box ($110 \times 31 \times 27$ Å^3) contains 1,536 molecules. The size of the initial block of ice along the direction perpendicular to the surface is 59 Å. (Courtesy of M. M. Conde and C. Vega.)

This model results from a slight modification of the parameters of the original TIP4P model (see Table 1.5) in order to reproduce the experimental melting point value. As is evident from the simulation, the initially oxygen-ordered (0 ns) surface develops a QLL as a function of time. The thickness increases with temperature and at 268 K it is about 1 nm thick. At a given temperature the QLL thickness is larger for the basal plane than for the primary prismatic plane, and for the primary prismatic plane it is larger than for the secondary prismatic plane. Other water models essentially give the same value for the liquid layer thickness; about 1 nm at temperatures up to 3–4 K below the melting point.

From the experimental side, different techniques have been used in order to characterize the QLL, leading to different thickness values. We mention some of them here. Dosch, Lied, and Bilgram (1995) determined by means of glancing-angle X-ray scattering (GAXS) the layer thickness (L) as a function of temperature, which follows the expression:

$$L = A \ln \left(\frac{T_s}{T_m - T} \right) \qquad (2.13)$$

where A represents a growth amplitude. For the Ih basal plane, the thickness is $\sim$50 nm at 272.7 K ($-0.3°$C) and $T_s = 259.5$ K. In this case ex situ grown Ih single crystals were used. Cutting the crystal to the required shape (e.g., with heated wires) in order to obtain mirrorlike surfaces and assembling it in the sample holder has to be performed in cold rooms and transferred to the experimental stations with adapted chambers. The correct preparation of the surface is crucial for such experiments.

Döppenschmidt, Kappl, and Butt (1998) determined the QLL thickness with an AFM. Upon approaching the cantilever tip to the surface, capillary forces induce the tip to jump into contact. The upper limit in thickness of the liquidlike layer varies from 12 nm at 249 K ($-24°$C) to 70 nm at 272.3 ($-0.7°$C). At about 240 K ($-33°$C) surface melting starts. Caution has to be taken when performing such experiments, because thermal gradients have to be avoided. The cantilever can be at a higher temperature because it is mechanically isolated from the ice surface and because of the detection system, which is based on a laser illuminating the back side of the cantilever. This effect can be reduced by performing the measurements in a cold room or using adapted freezers (students prefer by far the second option).

Bluhm et al. (2002) confirmed the existence of a liquidlike film at temperatures as low as 253 K ($-20°$C) by near-edge X-ray absorption fine-structure (NEXAFS) spectroscopy and estimated the QLL thickness in 2–3 nm close to the melting point. They observed that the intensity of the pre-edge feature at 535 eV, associated with the dangling OH at the surface (Nordlund et al. 2004), changed when going from ice to liquid water (more prominent for liquid water). This feature corresponds to the transition from the O1s core level to $4a_1$-like empty states (see Figures 1.6 and 1.22). The study of liquids with photoemission techniques is a breakthrough that deserves a more detailed comment. We already saw in Figure 1.20 the results from photoemission performed on droplets. The principal obstacle encountered with photoemission at elevated pressures is the intrinsic limited mean free path of electrons through gases, with an approximate value of 1 mm at a kinetic energy of 100 eV in 1 torr of water vapor. However, there is an increasing interest in being able to work at pressures higher than 5 torr for environmental science because the vapor pressure of water at the triple point is 4.6 torr at 273 K. After the seminal work by Siegbahn, Svensson, and Lundholm (1981) nowadays near ambient pressure X-ray photoelectron (NAPP) spectroscopy can be considered a mature technique. H. Siegbahn inherited the knowledge on photoemission from his father, K. Siegbahn, who was awarded the 1981 Nobel Prize in Physics for the development of the photoelectron spectroscopy technique (Siegbahn 1982). The pressure problem is overcome by means of differential pumping stages, where the sample is placed very close (about 1 mm) to the entrance aperture of a differentially pumped electrostatic lens system, thereby limiting the path length of the electrons in the high-pressure region while keeping the electron detector in UHV (Ogletree et al. 2002). The NAPP technique is increasingly being used in catalysis and atmospheric chemistry and an example of water on ionic materials is given in Section 3.5.

Premelting or surface melting is observed not only in water ice but also in other solids (Dash, Fu, and Wettlaufer 1995). It has been extensively studied in metal surfaces such as lead and aluminum as well as in semiconductors (van der Veen 1999). The (001) surface of silicon is particularly interesting inasmuch as it exhibits what has been termed incomplete melting. X-ray photoelectron diffraction performed in UHV conditions has shown that when such a surface is heated approaching $T_m = 1,685$ K, the bulk melting point of silicon, it shows premelting above $\sim$1,400 K forming a QLL layer about 0.2 nm thick that is stable up to close to T_m (Fraxedas et al. 1994).

We finish this section by briefly introducing two relevant examples where QLL plays an important role.

Crystal Growth from the Vapor Phase

Snow crystals (snowflakes) are single crystals of ice that grow from water vapor. They show a surprisingly rich morphology diagram depending on the growth temperature and supersaturation, crystallizing in the form of plates, columns, dendrites, and needles. Nakaya (1954) performed the first systematic laboratory studies of snow crystal growth back in the 1930s. He classified natural snow crystals under different meteorological conditions and was the first to grow synthetic crystals in controlled environments. A detailed description of the growth morphology has been published by Libbrecht (2005) and we briefly summarize the essential points here. At temperatures near $-2°C$, the growth is platelike, with thick plates at lower supersaturations, thinner plates at intermediate supersaturations, and platelike dendritic structures at higher supersaturations. For temperatures near $-5°C$, the growth is columnar, with stout columns at the lower supersaturations, hollow columns at intermediate supersaturations, and clusters of needlelike crystals at higher supersaturations. Near $-15°C$, the growth again becomes platelike. Finally, at the lowest temperatures the growth becomes a mixture of thick plates at low supersaturations and columns at higher supersaturations.

Within the mentioned temperature range water vapor condenses on the QLL, and not on crystalline solid surfaces, so that the well-known theories of vapor–solid homogeneous growth cannot be applied. In order to understand the kinetics of growth, Carignano, Shepson, and Szleifer (2005) performed MD simulations of ice growth

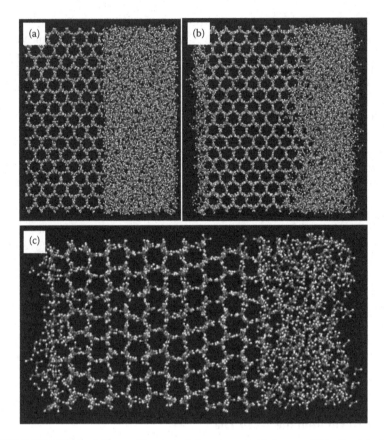

FIGURE 2.10 MD simulations of ice growth from supercooled water. Water/ice interface at the secondary prismatic plane, $(1\bar{2}10)$, viewed from the basal plane, (0001). Both the liquid and ice are the result of equilibration runs at 271 K. (a) $t = 0$ ns and (b) $t = 1.7$ ns. (c) Water/ice interface at the basal plane viewed from the secondary prismatic plane. Both the liquid and ice are the result of equilibration runs at 275 K after 4.2 ns. (Reproduced from M.A. Carignano, P.B. Shepson, and I. Szleifer, *Mol. Phys.*, vol. 103: 2957–2967, 2005. With permission.)

from supercooled water making use of the TIP6P water model (see Table 1.5). Figure 2.10 shows snapshots of the solid/liquid interfaces corresponding to both the secondary prismatic plane (a, b) and the basal plane (c). The simulated system is a quasi-2D layer, with two free surfaces facing a vacuum, which, upon molecular evaporation, results in water vapor. In the initial state, the system consists of a block of ice in contact with a water layer. In the final state the system is a block of ice with a symmetric QLL on both sides of the ice. The results show that ice grows in a layer-by-layer mode at the basal plane, whereas for the prismatic plane, a rough ice/water interface is generated. In addition, the growth rate of ice at the prismatic plane is approximately twice as fast as the growth rate at the basal plane.

In addition to temperature and supersaturation, impurities also play a relevant role in the growth morphology. It has been observed that the presence of acetic acid has a profound impact on the transition from needles to dendrites, which takes place at $\sim -10°C$, with the transition occurring at progressively lower temperatures as the gas phase acetic acid concentration increases (Knepp, Renkens, and Shepson 2009). Given the host of chemical species that exist in the atmosphere–stratosphere, we can understand how difficult it is to understand fully the role played by the QLL. Based on photoemission studies, Bluhm et al. (2002) concluded that premelting of water can be strongly enhanced by the presence of hydrocarbon contamination. We discuss next an example involving chemicals of extreme importance for us humans.

The Role of the Surface of Ice on Polar Ozone Depletion

The existence of the QLL on ice has tremendous implications in stratospheric chemistry. The stratosphere is the atmospheric layer characterized by an inverted temperature profile, rising from ~210 K at its base (10–15 Km altitude) to ~275 K at 50 km altitude. Ozone, which protects life on Earth against harmful energetic ultraviolet radiation, is produced in the upper part of the stratosphere by solar irradiation of molecular oxygen and destroyed by chemical processes involving free radicals. This ozone destruction occurs mainly in the early Antarctic spring, around September, and has been enhanced by human activity. The Nobel Prize in Chemistry 1995 was awarded jointly to P.J. Crutzen, M.J. Molina, and F.S. Rowland for their work in atmospheric chemistry, particularly concerning the formation and decomposition of ozone. It was Molina who pointed out the relevance that the QLL has concerning ozone depletion. An instructive introduction on the subject was given by Molina himself in his Nobel Lecture (Molina 1996).

The key chemical reaction is:

$$ClONO_2 + HCl \longrightarrow Cl_2 + HNO_3$$

which is greatly enhanced in the presence of ice particles in polar stratospheric clouds (Molina et al. 1987). Thus, the reservoir $ClONO_2$ + HCl species are transformed into the active species Cl_2 and HNO_3 catalyzed by the ice QLL. This reaction arises from the previous reactions $ClONO_2 + H_2O \longrightarrow HOCl + HNO_3$ and $HOCl + HCl \longrightarrow Cl_2 + H_2O$. Molecular chlorine becomes activated by irradiation giving rise to the free radical Cl, which transforms ozone into oxygen. Because the concentration of chlorinated molecules has increased in the atmosphere with the photolysis of gaseous chlorofluorocarbons (CFCs) and hydrochlorofluorocarbons (HCFCs) international treaties were established in order to reduce the consumption and production of such ozone-depleting substances. The most successful of such treaties are the Vienna Convention for the Protection of the Ozone Layer and its Montreal Protocol on Substances That Deplete the Ozone Layer (http://ozone.unep.org/) which have enabled reductions of over 97% of all global consumption of controlled ozone-depleting substances. This is a nice example of how fundamental research has helped humanity but one has to accept that the mechanisms that make the referred chemical reactions effective under the action of the QLL are far from being understood so that research on environmental science has to be pursued.

2.3 SUMMARY

- The water vapor/liquid interface has a finite width, of the order of the mean molecular diameter, resembling, to a certain extent, the interface formed between liquid water and a flat hydrophobic surface, the main difference arising from the unlayered character of the former. The finite width induces a dipolar double layer, where, on average, molecules closer to the vapor phase point their dipoles preferentially toward such a phase and molecules closer to the liquid phase point their dipoles toward the bulk.

- The surface tension of a liquid is defined as the work required to produce a new surface per unit area and depends on the surface curvature, temperature, and presence of contaminants. For a flat surface of pure water at room temperature, the surface tension is 72.75 mN m^{-1}.

- The basal and prismatic surfaces of ice Ih in vacuum should exhibit proton-order at low temperatures, according to theoretical calculations. An analogy to the XI phase can be established, although for the surface no impurities are needed to induce order. The predicted surface proton-order has not been conclusively demonstrated.

- The surface of pure water is able to structure monolayers of amphiphilic molecules, due to the intrinsic polar character, for sufficiently large surface densities, that is, when intermolecular lateral interactions are relevant. In addition, it can induce chiral supramolecular assemblies from achiral molecules.

- Below the bulk melting temperature the surface of ice is covered by a (quasi)liquid layer building a solid/liquid/vapor interface where the liquid is undercooled. The thickness of such a layer is of a few nm a few degrees below the bulk melting temperature and increases logarithmically when approaching such temperature. The presence of this layer explains the slippery character of ice, is responsible for the rich variety of shapes ice crystals can adopt when growing from the vapor phase, and has important consequences for polar ozone depletion.

REFERENCES

1. Ariga, K., Michinobu, T., Nakanishi, T., and Hill, J.P. 2008. Chiral recognition at the air–water interface. *Curr. Op. Colloid. Inter. Sci.* 13:23–30.
2. Arnett, E.M., Harvey, N.G., and Rose, P.L. 1989. Stereochemistry and molecular recognition in two dimensions. *Acc. Chem. Res.* 22:131–138.
3. Beattie, J.K. and Djerdjev, A.M. 2004. The pristine oil-water interface: Surfactant-free hydroxide-charged emulsions. *Angew. Chem. Int. Ed.* 43:3568–3571.
4. Benjamin, T.B. and Ursell, F. 1954. The stability of the plane free surface of a liquid in vertical periodic motion. *Proc. R. Soc. London A* 225:505–515.
5. Blodgett, K.B. 1935. Films built by depositing successive monomolecular layers on a solid surface. *J. Am. Chem. Soc.* 57:1007–1022.
6. Bluhm, H., Ogletree, D.F., Fadley, C.S., Hussain, Z., and Salmeron, M. 2002. The premelting of ice studied with photoelectron spectroscopy. *J. Phys. Condens. Matter* 14:L227–L233.

7. Braslau, A., Deutsch, M., Pershan, P.S., Weiss, A.H., Als-Nielsen, J., and Bohr, J. 1985. Surface roughness of water measured by X-ray reflectivity *Phys. Rev. Lett.* 54:114–117.

8. Bresme, F., Chacón, E., Tarazona, P., and Tay, K. 2008. Intrinsic structure of hydrophobic surfaces: The oil-water interface. *Phys. Rev. Lett.* 101:056102.

9. Buch, V., Groenzin, H., Li, I., Shultz, M.J., and Tosatti, E. 2008. Proton order in the ice crystal surface. *Proc. Natl. Acad. Sci. USA* 105:5969–5974.

10. Buch, V., Milet, A., Vácha, R., Jungwirth, P., and Devlin, J.P. 2007. Water surface is acidic. *Proc. Natl. Acad. Sci. USA* 104:7342–7347.

11. Butt, H.J. and Kappl, M. 2009. Normal capillary forces. *Adv. Colloid Interface Sci.* 146:48–60.

12. Butt, H.J., Graf, K., and Kappl, M. 2013. *Physics and Chemistry of Interfaces.* Weinheim: Wiley-VCH.

13. Carignano, M.A., Shepson, P.B., and Szleifer, I. 2005. Molecular dynamics simulations of ice growth from supercooled water. *Mol. Phys.* 103:2957–2967.

14. Conde, M.M., Vega, C., and Patrykiejew, A. 2008. The thickness of a liquid layer on the free surface of ice as obtained from computer simulations. *J. Chem. Phys.* 129:014702.

15. Cross, M.C., and Hohenberg, P.C. 1993. Pattern formation outside of equilibrium. *Rev. Mod. Phys.* 65:851–1112.

16. Dash, J.G., Fu, H., and Wettlaufer, J.S. 1995. The premelting of ice and its environmental consequences. *Rep. Prog. Phys.* 58:115–167.

17. Döppenschmidt, A., Kappl, M., and Butt, H.-J. 1998. Surface properties of ice studied by atomic force microscopy. *J. Phys. Chem. B* 102:7813–7819.

18. Dorsey, N.E. 1897. The surface tension of water and of certain dilute aqueous solutions, determined by the method of ripples II. *Phys. Rev.* 5:213–230.

19. Dosch, H., Lied, A., and Bilgram, J.H. 1995. Glancing-angle X-ray scattering studies of the premelting of ice surfaces. *Surf. Sci.* 327:145–164.

20. Du, Q., Superfine, R., Freysz, E., and Shen, Y.R. 1993. Vibrational spectroscopy of water at the vapor/water interface. *Phys. Rev. Lett.* 70:2313–2316.

21. Fan, Y., Chen, X., Yang, L., Cremer, P.S., and Gao, Y.Q. 2009. On the structure of water at the aqueous/air interface. *J. Phys. Chem. B* 113:11672–11679.

22. Faraday, M. 1831. On a peculiar class of acoustical figures; and on certain forms assumed by groups of particles upon vibrating elastic surfaces. *Phil. Trans. R. Soc. London* 121:299–340.

23. Faraday, M. 1859. On regelation. *Phil. Mag.* 17:162–166.

24. Fletcher, N.H. 1968. Surface structure of water and ice. II. A revised model. *Phil. Mag.* 18:1287–1300.

25. Fletcher, N.H. 1992. Reconstruction of ice crystal surfaces at low temperatures. *Phil. Mag. B* 66:109–115.

26. Fradin, C., Braslau, A., Luzet, D., Smilgies, D., Alba, M., Boudet, N. et al.. 2000. Reduction in the surface energy of liquid interfaces at short length scales. *Nature* 403:871–874.

27. Fraxedas, J., Ferrer, S., and Comin, F. 1994. Temperature dependent photoelectron diffraction of the Si(001) surface. *Surf. Sci.* 307–309:775-780.

28. Fraxedas, J., Verdaguer, A., Sanz, F., Baudron, S., and Batail, P. 2005. Water nanodroplets confined in molecular nanobeakers. *Surf. Sci.* 588:41–48.

29. Fraxedas, J. 2006. *Molecular Organic Materials.* Cambridge, UK: Cambridge University Press.

30. Furukawa, Y. and Nada, H. 1997. Anisotropic surface melting of an ice crystal and its relationship to growth forms. *J. Phys. Chem. B* 101:6167–6170.

31. Gavish, M., Popovitz-Biro, R., Lahav, M., and Leiserowitz, L. 1990. Ice nucleation by alcohols arranged in monolayers at the surface of water drops. *Science* 250:973–975.

32. Glebov, A., Graham, A.P., Menzel, A., Toennies, J.P., and Senet, P. 2000. A helium atom scattering study of the structure and phonon dynamics of the ice surface. *J. Chem. Phys.* 112:11011–11022.

33. Guo, P., Tang, R., Cheng, C., Xi, F., and Liu, M. 2005. Interfacial organization–induced supramolecular chirality of the Langmuir–Schaefer films of a series of PPV derivatives. *Macromolecules* 38:4874–4879.

34. Handel, R., Davidchack, R.L., Anwar, J., and Brukhno, A. 2008. Direct calculation of solid-liquid interfacial free energy for molecular systems: TIP4P ice-water interface. *Phys. Rev. Lett.* 100:036104.

35. Hardy, S.C. 1977. A grain boundary groove measurement of the surface tension between ice and water. *Phil. Mag.* 35:471–484.

36. Huang, X., Li, Ch., Jiang, S., Wang, X., Zhang, B., and Liu, M. 2004. Self–assembly spiral nanoarchitecture and supramolecular chirality in Langmuir–Blodgett films of an achiral amphiphilic barbituric acid. *J. Am. Chem. Soc.* 126:1322–1323.

37. Iglev, H., Schmeisser, M., Simeonidis, K., Thaller, A., and Laubereau, A. 2006. Ultrafast superheating and melting of bulk ice. *Nature* 439:183–186.

38. Ikeda-Fukazawa, T. and Kawamura, K. 2004. Molecular-dynamics studies of surface of ice Ih. *J. Chem. Phys.* 120:1395–1401.

39. Jarvis, N.L. and Scheiman, M.A. 1968. Surface potentials of aqueous electrolyte solutions. *J. Phys. Chem.* 72:74–78.

40. Ji, N., Ostroverkhov, V., Tian, C.S., and Shen, Y.R. 2008. Characterization of vibrational resonances of water-vapor interfaces by phase-sensitive sum-frequency spectroscopy. *Phys. Rev. Lett.* 100:096102.

41. Jungwirth, P. 2009. Ions at aqueous interfaces. *Faraday Discuss.* 141:9–30.

42. Kathmann, S.M., Kuo, I.W., and Mundy, C.J. 2008. Electronic effects on the surface potential at the vapor–liquid interface of water. *J. Am. Chem. Soc.* 130:16556–16561.

43. Kayser, R.F. 1986. Effect of capillary waves on surface tension. *Phys. Rev. A* 33:1948–1956.

44. Kestin, J., Sengers, J.V., Kamgar-Parsi, B., and Levelt Sengers, J.M.H. 1984. Thermophysical properties of fluid H_2O. *J. Phys. Chem. Ref. Data* 13:175–183.

45. Khossravi, D. and Connors, K.A. 1993. Solvent effects on chemical processes. 3. Surface tension of binary aqueous organic solvents. *J. Solution Chem.* 22:321–330.

46. Knepp, T.N., Renkens, T.L., and Shepson, P.B. 2009. Gas phase acetic acid and its qualitative effects on snow crystal morphology and the quasi-liquid layer. *Atmos. Chem. Phys.* 9:7679–7690.

47. Kühne, T.D., Pascal, T.A., Kaxiras, E., and Jung, Y. 2011. New insights into the structure of the vapor/water interface from large–scale first–principles simulations. *J. Phys. Chem. Lett.* 2:105–113.

48. Kuo, I.W. and Mundy, C.J. 2004. An *ab initio* molecular dynamics study of the aqueous liquid–vapor interface. *Science* 303:658–660.

49. Langmuir, I. 1917. The constitution and fundamental properties of solids and liquids. II. Liquids. *J. Am. Chem. Soc.* 39:1848–1906.

50. Libbrecht, K.G. 2005. The physics of snow crystals. *Rep. Prog. Phys.* 68:855–895.

51. Luo, S.-N., Strachan, A., and Swift, D.C. 2005. Deducing solid-liquid interfacial energy from superheating or supercooling: Application to H_2O at high pressures. *Modelling Simul. Mater. Sci. Eng.* 13:321–328.

52. Marinova, K.G., Alargova, R.G., Denkov, N.D., Velev, O.D., Petsev, D.N., Ivanov, I.B., and Borwankar, R.P. 1996. Charging of oil-water interfaces due to spontaneous adsorption of hydroxyl ions. *Langmuir* 12:2045–2051.

53. Mecke, K.R. and Dietrich, S. 1999. Effective Hamiltonian for liquid-vapor interfaces. *Phys. Rev. E* 59:6766–6784.
54. Moilanen, D.E., Fenn, E.E., Wong, D., and Fayer, M.D. 2009. Water dynamics in large and small reverse micelles: From two ensembles to collective behavior. *J. Chem. Phys.* 131:014704.
55. Molina, M.J. 1996. Polar ozone depletion (Nobel Lecture). *Angew. Chem. Intl. Ed. Engl.* 35:1778–1785.
56. Molina, M.J., Tso, T., Molina, L.T., and Wang, F.C.-Y. 1987. Antarctic stratospheric chemistry of chlorine nitrate, hydrogen chloride and ice: Release of active chlorine. *Science* 238:1253–1257.
57. Nagata, Y., Pool, R.E., Backus, E.H.G., and Bonn, M. 2012. Nuclear quantum effects affect bond orientation of water at the water-vapor interface. *Phys. Rev. Lett.* 109:226101.
58. Nakaya, U. 1954. *Snow Crystals: Natural and Artificial.* Cambridge, MA: Harvard University Press.
59. Nordlund, D., Ogasawara, H., Wernet, P., Nyberg, M., Odelius, M., Pettersson, L.G.M., and Nilsson, A. 2004. Surface structure of thin ice films. *Chem. Phys. Lett.* 395:161–165.
60. Ogletree, D.F., Bluhm, H., Lebedev, G., Fadley, C.S., Hussain, Z., and Salmeron, M. 2002. A differentially pumped electrostatic lens system for photoemission studies in the millibar range. *Rev. Sci. Instrum.* 73:3872–3877.
61. Paluch, M. 2000. Electrical properties of free surfaces of water and aqueous solutions. *Adv. Colloid Inter. Sci.* 84:27–45.
62. Pan, D., Liu, L.-M., Tribello, G.A., Slater, B., Michaelides, A., and Wang, E. 2010. Surface energy and surface proton order of the ice Ih basal and prism surfaces. *J. Phys. Condens. Matter* 22:074209.
63. Pittenger, B., Fain, S.C., Cochran, M.J., Donev, J.M.K., Robertson, B.E., Szuchmacher, A., and Overney, R.M. 2001. Premelting at ice-solid interfaces studied via velocity-dependent indentation with force microscope tips. *Phys. Rev. B* 63:134102.
64. Pockels, A. 1891. Surface tension. *Nature* 43:437–439.
65. Quincke, G. 1861. Über die Fortführung materieller Theilchen durch strömende Elektricität. *Ann. Phys. Chem.* 113:513–598.
66. Sadtchenko, V. and Ewing, G.E. 2002. Interfacial melting of thin ice films: An infrared study. *J. Chem. Phys.* 116:4686–4697.
67. Sedlmeier, F., Janecek, J., Sendner, C., Bocquet, L., Netz, R.R., and Horinek, D. 2008. Water at polar and nonpolar solid walls. *Biointerfaces* 3:FC23–FC39.
68. Siegbahn, K. 1982. Electron spectroscopy for atoms, molecules, and condensed matter. *Rev. Mod. Phys.* 54:709–728.
69. Siegbahn, H., Svensson, S., and Lundholm, M. 1981. A new method for ESCA studies of liquid-phase samples. *J. Electron Spectrosc. Rel. Phenom.* 24:205–213.
70. Srivastava, G.P. 1997. Theory of semiconductor surface reconstruction. *Rep. Prog. Phys.* 60:561–613.
71. Stillinger, F.H. 1973. Structure in aqueous solutions of nonpolar solutes from the standpoint of scaled-particle theory. *J. Solution Chem.* 2:141–158.
72. Stiopkin, I.V., Weeraman, C., Pieniazek, P.A., Shalhout, F.Y., Skinner, J.L., and Benderskii, A.V. 2011. Hydrogen bonding at the water surface revealed by isotopic dilution spectroscopy. *Nature* 474:192–195.
73. Teschke, O. and de Souza, E.F. 2005. Water molecular arrangement at air/water interfaces probed by atomic force microscopy. *Chem. Phys. Lett.* 403:95–101.
74. Thomson, W. 1871. On the equilibrium of vapour at a curved surface of liquid. *Phil. Mag.* 42:448–452.

75. Tolman, R.C. 1949. The effect of droplet size on surface tension. *J. Chem. Phys.* 17:333–337.

76. van der Veen, J.F. 1999. Melting and freezing at surfaces. *Surf. Sci.* 433–435:1–11.

77. van Oss, C.J., Giese, R.F., Wentzek, R., Norris, J., and Chuvilin, E.M. 1992. Surface tension parameters of ice obtained from contact angle data and from positive and negative particle adhesion to advancing freezing fronts. *J. Adhesion Sci. Technol.* 6:503–516.

78. Vega, C., Martín–Conde, M., and Patrykiejew, A. 2006. Absence of superheating for ice Ih with a free surface: A new method of determining the melting point of different water models. *Mol. Phys.* 104:3583–3592.

79. Viswanathan, R., Zasadzinski, J.A., and Schwartz, D.K. 1994. Spontaneous chiral symmetry breaking by achiral molecules in a Langmuir–Blodgett film. *Nature* 368:440–443.

80. Wagner, W. and Pruss, A. 1993. International equations for the saturation properties of ordinary water substance. Revised according to the international temperature scale of 1990. *J. Phys. Chem. Ref. Data* 22:783–787.

81. Wang, D., Duan, H., and Möhwald, H. 2005. The water/oil interface: The emerging horizon for self-assembly of nanoparticles. *Soft Matter* 1:412–416.

82. Wannier, G.H. 1950. Antiferromagnetism. The triangular Ising net. *Phys. Rev.* 79:357–364.

83. Wei, X. and Shen, Y.R. 2002. Vibrational spectroscopy of ice interfaces. *Appl. Phys. B* 74:617–620.

84. Wei, X., Miranda, P.B., and Shen, Y.R. 2001. Surface vibrational spectroscopic study of surface melting of ice. *Phys. Rev. Lett.* 86:1554–1557.

85. Wilson, K.R., Schaller, R.D., Co, D.T., Saykally, R.J., Rude, B.S., Catalano, T., and Bozek, J.D. 2002. Surface relaxation in liquid water and methanol studied by X-ray absorption spectroscopy. *J. Chem. Phys.* 117:7738–7744.

86. Wilson, M.A., Pohorille, A., and Pratt, L.R. 1988. Surface potential of the water liquid–vapor interface. *J. Chem. Phys.* 88:3281–3285.

87. Yamada, T., Okuyama, H., Aruga, T., and Nishijima, M. 2003. Vibrational spectroscopy of crystalline multilayer ice: Surface modes in the intermolecular-vibration region. *J. Phys. Chem. B* 107:13962–13968.

3 Water on Ideal Solid Surfaces

Behind him the water was melted gold
J. J. Fraxedas, The Lonely Crossing of Juan Cabrera

This chapter is devoted to the adsorption of water molecules on well-ordered (crystalline), defect-free surfaces, deserving the label ideal, spanning from the most ideal 0D case, a single water molecule on a surface, to the mono- and multilayer regime. It is strongly recommend to read the *Surface Science Reports* review articles of Thiel and Madey (1987), written before the irruption of the scanning probe techniques such as the STM and AFM, and those of Henderson (2002) and of Hodgson and Haq (2009) as well as the 2006 *Chemical Review* issue on the structure and chemistry at aqueous interfaces (volume 106, issue 4), in particular the review article by Verdaguer et al. (2006). In such articles detailed descriptions and discussions on many aspects related to water at interfaces can be found, with an extensive list of the surfaces that have been explored. Such works are indeed the sustaining pillars of the present chapter, which has been conceived to provide a global overview on the subject covering different topics in a simplified manner, such as electronic structure using frontier orbitals, substrate-induced structuring, confinement, RT ice, and the role of ions, to mention a few. The selected examples capture the essentials of the most relevant phenomena, avoiding a compendium of published water–substrate systems.

3.1 SINGLE WATER MOLECULES AND CLUSTERS

3.1.1 SINGLE WATER MOLECULES ON METALLIC SURFACES

Let us start at the most fundamental level with the study of the interaction of an individual water molecule, a monomer, with an ideal metallic surface. Can we ascertain the orientation of the molecule from very simple concepts? Let us try it. If we consider the water molecule as an electrical dipole on a metallic surface in the Fermi scenario (a sea of free electrons) then the water dipole will induce an image dipole on the metal, as represented in Figure 3.1a. The energetically most favorable distribution corresponds to a collinear alignment, indicating that the H–O–H plane should be perpendicular to the surface's plane. The maximum attraction between two dipoles occurs when they are in line, with an energy twice the value when both dipoles are aligned antiparallel to each other, as can be deduced from (1.27). However, if the molecule interacts through one of its lone-pairs in the pseudo-tetrahedral sp^3 scenario, one would expect that the H–O–H plane forms an angle close to $35.26°$ ($90 - 109.47/2$ or half of the dihedral angle) with the metal surface, provided the lone-pair points perpendicularly

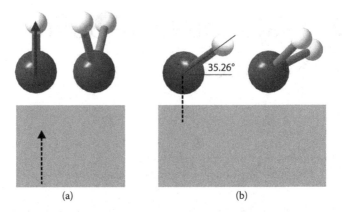

FIGURE 3.1 Scheme of the interaction of a water molecule with the surface of a metal considering: (a) dipole-induced dipole interaction and (b) a lone-pair pointing perpendicularly to the surface. The water dipole and the induced dipole are represented by a continuous and discontinuous arrow, respectively. In (b) the angle between the H–O–H plane and the metal surface is 35.26°, where the sp^3-like lone-pair is represented by a discontinuous line. Water molecules are represented with the H–O–H plane perpendicular to the page (left of each figure) and with a certain perspective (right of each figure).

to the surface (see Figure 3.1b). Then, does the water molecule lie nearly parallel or perpendicular to the surface? The dipole-induced model would be appropriate in the absence of chemical bonding (lone-pairs play no role), but ignoring the electronic structure of a metal when considering the interaction with a single molecule seems a rather crude approximation. Thus, when the relevant orbitals from the metal are considered, Figure 3.1b should be a better approximation.

Values of $\simeq 30°$ have been derived from high-resolution electron energy loss spectroscopy (HREELS) performed on Cu(100) and Pd(100) surfaces (Andersson, Nyberg, and Tengstål 1984), whereas lower values, about 23° and below 15°, have been computed for the Cu(H$_2$O) complex (Pápai 1995) and for close-packed surfaces of transition metals (Carrasco, Michaelides, and Scheffler 2009), respectively, as we show below. Such planar disposition means that the water dipole component perpendicular to the surface is small, which has important consequences for IR measurements.

Electronic and Vibrational Structure

We start this part by summarizing the conceptually simple approach of Hoffmann (1988) based on frontier orbitals considering only metallic surfaces for easier understanding. Disregarding vdW interactions, the molecule–surface interaction can be described as follows. Figure 3.2a shows a typical molecule–molecule interaction diagram (we have used such diagrams in Section 1.2.2 when deriving the MOs of water) and in Figure 3.2b a molecule–metallic surface interaction diagram is schematized, assuming that molecule–metal interactions can be explained in terms of the interaction between frontier orbitals, that is, MOs (water) and energy bands (metal) close to E_F. The most relevant interactions are expected to be the two-orbital, two-electron stabilizing interactions **1** and **2**. For the molecule–metal case (**1**), the energy levels

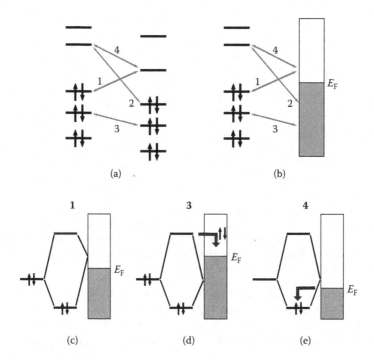

FIGURE 3.2 Schematic diagrams of (a) molecule–molecule orbital interactions and (b) molecule–metallic surface interactions. Cases (c), (d), and (e) represent the 1, 3, and 4 molecule–metallic surface interactions, respectively. (Adapted from R. Hoffmann, *Rev. Mod. Phys.* 60: 601–628, 1988. With permission.)

combine to give a lower energy two-electron MO and a higher energy unoccupied MO (Figure 3.2c). Depending on the relative MO energies and on the degree of overlap, both interactions will involve charge transfer from one system to the other. Interaction **3** is of the two-orbital, four-electron type, which for the molecule–molecule system is destabilizing and repulsive. However, in the molecule–surface system (Figure 3.2d), **3** may become attractive if the antibonding component is located above E_F. In this case electrons will transfer to the solid and the system become stabilized. Interaction **4** is expected to have no effect on molecule–molecule interactions, because both orbitals are empty, but in the molecule–surface case (Figure 3.2e) it may contribute significantly if the bonding level lies below E_F because in this case charge transfer from the solid to the molecule is expected, leading again to an attractive molecule–surface interaction.

According to DFT studies water monomers adsorb intact (no dissociation) on atop sites lying almost parallel to the close-packed (111) surfaces of transition metals such as Cu, Rh, Pd, Ag, Pt, and Au as well as the (0001) surface of Ru (Michaelides et al. 2003; Meng, Wang, and Gao 2004; Carrasco, Michaelides, and Scheffler 2009). The adsorption geometry is illustrated in Figure 3.3, and Table 3.1 summarizes the calculated adsorption parameters. The three angles, α (the tilt formed between the molecular plane and the metal surface), δ (between the surface normal and the direction defined by the oxygen and atop metal atoms), and θ (internal H–O–H) are given together with oxygen–metal (d_{O-M}) and oxygen–hydrogen (d_{O-H}) distances

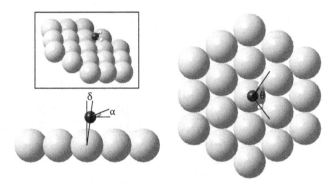

FIGURE 3.3 Side (left) and top (right) views of a water monomer adsorbed on a close-packed metal surface. The angle α describes the tilt angle formed by the water molecular plane and the metal surface, with δ the angle between the surface normal and the direction defined by both the oxygen and the atop metal atom. The H–O–H angle is represented by θ. The inset shows a perspective view of the molecule–surface system.

and the adsorption energies E_{ads}. Note that $0 < \alpha < 15°$, a constraint with important implications in terms of the $3a_1$ and $1b_1$ frontier MOs, as we show below. Because $\alpha > 0$ a net positive water dipole perpendicular to the surface is obtained ($\mu \sin \alpha$). The computed θ and d_{O-M} values are very close to the gas phase values, indicat-

TABLE 3.1
Calculated Adsorption Parameters for a Water Monomer on Metal Surfaces

	E_{ads} eV	d_{O-M} Å	d_{O-H} Å	θ Degrees	α Degrees	δ Degrees	Reference
Au(111)	0.13	3.02	0.97	105	13		[a]
Ag(111)	0.15	2.67	0.98	104.2	0.5	7.6	[b]
	0.18	2.78	0.97	105	9		[a]
Cu(111)	0.19	2.36	0.98	104.5	8.7		[c]
	0.24	2.25	0.98	106	15		[a]
Pd(111)	0.27	2.33	0.98	104.8	5.1	5.4	[b]
	0.33	2.28	0.98	105	7		[a]
Pt(111)	0.35	2.36	0.98	106	7		[a]
	0.30	2.34		105.7	8		[d]
Rh(111)	0.36	2.34	0.99	104.6	4.2	4.1	[b]
	0.42	2.31	0.98	106	9		[a]
Ru(0001)	0.38	2.29	0.98	106	6		[a]
	0.42	2.31	0.99	105.3	11.5	3.0	[b]

[a] (Michaelides et al. 2003), [b] (Carrasco, Michaelides, and Scheffler 2009), [c] (Tang and Chen 2007), [d] (Árnadóttir et al. 2010)

E_{ads} stands for the adsorption energy, d_{O-M} and d_{O-H} for the oxygen–metal and oxygen–hydrogen distances, respectively, and θ, α and δ for the internal H–O–H angle, the tilt formed between the molecular plane and the metal surface and the angle between the surface normal and the direction defined by the oxygen and atop metal atoms, respectively. See Figure 3.3 for a scheme.

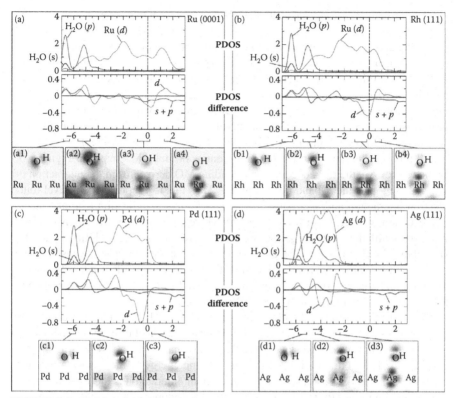

FIGURE 3.4 PDOS and PDOS difference plots (in electrons/eV units) projected onto $s + p$ and d states of the atop metal atom upon which the water is adsorbed. Negative (positive) values stand for depletion (creation) of metal states. E_F sets the binding energy reference. The insets display electron density contour plots. (Reprinted from J. Carrasco, A. Michaelides, and M. Scheffler, *J. Chem. Phys.* 130:184707, 2009, American Institute of Physics. With permission.)

ing a weak molecule–surface interaction. The metal–oxygen distances lie below 3 Å and E_{ads} ranges from 0.13 to 0.42 eV, that is, of the order of the H-bonding energy. Thus we can anticipate that the balance between the in-plane (H-bonding) and out-of-plane (molecule–surface) interactions may become quite complex when an increasing number of water molecules are involved.

As mentioned above, the major contribution to the bonding should arise from the orbitals $3a_1$ and $1b_1$. The fact that the water molecule lies nearly flat on the surface is telling us that $1b_1$ plays the leading role, because it is perpendicular to the molecular plane (see Figures 1.5 and 1.6). This can be analyzed by inspecting the partial density-of-states (PDOS) of the joint water–surface system. Figure 3.4 shows the calculated PDOS and PDOS difference plots of water adsorbed on the $4d$ metal surfaces Ru(0001), Rh(111), Pd(111), and Ag(111) (Carrasco et al. 2009). The peaks closest in energy to E_F in the PDOS plots, the zero in the energy scale, with binding energies around 4–6 eV, correspond to $1b_1$.

The $3a_1$ features are located at about 6 eV. The depletion regions with respect to the clean surface in the PDOS difference plots (negative values) indicate states involved in the bonding. The prominent d depletion around E_F is related to the interaction of the $1b_1$ orbital with the substrate, which becomes larger when moving from Ru to Pd. This portion of the d-band is formed essentially by d_{z^2}-like orbitals that are pointing perpendicularly to the surface, thus enhancing the effective overlap with the $1b_1$ orbital. Because the d orbitals involved in the bonding are initially located very close to E_F it implies that the antibonding states appear above it conferring a stabilizing character (as in Figure 3.2d). Note that Ag behaves differently.

All this is schematized in Hoffmann's frontier orbital diagrams from Figure 3.5. In the case of Ag, the fully occupied antibonding MO derived from the interaction of $1b_1$ and the top of the d-band lies below E_F. Thus this is a repulsive two-level four-electron interaction and as a consequence no covalent stabilization can take place. The situation is different for Ru, Rh, and Pd, where the antibonding component of the interaction with the most energetic d metal states rises totally (Ru and Rh) or partially (Pd) above E_F (Figures 3.5a–c), contributing to the stabilization of the system. The energy gain produced by the orbital mixing, that is, the energy difference between the $1b_1$ level and the corresponding new bonding state formed, decreases when moving Ru to Pd. This is why $E_{ads}(Ru) > E_{ads}(Rh) > E_{ads}(Pd)$.

Concerning the $3a_1$ MO the PDOS plots in Figure 3.4 show that it interacts mainly with a narrow region of the d-band, leading to the formation of fully occupied bonding and antibonding states, a two-level four-electron scenario which is destabilizing (see Figure 3.5). This is due to the fact that the binding energy of the $3a_1$ state is larger than that of $1b_1$ and, as a consequence, the resultant interaction is too weak to enable the antibonding states to rise above E_F and become depopulated. The MO description seems more appropriate than the sp^3 one but just remember that both descriptions are equivalent (see discussion in Section 1.4.2).

Thus far we have seen that theoretical calculations provide structural and energetic information on the adsorption of individual water molecules on ideal metallic surfaces, but can such monomers be imaged? The answer is yes and the technique that permits such a direct observation has the generic name of scanning probe microscopy (SPM). But in order to measure individual molecules one has to work in UHV, avoiding contamination that would complicate the identification of the monomers, at low temperatures, avoiding surface diffusion, and indeed at extremely low water coverages, in order to avoid molecule–molecule lateral interactions. The STM technique demands the use of metallic substrates, because the tunneling current between the sample and probe (a metallic tip) has to be measured, but insulating materials can be measured with AFM in the frequency modulation operation mode. We show a few examples next but we anticipate that the SPM technique allows not only the visualization of the molecules but performing spectroscopy as well.

Water monomers appear as simple protrusions in the STM images with apparent heights that depend on the experimental conditions (chemical nature of the surface, temperature, and applied voltage and tunneling current). If the interaction with the surface is weak, the images reveal the HOMO and LUMO of the molecule, which are rather structureless inasmuch as the water molecule is so small. However, for larger molecules it is possible to probe the real-space distribution of the MOs and several examples can be found in the literature [see, e.g., Repp et al. (2005)

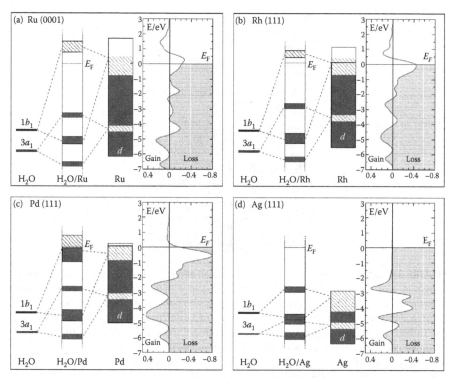

FIGURE 3.5 Schematic molecular-level interaction diagrams for water monomer adsorption on (a) Ru(0001), (b) Rh(111), (c) Pd(111), and (d) Ag(111). The principal interactions between the $1b_1$ and d states of the substrate are indicated. The graphs on the right of each case are d projected PDOS difference plots (same as those in Figure 3.4). (Reprinted from J. Carrasco, A. Michaelides, and M. Scheffler. *J. Chem. Phys.* 130:184707, 2009, American Institute of Physics. With permission.)

and Fernández-Torrente et al. (2007)]. The protrusions are more or less broadened depending on the available configurations, for example, the different ways the water molecule can accommodate on the surface without diffusion. For the metallic surfaces discussed above there are six available positions leading to a sixfold configuration (see Figure 3.3). Figure 3.6a shows STM images taken at 6 K of isolated water molecules on a Ru(0001) surface (Mugarza et al. 2009). Some of the molecules are bound to a Ru atom (I in figure) and some are also bound to a carbon atom (II in figure). The carbon atoms appear as 40-Pm deep depressions. Type I protrusions appear brighter with an apparent height of 0.055 nm and type II exhibit apparent heights of 0.030 nm.

By varying the applied tip voltage in a controlled manner, the different vibrational modes of the molecule can be excited, opening up the extraordinary possibility of performing vibrational spectroscopy with single molecules on surfaces, a major achievement because the vibrational fingerprint of an adsorbate allows its chemical identification (Stipe, Rezaei, and Ho 1998). When an isolated water molecule is bound nondissociatively to a surface its nine normal modes transform to those shown in Figure 3.7. In addition to the three modes (symmetric and asymmetric stretching

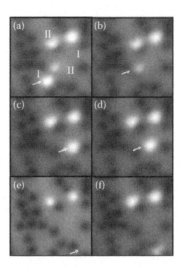

FIGURE 3.6 STM images (5 × 5 nm) taken with an applied bias of 50 mV and a tunneling current of 0.2 nA at 6 K of a Ru(0001) surface covered with water monomers. The sequence shows a series of manipulation experiments by application of voltage pulses. The arrows point to the location of the STM tip where a voltage pulse of 3 s will be applied. (a) Image containing two isolated (I) (brightest spots) and two carbon-bonded (II) water molecules. (b) The type I molecule pointed to by the arrow in (a) has been desorbed by a 500 mV pulse. (c) The type II molecule pointed to by the arrow in (b) is moved and the complex converted to type I by a 550 mV pulse. (d) Type I molecule in (c) diffused by one Ru lattice site after a 450 mV pulse. (e) Type I molecule was transferred to the tip, indicated by its disappearance and by the change in image contrast. (f) The water molecule on the tip is transferred back to the surface by a 550 mV pulse becoming type II. (Reproduced from A. Mugarza, et al. *Surf. Sci.* 603:2030–2036. 2009. With permission from Elsevier.)

and bending) discussed for the isolated molecule, three frustrated translations along the x-, y-, and z-axes and three frustrated rotations about the same axes are obtained. The surface-oxygen stretch indicated in the figure can be understood as a frustrated translation along the z-axis. The hindered rotations (librations) are termed rocking (x), wagging (y), and twisting (z). The modes are classified in terms of dipole-allowed and dipole-forbidden transitions and labeled according to the irreducible representations within the C_{2v} point group. Note that this is only valid in the ideal case when the molecular plane is perpendicular to the flat surface.

The excitation induced by the tunneling electrons can lead to the diffusion, desorption, association, or dissociation of the species in a selected manner (Pascual et al. 2003) and Figure 3.6 provides an example. Above ~450 mV pulses, a value close to the O–H stretching mode (450 meV = 3,630 cm^{-1}), diffusion, desorption, or transfer to the tip is induced as revealed by the displacement or the disappearance of the water molecule in subsequent STM images. Between Figures 3.6a and b the type I molecule near the bottom was desorbed. A pulse applied to the type II molecule in (b) results in the displacement of the molecule to a carbon-free region, thus becoming a type I monomer (see Figure 3.6c). Another pulse to the same molecule in (c) moves it

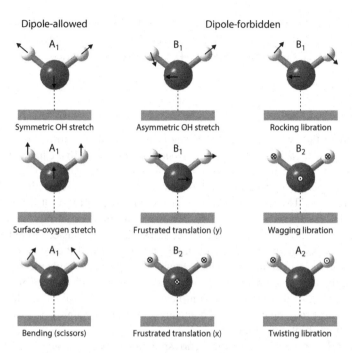

Dipole-allowed

Dipole-forbidden

A_1	B_1	B_1
Symmetric OH stretch	Asymmetric OH stretch	Rocking libration
A_1	B_1	B_2
Surface-oxygen stretch	Frustrated translation (y)	Wagging libration
A_1	B_2	A_2
Bending (scissors)	Frustrated translation (x)	Twisting libration

FIGURE 3.7 Normal modes of a water monomer adsorbed on a flat surface indicating their symmetry. (Adapted from P.A. Thiel, F.M. Hoffmann, and W.H. Weinberg, *J. Chem. Phys.* 75: 5556–5572, 1981. With permission.)

one lattice constant to the left (see Figure 3.6d). Other examples showing induced diffusion involving scissoring and stretching modes can be found in the literature (Morgenstern and Rieder 2002; Fomin et al. 2006). Water dissociation is found for voltages above 1 eV. This is a serious drawback to the determination of the electronic structure of water on metallic surfaces with STM, because MO are separated by more than 1 eV and thus local (electro)chemistry is induced for applied voltages >1 eV.

A clear spectroscopic signature of the absence of H-bonding is the bending mode. If the measured values are very close to those corresponding to isolated molecules (1,595 cm^{-1}) this implies that the scissoring movement is not stiffened by neighboring molecules. Andersson, Nyberg, and Tengståhl (1984) were the first to characterize the adsorbed monomer on Cu(100) and Pd(100) surfaces at 10 K using HREELS. The scissoring mode was found at 1,589 and 1,597 cm^{-1} for Cu(100) and Pd(100), respectively, close to the free-molecule value. In addition, translational/rotational modes at 230 and 335 cm^{-1} for Cu(100) and Pd(100), respectively, were also detected. Monomers have also been identified using reflection absorption infrared spectroscopy (RAIRS). Most of the studies have been performed with D_2O in order to improve the signal-to-noise ratio of the data. On Pt(111) at 25 K with a coverage of 0.18 ML, two bands at 2,706 and 2,465 cm^{-1} have been assigned to the asymmetric and symmetric OD stretches of the monomer, respectively, with the corresponding scissoring mode at 1,166 cm^{-1} (Ogasawara, Yoshinobu, and Kawai 1999). Bending modes at

1,157 cm^{-1} on Ru(0001) at 25 K below 0.36 ML (Thiam et al. 2005), at 1161 on Rh(111) at 20 K below 0.16 ML (Yamamoto et al. 2005), and at 1161 on Ni(111) at 20 K at 0.03 ML (Nakamura and Ito 2004) have also been reported.

3.1.2 SMALL CLUSTERS ON SURFACES

What happens then when a second water molecule approaches sufficiently? The second molecule will vacillate between the possibility of bonding to the metal, just as the first arrived molecule, and the formation of H-bonding. The reason is simply that the involved energies are very similar (0.2–0.3 eV). If both molecules are directly bonded to the metal, then STM images would show a dimer formed by two protrusions roughly separated by the distance between two metallic near-neighbor surface atoms. However, what is observed for close-packed metallic surfaces is a cluster formed by seven protrusions! This has been nicely illustrated for the particular case of the Pt(111) surface (Motobayashi et al. 2008) where monomers appear as single protrusions, and can be simply interpreted with a dimer where a central molecule is fixed and the other is freely rotating over the six equivalent nearest-neighbor metal sites. The hopping frequency of the moving molecule is much higher than the scanning frequency, hence the apparent multiplicity of the image.

Figure 3.8 compares the structures corresponding to both a monomer and a dimer on the Cu(111) surface according to DFT calculations (Michaelides 2007). Both molecules in the dimer lie at different heights, in a buckled geometry, as observed in Figure 3.8d. The buckled geometry on surfaces is very important inasmuch as it opens the door to the formation of bilayers and thus the heterogeneous formation of hexagonal ice externally induced by surfaces. The computed adsorption energy of this dimer is 0.32 eV per water molecule. In this structure both molecules adsorb above atop sites with the D molecule closer to the surface than the A molecule. The shortest Cu–O bond lengths for the D and A molecules are 2.20 and 3.00 Å, respectively. In

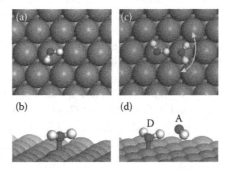

FIGURE 3.8 Top and side views of the lowest energy configurations of an adsorbed H_2O monomer, (a) and (b), and a H_2O dimer, (c) and (d), on Cu(111). Donor and acceptor molecules are represented by D and A, respectively. In (c) the arrow illustrates the facility of rotation of the acceptor molecule. (Reprinted from A. Michaelides, *Faraday Discuss.* 136:287–297, 2007. With permission of The Royal Society of Chemistry.)

this configuration the O–O distance is 2.74 Å. The essentially free rotation of the acceptor molecule is illustrated with an arrow in the figure. In the case of the Pd(111) surface, DFT calculations (Ranea et al. 2004) show that the low-lying water molecule (D) is 0.50 Å closer to the substrate than the high-lying one (A). Compared to the water monomer, the donor water molecule is 0.10 Å closer to the surface and interacts strongly with it, whereas the acceptor molecule interacts weakly with the substrate at 2.90 Å from it. The estimated rotation barrier is very low, of only 0.02 eV, hence the observed rotation. The ease of rotation of the high-lying water molecule around the axis centered at the donor molecule triggers the dimer diffusion. If the A molecule binds to the metallic atom, then the initial D and A molecules can exchange their roles leading to a net translation of the dimer by one lattice spacing.

Adding more water molecules one by one leads to the formation of trimers, tetramers, pentamers, hexamers, and so on. On Cu(111) two-trimer structures, both with 0.37 eV adsorption energy per water molecule, have been predicted (Michaelides 2007). They are bent structures with only two H-bonds connecting the three water molecules, differing from the trimers in the gas phase. In one of the configurations one water molecule donates two H-bonds. The lowest energy structure identified for the water tetramer resembles one of the low-energy trimers but with a fourth water added as a H-bond donor to the central water. The binding energy of this cluster is 0.41 eV per molecule. Again the adsorbed structure differs significantly from the lowest energy gas phase isomer which is, like the trimer, a cyclic structure and is comprised of four H-bonds. The most stable pentamer structure is cyclic, where each water acts as a single H-bond donor and a single H-bond acceptor, rather similar to the low-energy structure of the gas phase pentamer. The adsorption energy is 0.44 eV per water molecule.

The case of cyclic hexamers is of great interest because an hexamer constitutes "the smallest piece of ice", an expression coined by Nauta and Miller (2000). The cyclic hexamers can be thought of as building blocks that can lead to 2D mono- or bilayers in a LEGO-like way. In this context a bilayer is a buckled monolayer (ML). This is a rather naïve view but in the case of benzene, it rationalizes in a pedagogical way the host of 2D and 3D arrangements of carbon (linear and planar molecules, nanotubes, graphene, etc.) described in Fraxedas (2006). Figure 3.9a shows a high-resolution STM image of an hexamer on Cu(111) at 17 K (Michaelides and Morgenstern 2007). The inset shows the atomically resolved surface indicating that the hexagon is approximately aligned with the close-packed directions of the substrate. The isolated protrusion at the bottom of the image corresponds to a water monomer. Note that the cluster is formed by six protrusions and not seven as in the case of the dynamic dimer. DFT calculations predict that on Cu(111) and Ag(111) the lowest energy six-member cluster is a cyclic hexamer with each water molecule acting as a single H-bond donor and single H-bond acceptor and located close to atop sites, as depicted in Figure 3.10a. This explains why the hexamer is in a static configuration (at very low temperatures). The DFT structure of the hexamer exhibits a noticeable buckling in the heights of adjacent water molecules. The vertical displacement between adjacent water molecules is about 0.76 Å on Cu and 0.67 Å on Ag. Furthermore, the six nearest-neighbor O–O distances are not equal: they alternate between two characteristic values of 2.76 and 2.63 Å on Cu and 2.73 and 2.65 Å on Ag. These values have to be compared to the 2.75 Å found in Ih ice at 10 K (Röttger et al. 1994). The values of the lattice

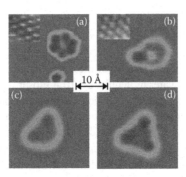

FIGURE 3.9 High–resolution STM images of adsorbed water clusters. (a) H_2O hexamer on Cu(111) taken with a bias voltage of 20 mV and a tunnel current of 11 pA. (b) D_2O heptamer on Ag(111) (11 mV and 2 nA). Images of the substrates with atomic resolution are shown in the corresponding insets. (c) D_2O octamer on Ag(111) (−21 mV and 2 nA) and (d) D_2O nonamer on Ag(111) (11 mV and 2 nA). (Reprinted from Macmillan, *Nature Materials* 6:597–601, 2007. With permission.)

constants extrapolated to 0 K for Cu and Ag are 3.6024 and 4.0690 Å, respectively (Giri and Mitra 1985). Thus, in the (111) plane, the surface lattice constants are 4.412 and 4.983 Å, respectively, which result from multiplying the nearest-neighbor distance (2.547 and 2.877 Å) by $\sqrt{3}$. The comparison of such surface lattice constants with those of the basal plane of ice Ih at 10 K, gives mismatches of about 2% (Cu) and −11% (Ag), which explains the differences in buckling.

Addition of one, two, and three water molecules leads to the formation of heptamers, octamers, and nonamers, respectively (Figures 3.9b–d and 3.10b–d). In the case of heptamers (Figures. 3.9b and 3.10b), the extra water molecule (H-bond acceptor) can be allocated in six equivalent lowest-energy positions. Once the water molecule is incorporated in the cluster, additional molecules will be allocated in

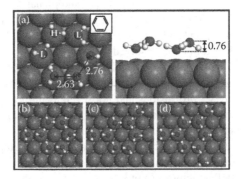

FIGURE 3.10 (a) Top and side views of the equilibrium cyclic hexamer optimized structures obtained from DFT on Cu(111). High- and low-lying molecules are indicated by H and L, respectively. The inset shows a schematic diagram of the Kekulé structure of benzene. Top views of (b) an heptamer, (c) an octamer, and (d) a nonamer. (Reprinted from Macmillan, *Nature Materials* 6:597–601, copyright 2007. With permission.)

the free equivalent positions, as H-bond acceptors, as shown in Figure 3.10c for the octamer and in Figure 3.10d for the nonamer.

3.2   SUBSTRATE-INDUCED STRUCTURING OF MONO- AND BILAYERS

Given the formation of hexamers discussed above, one can consider the possibility of artificially ordering or structuring water in its hexagonal Ih phase with the help of foreign bodies such as crystalline surfaces. If this is feasible, we would then go from the smallest pieces of ice (hexamers) to real ice. It seems reasonable to consider first materials that can expose surfaces with hexagonal symmetry, provided that their lattice dimensions are not that different from those of the basal plane of ice Ih. In addition, the water–surface interaction should be sufficiently weak in order to prevent water dissociation but sufficiently strong in order to allow wetting of the surface. It is thus a matter of a balance. An estimation of the value of E_{ads} can be obtained from the STM results shown above for metals, where molecules remain intact below ~ 1 eV. But before considering the formation of water MLs on suitable surfaces, let us make a point regarding water dissociation. In the liquid state water dissociates according to the reversible reaction:

$$2H_2O(l) \rightleftharpoons H_3O^+(aq) + OH^-(aq) \qquad (3.1)$$

where l and aq stand for liquid and aqueous, respectively. This reaction is usually simplified into the formal expression

$$H_2O(l) \rightleftharpoons H^+(aq) + OH^-(aq)$$

understanding that (3.1) is the correct one. The dissociation of a molecule, which is due to local electric field fluctuations generated by the surrounding molecules, is also termed autoionization because no external driving force is at hand. The mechanisms leading to autoionization of water are of short–range order, according to MD calculations, involving mainly the first and second hydration shells (Reischl, Köfinger, and Dellago 2009). For pure water at 298 K, the concentration of $H^+(aq)$ and of $OH^-(aq)$ are 10^{-7} moles L^{-1} and because for intact water the concentration is 55.55 moles L^{-1} we obtain a concentration ratio between dissociated and undissociated water molecules of 1.8×10^{-9}: 9 orders of magnitude separating both physical states. Such a small number is telling us that the role of H^+ and OH^- should be, in principle, negligible on nonreactive surfaces. As we show below, this is the case. However, we anticipate that when E_{ads} increases, the role of OH^- becomes relevant and in fact determines the wetting properties of the surface.

3.2.1   INTERFACIAL REGISTRY

The relative ordering or registry that an ordered ML can achieve on an underlying crystalline substrate surface can be described in a generalized way using matrix formulations (Wood 1964; Hooks, Fritz, and Ward 2001). The recommended nomenclature for surfaces by the International Union of Pure and Applied Chemistry (IUPAC) can be found in Bradshaw and Richardson (1996). The substrate surface can be described by the lattice parameters a_s, b_s and the angle α_s ($\alpha_s > 0$) formed between the corresponding vectors a_s and b_s. Analogously, the lattice parameters of the ML can be represented by a_{ML}, b_{ML} and the angle α_{ML} ($\alpha_{ML} > 0$) between both vectors a_{ML} and b_{ML}. Finally, both lattices are related by the azimuthal angle θ between the lattice vectors a_s and a_{ML}. Figure 3.11a shows the referred parameters for two hexagonal lattices.

For a given azimuthal orientation θ, the substrate and ML lattice vectors are related through the general expression:

$$\begin{pmatrix} R_{11} & R_{12} \\ R_{21} & R_{22} \end{pmatrix} \begin{bmatrix} a_s \\ b_s \end{bmatrix} = \begin{bmatrix} a_{ML} \\ b_{ML} \end{bmatrix} \tag{3.2}$$

where the matrix coefficients R_{ij} (from registry) with $i, j = 1, 2$ are defined as:

$$R_{11} = \frac{a_{ML}}{a_s} \frac{\sin(\alpha_s - \theta)}{\sin \alpha_s} \tag{3.3a}$$

$$R_{12} = \frac{a_{ML}}{b_s} \frac{\sin \theta}{\sin \alpha_s} \tag{3.3b}$$

$$R_{21} = \frac{b_{ML}}{a_s} \frac{\sin(\alpha_s - \alpha_{ML} - \theta)}{\sin \alpha_s} \tag{3.3c}$$

$$R_{22} = \frac{b_{ML}}{b_s} \frac{\sin(\alpha_{ML} + \theta)}{\sin \alpha_s} \tag{3.3d}$$

Registry is classified according to the R_{ij} values as commensurate, coincident, and incommensurate. Commensurism is achieved when all R_{ij} are integers ($R_{ij} \in \mathbb{Z}$). In this case all the ML lattice points coincide with symmetry-equivalent substrate lattice points. For this reason, this registry is also known under the term *point-on-point* coincidence. If among R_{ij} there are at least two integers confined to a single column of the transformation matrix, registry is called type I coincidence. Every lattice point of the ML lies at least on one primitive lattice line of the substrate, a condition that has been described as *point-on-line* coincidence. Type II coincidence is achieved when $R_{ij} \in \mathbb{Q}$, $R_{ij} \notin \mathbb{Z}$. Registry is incommensurate when at least one R_{ij} is irrational and neither column of the translation matrix consists of integers. Examples involving organic molecules on inorganic substrates can be found in Fraxedas (2006). Note that this is a strictly geometrical model, thus ignoring the underlying chemistry.

We next explore the most usual and simplest cases of substrates with hexagonal, cubic, and rectangular surface lattices with regard to their registry with the basal plane of ice Ih. This has to be taken as a systematic way to identify potential candidates that might structure interfacial water and not as a complete exhaustive list of known surfaces.

3.2.2 INORGANIC SURFACES WITH HEXAGONAL SYMMETRY

In the particular interesting case of ice Ih on a substrate with hexagonal symmetry we have $a_s = b_s \equiv a_s$, $\alpha_s = 60°$ and $a_{Ih} = b_{Ih} \equiv a_{Ih}$, $\alpha_{Ih} = 60°$, where the Ih subscript

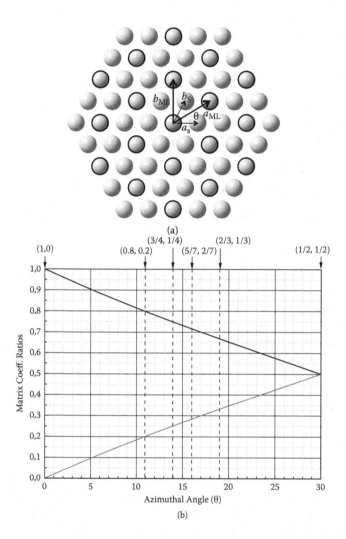

FIGURE 3.11 (a) Scheme of two hexagonal lattices in registry. The lattice vectors of the lattice represented by light gray balls (substrate) are a_s and b_s ($\alpha_s = 60°$) and the overlayer is represented by black circles with a_{ML} and b_{ML} ($\alpha_{ML} = 60°$). θ represents the azimuthal angle between a_s and a_{ML} (and equivalently between b_s and b_{ML}). (b) $\sin(60 - \theta)/\sin(60 + \theta)$ (black continuous line) and $\sin\theta/\sin(60 + \theta)$ (gray continuous line) as a function of θ. The corresponding ratios are: $(1, 0)$ for $\theta = 0°$, $(0.8, 0.2)$ for $\theta = 10.9°$, $(3/4, 1/4)$ for $\theta = 13.85°$, $(5/7, 2/7)$ for $\theta = 16.1°$, $(2/3, 1/3)$ for $\theta = 19.1°$, and $(1/2, 1/2)$ for $\theta = 30°$ and are indicated by discontinuous vertical dark gray lines.

refers to the hexagonal lattice of ice and substitutes the ML subscript. In this case $a_{Ih} \simeq 4.5$ Å and (3.3) simplifies to:

$$R_{11} = \frac{na_{Ih}}{a_s} \frac{\sin(60 - \theta)}{\sin 60} \qquad (3.4a)$$

$$R_{12} = \frac{n a_{Ih}}{a_s} \frac{\sin \theta}{\sin 60} \tag{3.4b}$$

$$R_{21} = -\frac{n a_{Ih}}{a_s} \frac{\sin \theta}{\sin 60} \tag{3.4c}$$

$$R_{22} = \frac{n a_{Ih}}{a_s} \frac{\sin(60 + \theta)}{\sin 60} \tag{3.4d}$$

where $n \in \mathbb{N}$ ($n = 1, 2, 3, \ldots$). The R-matrix can be expressed as:

$$\begin{pmatrix} R_{11} & R_{12} \\ R_{21} & R_{22} \end{pmatrix} = \frac{n a_{Ih}}{a_s} \frac{1}{\sin 60} \begin{pmatrix} \sin(60 - \theta) & \sin \theta \\ -\sin \theta & \sin(60 + \theta) \end{pmatrix} \tag{3.5}$$

and its determinant, $D_R = R_{11} R_{22} - R_{12} R_{21}$, turns out to be independent of θ:

$$\sqrt{D_R} = \frac{n a_{Ih}}{a_s} \tag{3.6}$$

as can be deduced from simple trigonometrics. Inasmuch as we are mostly interested here in commensurate registry, we have to find those θ values that make $R_{ij} \in \mathbb{Z}$. Thus, once the integer R_{ij} numbers are found, we have to calculate a_s and find, with the help of crystal structure databases, the potential surfaces capable of inducing hetero(quasi)epitaxy with ice Ih. One strategy directed toward finding combinations of integer R_{ij} numbers consists in plotting the $\sin(60-\theta)/\sin(60+\theta)$ and $\sin \theta / \sin(60+\theta)$ ratios as a function of θ, as depicted in Figure 3.11b, and look for simultaneous combinations of simple ratios such as 0, 1/2, 1/3, 1/4, 1, and so on.

The most representative cases are listed next within the $0 \leq \theta \leq 30°$ range.

(i) $\theta = 0°$, $\begin{pmatrix} 1 & 0 \\ 0 & 1 \end{pmatrix}$

This corresponds to the identity matrix, so that $a_s = a_{Ih}$ because $D_R = 1$. Apart from the trivial homoepitaxy case of ice Ih on the basal plane of ice Ih, we have several materials for which $a_s \simeq a_{Ih}$. The (111) surface of W ($Im\bar{3}m$) has $a_s = 4.48$ Å, so that tungsten (or wolfram) seems to be an ideal material for the heteroepitaxial growth of ice Ih. However, this surface turns out to be extraordinarily efficient in dissociating water. This is an example showing that registry might not be a sufficient condition for heteroepitaxy. Geometry does not account for the chemical state of the surface. The β-phase of AgI ($P6_3mc$), with $a_s = 4.592$ Å for the basal plane, is another candidate, as well as BaF$_2$ ($Fm\bar{3}m$), whose (111) surface exhibits $a_s = 4.38$ Å. We discuss both materials in detail below. Some other interesting surfaces are the basal planes of LiI with $a_s = 4.48$ Å ($P6_3/mmc$), Bi with $a_s = 4.53$ Å ($R\bar{3}m$), PbI$_2$, with $a_s = 4.56$ Å ($P6_3mc$), and VSi$_2$ with $a_s = 4.56$ Å ($P6_222$).

(ii) $\theta = 10.9°$, $\begin{pmatrix} 4 & 1 \\ -1 & 5 \end{pmatrix}$

In this case $a_s = n a_{Ih}/\sqrt{21}$. For $n = 3$, then $a_s = 2.94$ Å, which is close enough to the surface lattice constant of MgO(111) ($Fm\bar{3}m$), with $a_s = 2.97$ Å. This surface has not been explored because the cleavage face of this material is the (100), which is discussed in Section 3.2.3. Additional candidates belonging to the hexagonal systems are the basal planes of Ti ($P6_3/mmc$) and TiB$_2$ ($P6/mmm$) with $a_s = 2.95$ and $a_s = 3.03$ Å, respectively. The case of titanium is interesting, because the pure metal is extremely reactive, so there is no hope of building ordered ice Ih layers on top of it.

(iii) $\theta = 13.8°$, $\begin{pmatrix} 3 & 1 \\ -1 & 4 \end{pmatrix}$

In this case $a_s = na_{Ih}/\sqrt{13}$ and for $n = 2$, then $a_s = 2.5$ Å. The hexagonal diamond modification known as lonsdaleite, with the wurtzite structure and mentioned in Section 1.4.2, is a potential candidate because $a_s = 2.49$ Å as well as the (111) surface of cubic (zinkblende) diamond (111) ($Fd\bar{3}m$), with $a_s = 2.52$ Å. However, such surfaces have not been explored for the interfacial structuration of ice Ih. For $n = 3$, $a_s = 3.74$ Å, which is close to the surface lattice constant of the basal plane of covellite (CuS) and wurtzite (ZnS), 3.79 and 3.82 Å, respectively.

(iv) $\theta = 16.1°$, $\begin{pmatrix} 5 & 2 \\ -2 & 7 \end{pmatrix}$

In this case $a_s = na_{Ih}/\sqrt{39}$. For $n = 4$, then $a_s = 2.88$ Å, a value above the surface lattice constant of Pt(111) ($Fm\bar{3}m$), $a_s = 2.77$ Å. This corresponds to the rather involved $\sqrt{39} \times \sqrt{39}R16.1°$ ordering and a symmetry–related example can be found in Figure 3.19b (see discussion below for negative θ values).

(v) $\theta = 19.1°$, $\begin{pmatrix} 2 & 1 \\ -1 & 3 \end{pmatrix}$

In this case $a_s = na_{Ih}/\sqrt{7}$ and for $n = 2$, then $a_s = 3.4$ Å. Potential candidates are cobalt, iron, and titanium sulphides ($P6_3/mmc$) with lattice a values of 3.38, 3.44, and 3.30 Å, respectively, as well as the basal plane of NbSe$_2$ ($P\bar{6}m2$) with $a_s = 3.44$ Å.

(vi) $\theta = 30°$, $\begin{pmatrix} 1 & 1 \\ -1 & 2 \end{pmatrix}$

In this case $a_s = na_{Ih}/\sqrt{3}$. Note that, according to (3.4a) and (3.4b), $\sin(60 - \theta) = \sin\theta$ because $R_{11} = R_{12}$. For $n = 1$, then $a_s = 2.6$ Å, which is close enough to the surface lattice constant of Cu(111) ($Fm\bar{3}m$), $a_s = 2.56$ Å. Remember that water hexamers accomodate quite well on this hydrophobic surface, as shown in Figure 3.10, but at very low temperatures due to the low interaction energy. At higher temperatures water diffuses, ignoring this surface. Such a registry is the well-known $\sqrt{3} \times \sqrt{3}R30°$ reconstruction found in several interfaces not only involving water. For $n = 2$, then $a_s = 5.2$ Å, quite close to the lattice constants of the (0001) surfaces of LiNbO$_3$ and LiTaO$_3$, both belonging to the $R3c$ space group, with $a_s = 5.148$ and $a_s = 5.154$ Å, respectively.

The registries discussed above correspond to the most relevant ones for $30° \geq \theta \geq 0°$. By symmetry one can build the equivalent angles. The matrices corresponding to negative values of θ can be simply obtained by interchanging R_{11} and R_{22} as well as the signs of R_{12} and R_{21}, as can be readily demonstrated from (3.5). Following the mathematical formalism, the $\theta < 0$-matrices are the inverse matrices of those corresponding to $\theta > 0$ multiplied by their determinant. If we take $\theta = -16.1°$ and $\theta = -25.3°$ the resulting transformation matrices would be, respectively:

$$\begin{pmatrix} 7 & -2 \\ 2 & 5 \end{pmatrix} \quad \text{and} \quad \begin{pmatrix} 7 & -3 \\ 3 & 4 \end{pmatrix}$$

which correspond to the $\sqrt{39} \times \sqrt{39}R16.1°$ and $\sqrt{37} \times \sqrt{37}R25.3°$ arrangements illustrated in Figures 3.19a and b, respectively.

With the perspective of the geometrical arrangements leading to commensurate registry, we next discuss particular materials classified according to the adsorption energy of water on top of them. Those that are susceptible to form ice at temperatures around its melting point are included in the ambient conditions realm, which corresponds to $E_{ads} > 0.5$ eV, and those involving considerably lower temperatures, (e.g., below 200 K) are included in the cryogenic conditions part, corresponding to $E_{ads} < 0.5$ eV.

Ambient Conditions

It was B. Vonnegut (1947) who predicted, within a pure geometrical scheme as described above, that nearly lattice matching may induce nucleation of ice in ambient, not cryogenic, conditions. This may have extraordinary consequences because it would allow, in principle, control of the weather to a certain extent, a point that is discussed in detail in Chapter 5. A material that efficiently condenses water may trigger the formation of rain droplets, artificially avoiding droughts and heavy rain, as well as other atmospheric phenomena. Hence, its practical importance.

Table 3.2 summarizes potential candidates as ice nucleating agents classified according to the E_{ads} values of water on them and showing their lattice mismatch with ice Ih, referred to the 4.5117 Å lattice parameter at 223 K (Section 1.4.2). Given that the mismatch can be taken as a guess parameter, we take the liberty to compare lattice parameters at different temperatures. This is not a major problem for the sake of comparison. Many of the structures have been determined at RT, therefore it would be a real problem to find the structure of ice Ih at RT.

A relevant candidate is β-AgI (see Figure 3.12), the wurtzite-type polymorph of AgI, with lattice constants $a = 4.592$ and $c = 7.510$ Å (Burley 1963) and a resulting lattice mismatch of ~1.8 %.

Along the c-axis the crystal structure is formed by alternate atomic planes of either silver or iodine ions separated by 0.938 and 2.814 Å, respectively. This material has been barely investigated experimentally at the fundamental level (most of the studies have been at the application level) and the few available works are theoretical. According to computer simulations based on the Monte Carlo method, a ML of water is formed mimicking the substrate hexagonal symmetry, as illustrated in Figure 3.13 (Shevkunov 2007). The water molecules are arranged above the iodine ions. One hydrogen atom participates in the formation of a H-bond with the adjacent molecule, and the other one either remains free or oriented toward the iodine atom lying below. The dipole moments of the molecules are oriented almost parallel to the substrate plane, a geometry that does not favor the formation of ice. This is quite common

TABLE 3.2
Inorganic Materials Exhibiting Crystal Faces with Hexagonal Symmetry Ordered by Decreasing Adsorption Energy (E_{ads}) Computed for Single Molecules

Material	Crystal Face	Space Group	Surface Lattice Constant (a_s)	Mismatch	E_{ads}
			Å	%	eV
α-Al$_2$O$_3$	(0001)	$R\bar{3}c$	4.75		1.5[i]
kaolinite	(001)	$C1$	5.154–5.168[b]	~ 14	0.65[f]
W	(111)	$Im\bar{3}m$	4.48	−0.7	0.53[h]
β-AgI	(0001)	$P6_3mc$	4.593[a]	1.8	
CaF$_2$	(111)	$Fm\bar{3}m$	3.845	−14.8	0.51[e]
BaF$_2$	(111)	$Fm\bar{3}m$	4.384	−2.8	0.49[e]
muscovite	(001)	$C2/c$			0.34–0.46[d]
Ru	(0001)	$P6_3/mmc$	2.706	−4	0.40[g]
Pt	(111)	$Fm\bar{3}m$	2.774	6.5	0.35[g]
Cu	(111)	$P6_3/mmc$	2.556	−2	0.24[g]

[a] (Bührer, Nicklow, and Brüesch 1978), [b] (Neder et al. 1999), [c] (Tunega, Gerzabek, and Lischka 2004), [d] (Odelius, Bernasconi, and Parrinello 1997), [e] (Foster et al. 2009), [f] (Hu and Michaelides 2007), [g] (Michaelides et al. 2003), [h] (Chen, Musaev, and Lin 2007), [i] (Thissen et al. 2009).
a_s corresponds to the surface lattice constant determined at RT. The lattice mismatch is referred to the lattice constant of the basal plane of Ih ice: 4.5117 Å at 223 K (Röttger et al. 1994).

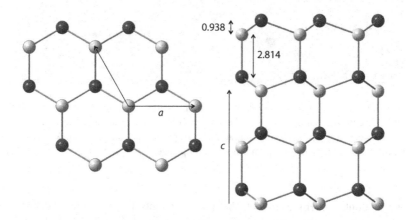

FIGURE 3.12 RT crystal structure of β-AgI along the c- (left) and b-axis (right). Crystallographic data taken from Burley (1963). Both lattice parameters $a = 4.592$ Å and $c = 7.510$ Å are indicated as well as the separation between adjacent atomic planes at distances 0.938 and 2.814 Å, respectively. Note that because of the tetrahedral arrangement, $0.938 = 1/3 \times 2.814$. Silver and iodine atoms are represented by light and dark gray balls, respectively.

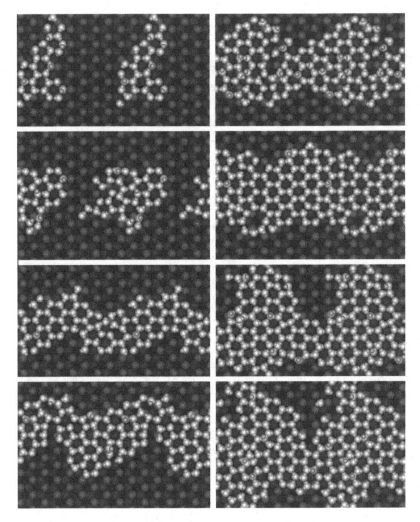

FIGURE 3.13 Consecutive stages of the growth of a water monomolecular film at 260 K on the defect-free infinite surface of a β-AgI crystal parallel to the basal face. The ST2 model has been used to describe the molecular interactions. (Reproduced from S.V. Shevkunov, *Colloid J.* 69: 360–377. 2007. With kind permission from Springer Science+Business Media B.V.)

and we encounter several examples: the first ML adopts a highly ordered distribution but does not help in the formation of an ordered second layer. The 2D character of the growth indicates that there is a high free energy barrier for the formation of the second and next layers (3D growth). Water seems to build a protective ML against further incorporation of water. Thus, the ideal defect-free basal plane of β-AgI is indicated for the formation of a well-ordered water ML but not for the formation of ice Ih. However, nano- and microparticles have shown their ability to condense water at temperatures below freezing. Two important issues have to be considered. First,

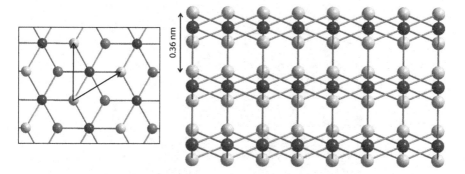

FIGURE 3.14 Views along the [111] (left) and the [1$\bar{2}$1] (right) directions of the crystal structure of BaF$_2$ ($Fm\bar{3}m$, $a = 0.62$ nm). The surface lattice vectors (0.44 nm) are indicated in the left figure. The separation between trilayers is 0.36 nm. Fluorine and barium atoms are represented by light and medium gray and dark gray balls, respectively.

the presence of defects, such as steps, cracks, contaminants, and the like, help in the 3D growth of water layers. On the other hand, nanoparticles, even electrically neutral, have a large polarization field. As mentioned above, β-AgI exhibits alternating atomic planes of iodine and silver ions along the c-direction. Because the material is ionic, the exposed surfaces have a given charge (e.g., positive if silver-terminated and negative if iodine-terminated), leading to an electric field that tends to orient the water molecules and induce condensation (Shevkunov 2009). It is clear that in the case of a large flat surface, the electric field will be weak and this effect will contribute negligibly.

Another candidate to grow ice Ih MLs from Table 3.2 is the alkaliearth fluoride BaF$_2$, which belongs to the $Fm\bar{3}m$ space group, because its (111) surface exhibits a lattice mismatch of -2.8%. Along the [111] direction, BaF$_2$ is formed by stacked F–Ba–F trilayers as shown in Figure 3.14. The distance between adjacent trilayers is 3.6 Å. Water adsorbs intact on top of the cation sites with a tilt due to H-bonding with a neighboring fluorine, according to detailed calculations (Nutt and Stone 2002; Foster, Trevethan, and Shluger 2009). The H–F attractive interaction makes the molecular plane point toward the surface plane resulting in a net negative water dipole perpendicular to the surface. In this case E_{ads} lies between 0.4 and 0.5 eV. Inasmuch as there are three symmetric fluorine ions around each cation, the water molecule can occupy three equivalent configurations with energy barriers of only $\sim$0.1 eV.

The first question to be answered is if water adsorbs intact or not on such a surface. According to Wu et al. (1994) water adsorbs dissociatively based on XPS studies performed on single crystals both at RT and at $\sim$130 K in UHV. The dissociation is revealed by the presence of signatures of hydroxyl groups with a binding energy of 531.6 eV and it might be related to the presence of defects induced by the preparation of the clean surface, which is mechanically performed (cleavage), and to irradiation with the formation of color centers (e.g., fluorine displacement). Unfortunately, NAPP experiments are not yet at hand to compare with but broad peaks should be expected due to surface charging caused by the insulating character of the material, a fact that would hinder the extraction of relevant information (see discussion in Section 3.5).

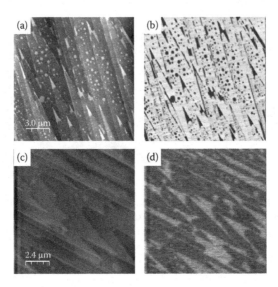

FIGURE 3.15 AM-AFM images taken at RT and 50% RH of a freshly cleaved BaF$_2$(111) surface: (a) topography, (b) phase, (c) SPFM, and (d) KPFM. (Reprinted from M. Cardellach, A. Verdaguer, J. Santiso, and J. Fraxedas, *J. Chem. Phys.* 132:234708, 2010. With permission.) American Institute of Physics and from M. Cardellach, A. Verdaguer, and J. Fraxedas, *Surf. Sci.*, 605: 1929–1933, (2011). With permission from Elsevier.) The WS×M free software has been used for image treatment: I. Horcas, et al. *Rev. Sci. Instrum.* 78:013705, 2007. With permission.)

When freshly cleaved BaF$_2$(111) surfaces are exposed to water vapor in a controlled way, the terraces and the steps become decorated, as illustrated in Figures 3.15b and c. Both figures correspond to the topography (a) and phase (b) images taken in AM-AFM mode at RT. In general water adsorbs preferentially at the lower terrace of the steps, although in some cases water accumulates on the higher terrace. In the case of V-shaped steps water becomes confined in the acute angles formed by steps of the same terrace. When water accumulates in the lower terrace of the steps, it forms 2D menisci. The accumulation of water at step edges is evidence of the high diffusion of water on the BaF$_2$(111) surface at RT (Cardellach et al. 2010). Foster, Trevethan, and Shluger (2009) computed (using DFT) the diffusion barriers to be 0.2 eV in the ideal case (water diffusing on defect-free surfaces) and below 1 eV when different kinds of vacancies are considered, concluding that molecular water should be extremely mobile on this surface, as well as on the (111) surfaces of CaF$_2$ and SrF$_2$. However, the 2D menisci are not formed on CaF$_2$(111), indicating the importance of lattice matching, because for this particular surface it amounts to -14.8% (Cardellach, Verdaguer, and Fraxedas 2011; see Table 3.2).

The observation of water patches in AM-AFM mode is surprising to a certain extent because the perturbation induced by the tip is in general considerable and water at surfaces is usually swept away. However, under certain experimental conditions the water layers can be imaged, leading to the so-called true noncontact operation mode.

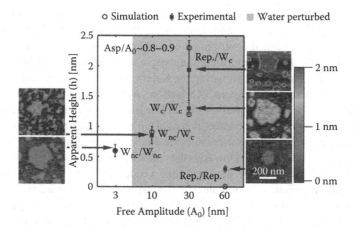

FIGURE 3.16 Experimental versus simulation values for the apparent height (h) of water patches on a $BaF_2(111)$ sample displaying both wet and dry regions. The values of apparent heights (nm) corresponding to free amplitudes (A_0) of 3, 10, 30, and 60 nm, respectively, are shown. Filled squares correspond to experimental data taken with a resonance frequency of 270 kHz, a cantilever force constant of 35 N m^{-1}, and a Q factor of about 400. Outlined circles correspond to simulations. W_c and W_{nc} represent perturbed (contact) and nonperturbed (noncontact) water layers and Rep. indicates a repulsive regime. (Reproduced from S. Santos, et al. *Nanotechnology* 22:465705, 2011. Institute of Physics. With permission.)

When both free amplitude (A_0) and setpoint (A) are systematically varied the apparent heights, as measured with an AFM, can vary substantially. Figure 3.16 shows both experimental and simulation data of the apparent height (h) of water patches on a $BaF_2(111)$ surface corresponding to A_0 values of 3, 10, 30, and 60 nm, respectively (Santos et al. 2011). Details are given in Section 4.3.2, when discussing the artifacts that water can induce when determining the height of objects with an AFM. As an appetizer we just mention that the interactions between a hydrated tip and hydrophilic and hydrophobic surfaces can be *grosso modo* classified in contact and noncontact regimes. In the pure (true) noncontact regime water remains unperturbed. However, in the contact regime water menisci can be formed or even mechanical contact with the solid surface through the water layers entering the repulsive mode. It seems clear that depending on the actual situation, the height of water patches should vary, and in fact large topographic variations are found, as demonstrated in Figure 3.16, where a dispersion in the 0.3–2.3 nm range for the same objects is observed, corresponding to different interaction regimes. As mentioned before, this is discussed in detail in Section 4.3.2 because we show that in AM-AFM mode the true heights of objects when regions of different affinity to water are involved cannot be determined with AFM in amplitude modulation.

Although the structure of such accumulated water is unknown, it shows a certain degree of orientation, as evidenced by scanning polarization force microscopy (SPFM) experiments (Figures 3.15c and d). SPFM is a noncontact electrostatic mode in which a conductive tip is set at about 10–20 nm from the surface and biased to a few volts and is ideal to obtain topographical and electrostatic properties of liquid films and droplets on surfaces (Hu, Xiao, and Salmeron 1995). Figure 3.15c shows an image resulting from the combined contribution of topography and dielectric response. In this case it is difficult to distinguish among step and water film edges because they show similar apparent heights. However, the Kelvin probe force microscopy (KPFM) image taken simultaneously (Figure 3.15d) shows a clear contrast between the water films (white color) and the dry $BaF_2(111)$ surface (dark color). The contrast between both regions (about +60 meV contact potential) in the KPFM images is induced by the averaged orientation of the water molecule dipoles pointing up from the surface at ambient conditions. However, at low RHs the contact potential is negative (Verdaguer, Cardellach, and Fraxedas 2008), demonstrating a cooperative and irreversible flipping of the preferential orientation of water dipoles, from pointing toward the surface at low coverages and evolving in the opposite direction at higher coverages. Under different conditions (below 165 K in UHV) water builds a long range order 1×1 phase as has been identified by means of HAS and assigned to a bilayer of water (Lehmann et al. 1996; Vogt 2007). The observation of the heteroepitaxial 1×1 phase is in contradiction to theoretical analysis of this system (Nutt and Stone 2002), which predicts that the bilayer should be unstable. Thus, at ambient conditions water does not wet the $BaF_2(111)$ surface and accumulates at steps and on terraces after diffusion. Near lattice matching contributes to the partial orientation of water molecules, with their dipoles pointing down (on average) at low RH and up above approximately 30% RH.

We come now to structurally more complex materials exposing surfaces with hexagonal symmetry that are known as good ice nucleators. They belong to the family of the phyllosilicates, layered materials built from silicate tetrahedra with Si_2O_5 (or 2:5) stoichiometry. We explore here the two more relevant representatives: kaolinite and muscovite. Kaolinite is a clay mineral with chemical formula $Al_2Si_2O_5(OH)_5$. In spite of the apparent complexity its crystal structure is relatively simple. Figure 3.17 shows the crystallographic structure along the b-axis (a) and a view of the (001) face (b). From (a) the layered structure becomes evident justifying why this material can be easily exfoliated. Each layer is built from silica tetrahedra (SiO_4) and aluminum octahedra (AlO_6) sharing oxygen atoms in a 1:1 ratio. The layers have two different terminations: oxygen atoms from the silica side and hydroxyl groups from the aluminum side implying that upon cleavage two chemically inequivalent surfaces are generated. The hydroxyl-terminated side has a pseudo-hexagonal distribution, as shown in Figure 3.17b. Shown in the figure is a circle with a radius of 2.78 Å centered in one hydroxyl group, which is indicated by an arrow. This distance corresponds to the shortest hydroxyl–hydroxyl one. The circle makes patent that the hexagonal symmetry is not perfect (see the resulting distorted hexagon with discontinuous line). The largest hydroxyl–hydroxyl separation is 3.43 Å. Thus the surface lattice parameter is either 5.15 or 5.17 Å, which results in a ~14 % mismatch. The nonperfect hexagonal symmetry that should in principle hinder the formation of MLs with ice Ih structure is compensated by the rotation of the hydroxyl groups, which allow a certain flexibility. Water can interact with the hydroxyl-terminated surface via H-bonds and the surfaces turn out to be amphoteric: it can act as H-donor or acceptor (Tunega, Gerzabeck, and Lischka 2004). This seems an ideal scenario for the formation of stable ice MLs.

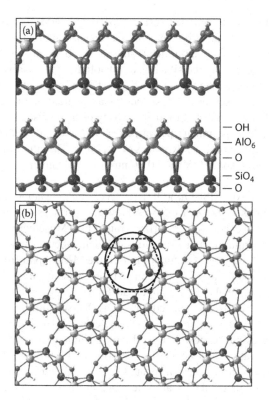

FIGURE 3.17 RT crystal structure of kaolinite. Chemical formula $Al_2Si_2O_5(OH)_5$, space group $C1$, lattice parameters $a = 5.154$ Å, $b = 8.942$ Å, $c = 7.401$ Å, $\alpha = 91.69°$, $\beta = 104.61°$, $\gamma = 89.82°$ (Neder et al. 1999). Aluminum, silicon, oxygen, and hydrogen atoms are represented by light gray, dark gray, medium gray, and white balls, respectively. (a) View along the b-axis and (b) of the (001) face. Shown in (b) is a circle with a radius of 2.78 Å centered in one hydroxyl group, indicated by an arrow. A distorted hexagon with a discontinuous line is also shown. The surface lattice parameter is either 5.15 or 5.17 Å.

However, DFT calculations (Hu and Michaelides 2007) predict a flat hexagonal ML (no bilayer), which should not be appropriate for 2D ice growth. The most stable overlayer is a H-down bilayer with each water molecule located approximately above a surface OH group. The E_{ads} of the periodic H-down overlayer is -0.65 eV, which is essentially identical to the cohesive energy of ice Ih (-0.66 eV). Every water molecule would be fully coordinated with four H-bonds and its stability matches that of ice, exceeding the predicted stability of similar overlayers on many other substrates. Its stability is not a consequence of a favorable lattice match with ice but rather because the substrate is amphoteric. Again, the water ML passivates the original surface preventing it from further H-bonding. As a consequence, the resulting surface has no affinity to water. However, kaolinite is a good nucleator as discussed in Chapter 5.

The experimental characterization of the kaolinite surfaces is quite difficult because the crystals are rather small (about 1 micron), and are aggregated, making the handling of individual crystals rather involved. However, Gupta et al. (2010) have

developed an appropriate technique to orient the kaolinite crystals, selectively exposing the silica and alumina faces by depositing a suspension of kaolinite on mica and
fused alumina substrates. The kaolinite crystals attach to the mica substrate with the
alumina face down, exposing the silica face of kaolinite; the positively charged alumina face of kaolinite is attached to the negatively charged mica substrate. The alumina
face of kaolinite is exposed on the fused alumina substrate; the negatively charged
silica face of kaolinite is attached to the positively charged fused alumina substrate.
Once oriented, AFM experiments performed under ambient conditions reveal the
hexagonal surface lattice of these two faces of kaolinite. The silica and alumina faces
exhibit periodicities of about 0.50 and 0.36 nm, in good agreement with the crystal
structure.

Muscovite mica, $KAl_2(AlSi_3O_{10})(OH)_2$, is very popular in laboratories using AFM
in ambient conditions because it exposes large flat surfaces (several microns) by
cleavage (with adhesive tape) and the prize is rather low. Muscovite is a 2:1 alumino–
silicate, with an aluminum octahedra layer sandwiched between two silicate layers.
Figure 3.18 shows the crystal structure of muscovite Mica. Apart from the 2:1 structure as compared to kaolinite, a key difference is the presence of layers of potassium,
which define the cleavage plane and compensate for the charge induced by the negative
charge on the tetrahedrally coordinated Al^{3+} ions that substitute Si^{4+} in the silicate
tetrahedra. This is a very important and characteristic property because, on average, cleaved surfaces will expose half of the potassium atoms, and water will have
the tendency to solvate such ions giving rise to a hydrophilic character. Note that
the lattice parameters of the ideal truncated (unrelaxed) surface, $a = 5.1988$ and
$b = 9.0266$ Å, do not match the basal plane of ice Ih. However, b is almost twice the
surface lattice parameter of ice Ih, with a mismatch below 0.05%.

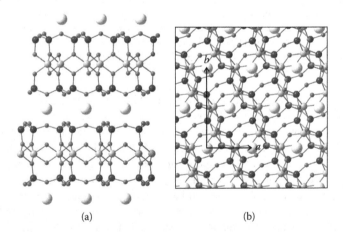

(a)   (b)

FIGURE 3.18 RT crystal structure of muscovite mica. $KAl_2(AlSi_3O_{10})(OH)_2$, space group
$C2/c$, lattice parameters $a = 5.1988$ Å, $b = 9.0266$ Å, $c = 20.1058$ Å, $\beta = 95.782°$
(Richardson and Richardson 1982). Aluminum, silicon, oxygen, and potassium atoms are
represented by light gray, dark gray, medium gray, and white balls, respectively. Hydrogen
atoms are not represented. (a) View along the [110] direction and (b) of the (001) face.

According to SPFM experiments performed at RT, water films exhibit an heteroepitaxial registry with the substrate (Hu et al. 1995). Islands ($\sim$2 Å high) intentionally produced by the action of the tip show polygonal structures with angles of $\sim$120°. The $\sim$2 Å thickness agrees with prior ellipsometry (Beaglehole and Christenson 1992) and IR (Cantrell and Ewing 2001) determinations. Odelius, Bernasconi, and Parrinello (1997), using MD simulations, found that at the ML coverage, water forms a fully connected 2D H-bonded network in commensurate registry with the mica lattice. Half of the water molecules in the unit cell are bound directly to basal oxygen atoms and partial solvation of potassium further stabilizes H-bonding. The water molecules are in two planes; together they form a periodic corrugated 2D structure that covers the surface. The simulations also predicted that no free OH bonds stick out from the surface, predictions that were confirmed both by SFG (Miranda et al. 1998) and SPFM experiments (Bluhm, Inoue, and Salmeron 2000). The SFG spectra showed little or no signal in the free OD stretch region near 2,740 cm^{-1} below 90% RH (D$_2$O was used in order to differentiate from the OH from muscovite). The SPFM experiments showed that the contact potential of the mica substrate decreased by about 400 mV from its value under dry conditions ($<$10% RH) when RH increased to 30% and that the potential remained nearly constant up to 80% RH. This is consistent with the formation of a structured water layer with the water dipoles pointing toward the surface.

To summarize this part, we must accept that the knowledge at the fundamental level of the wetting and ice Ih growth on the surfaces discussed thus far is only partial and far from complete. Water MLs seem to induce a protective wetting, hindering the growth of ice in a layer-by-layer mode because of unfavorable H-bonding architectures. The study of ideal surfaces is of little help for real applications, because what seems to define the ability to nucleate ice is the density and nature of defects (vacancies, steps, cracks, etc.; Hallett 1961) and the shape (nanoparticles) rather than the geometry. On the other hand we may ask ourselves if there are in nature more inorganic materials that expose chemically stable atomic planes with hexagonal symmetry which could allocate water films with the ice Ih structure at ambient conditions. Probably such materials exist but the fact that the crystallographic structures of most of the materials are known (to different degrees of accuracy) give limited space for surprises. Thus, other strategies are needed and one of them is the design of new (artificial) materials with the experience accumulated with the known materials (role of mismatch, H-bonding, electric fields, defects, etc.). This belongs to the field of materials engineering and flexibility should be one of the main issues. Flexibility in design for inorganic materials is difficult and organic molecules are the natural alternative.

Cryogenic Conditions

When the adsorption energies are small ($E_{ads} < 0.5$ eV as discussed here) and of the order of the intermolecular interaction, then the competition between molecule–molecule and molecule–substrate interactions usually leads to a rich variety of molecular distributions at the interface as a function of coverage and temperature. In general, the resulting phase diagrams are complex and have only been explored in detail in a few cases. A remarkable example is given for molecular nitrogen on highly-oriented pyrolytic graphite (HOPG), which exhibits a tremendous complexity, with commensurate and incommensurate orientationally disordered and ordered registries in addition to fluid phases (Marx and Wiechert 1996). Thus, an inert diatomic molecule already becomes a headache on an inert surface with hexagonal symmetry. The water molecule is triatomic and prone to dissociate, so that the registry issue is far from trivial. For selected well-defined substrates, complete phase diagrams of interfacial water in the ML regime are hard to find in the literature, most probably nonexistent, due to the experimental difficulty. In the vast majority of cases, selected coverage values and temperatures are used, and here many examples are at hand. This partial information of the structure of ordered interfacial water is to be compared to the well-characterized and widely accepted phase diagram of solid water, which is shown in Figure 1.16a.

Surfaces exhibiting low E_{ads} values, typically below 0.2 eV (see Table 3.1), are not wetted by water because the water–surface interaction is so weak. To give an idea of the interfacial strength note that such value corresponds to the characteristic cohesion energy of noble gas crystals, which are considered (by physicists) weakly interacting ordered gases. At very low temperatures 3D clusters are formed (Mehlhorn and Morgenstern 2007), the expected Volmer–Weber mechanism of growth for weak interactions. Remember that on Cu(111) clusters have been observed for $T < 20$ K (Michaelides and Morgenstern 2007). Nonwetting means hydrophobicity, a point discussed on such a surface in Section 4.3.2. Ag(111) and Au(111) behave similarly. When $E_{ads} > 0.2$ the water–surface interaction is sufficiently strong for the formation of MLs. The surfaces become wetted but water adsorbs non-dissociatively as is the case of the hexagonal (111) surfaces of certain transition metals (see Table 3.1), with the exception of Ru(0001), a point addressed below. Here we summarize some relevant issues concerning the formation of water layers on surfaces of transition metals at targeted registries. A detailed and extensive discussion on water adsorption and wetting on metal surfaces can be found in the review article from Hodgson and Haq (2009).

On Pd(111) surfaces water forms, at 40 K, small commensurate hexagonal rings that aggregate into larger clusters made of side-sharing hexagons in a $\sqrt{3} \times \sqrt{3}R30°$ network (Mitsui et al. 2002). XPS measurements show that water adsorbs and desorbs intact on such a surface (Gladys et al. 2008). On Pt(111) surfaces water also adsorbs intact, as has been shown by UPS experiments (Langenbach, Spitzer, and Lüth 1984), where the MO features are located at 12.6, 14.7, and 18.6 eV (referred to the vacuum level), respectively, in excellent agreement with the DFT and UPS results shown in Figure 1.6. Firment and Somorjai (1976) observed $\sqrt{3} \times \sqrt{3}R30°$ low-energy electron diffraction (LEED) patterns for films grown in the ~125–155 K range. However, it was later shown that such a LEED pattern does not correspond to the pristine water/Pt(111) interface but that it was in fact originated by electron damage caused by the LEED beam as well as by the presence of contaminants such as oxygen. Thus, the $\sqrt{3} \times \sqrt{3}R30°$ registry is associated with the formation of OH. In the absence of impurities water adopts a rather complex commensurate structure on Pt(111). Glebov et al. (1997) observed by means of HAS experiments two highly ordered water phases: $\sqrt{37} \times \sqrt{37}R25.3°$ and $\sqrt{39} \times \sqrt{39}R16.1°$, the latter replacing the former as the bilayer saturates. Both structures are shown in Figure 3.19, which consist

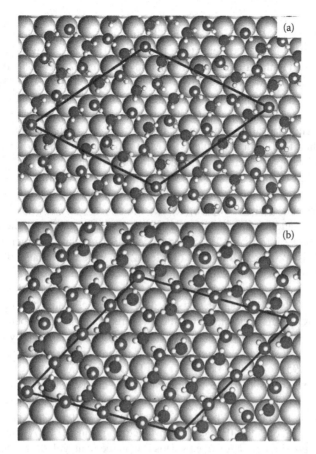

FIGURE 3.19 Structural model showing one possible lateral arrangement of water molecules in (a) the $\sqrt{37} \times \sqrt{37}R25.3°$ and (b) $\sqrt{39} \times \sqrt{39}R16.1°$ unit cell. Water is shown here in the H–up configuration, although the accepted arrangement corresponds to H–down. (Reprinted from A. Hodgson, and S. Haq, *Surf. Sci. Rep.* 64: 381–451, 2009. With permission from Elsevier.)

of a conventional ice bilayer rotated with respect to the Pt(111) axis. Because of the rotation, only a few oxygen atoms are in atop sites. The corresponding transformation matrices can be found in Section 3.2.2.

The actual structural arrangement of water molecules in such a complex unit cell is still a matter of debate, because 32 water molecules are involved. Proton disorder is an unsolved issue and Feibelman (2003) points toward the formation of H_3O^- and OH^- surface species. However, it is well established that water has a negligible amount of uncoordinated OH pointing into the vacuum (H-up) (note that in Figure 3.19 water molecules adopt a H–up configuration). This has been shown by XAS (Ogasawara et al. 2002) and IR (Haq, Harnett, and Hodgson 2002) experiments. In the XAS experiments the characteristic signature corresponding to the H–up structure is located at 536.5 eV, and the corresponding IR band is found near $2,720\ cm^{-1}$. Both features turn out to be very weak. The bilayer frustrated translational modes have wave numbers 133 and 266 cm^{-1} and the librations are found at 524 and 677 cm^{-1}, according to HREELS experiments (Jacobi et al. 2001). The 133 (266) cm^{-1} mode has been assigned to the stretch mode normal to the surface for H-bonded molecules corresponding to the lower (upper) layer of the bilayer (see Figure 3.7). The H–down configuration does not provide the best scenario for 2D growth. Half of the molecules bind to the surface through an oxygen lone-pair, half have a hydrogen atom pointing toward the metal, and all the molecules form H-bonds to three neighboring water molecules. Therefore, each molecule in the ML forms four bonds, leaving no dangling OH-bonds or lone-pair electrons protruding into the vacuum. The wetted surface is thus hydrophobic (Kimmel et al. 2005) so that multilayers will grow according to the Stranski–Krastanov mechanism. This is a further example of the surprising effect of how water can make a surface hydrophobic.

Ruthenium is another interesting material because it crystallizes in the $P6_3/mmc$ space group, the same as for Ih ice, and because it exhibits a larger E_{ads} value, about 0.4 eV, as compared to Pt(111) (Table 3.2). For the basal plane, (0001), $a_s = 2.71$ Å, so that a $\sqrt{3} \times \sqrt{3}R30°$ should be expected, because $a_{Ih}/a_s \simeq \sqrt{3}$. This is the case extensively studied by Held and Menzel (1995). According to RAIRS measurements (Clay, Haq, and Hodgson 2004) an intact water bilayer is formed below $\sim$150 K, because the characteristic OH stretch and H–O–H bending modes are observed. However, above such temperature the bilayer is partially dissociated because such modes disappear and only out-of-plane bending modes remain. Held and Menzel (1995) found that both H$_2$O and D$_2$O formed a diffuse $\sqrt{3} \times \sqrt{3}R30°$ LEED pattern when adsorbed at low temperature that sharpened up at 150 K for D$_2$O, indicating an extended $\sqrt{3} \times \sqrt{3}R30°$ structure, but not for H$_2$O, where additional spots around the integer order positions appeared. It was suggested that all oxygen atoms are almost coplanar, the vertical distance being just 0.10 Å compared to a buckling of approximately 1 Å in ice Ih. This is in agreement with the model proposed by Feibelman (2002), who pointed out that the energetically most favorable water MLs consist of half-dissociated–half-intact molecules. According to such a model, intact $\sqrt{3} \times \sqrt{3}R30°$ ice bilayer structures, with water adsorbed either H–up or H–down, have very similar binding energies (0.52 eV).

The intactness or not of water at interfaces opens the question of whether water dissociation is a necessary condition for wetting. Clearly, the presence of hydroxides enhances the accommodation of more water molecules, but can water wet a surface preserving its chemical integrity? It is clear that the higher E_{ads} is, the easier water dissociates, so that when surfaces become wetted at ambient conditions at least a considerable part of interfacial water should be dissociated. However, for surfaces with sufficiently low adsorption values, when wetting is achieved only at cryogenic temperatures, this is not that evident. In fact the experimental detection of OH species may be induced by the experimental technique used. This is quite clear for photoemission experiments, with excitation energies much larger than the electron binding energies. Theoretical calculations tend to be in favor of water dissociation so that the question remains open. This is another example of a problem related to water that is apparently simple but that in reality is not.

3.2.3 SURFACES WITH NON-HEXAGONAL SYMMETRY

Cubic Surface Lattice

Let us first consider the case of ice Ih on a substrate with cubic symmetry: $a_s = b_s \equiv a_s$, $\alpha_s = 90°$ and $a_{Ih} = b_{Ih} \equiv a_{Ih}$, $\alpha_{Ih} = 60°$. In this case (3.3) simplifies to:

$$R_{11} = \frac{n a_{Ih}}{a_s} \cos \theta \tag{3.7a}$$

$$R_{12} = \frac{n a_{Ih}}{a_s} \sin \theta \tag{3.7b}$$

$$R_{21} = \frac{n a_{Ih}}{a_s} \sin(30 - \theta) \tag{3.7c}$$

$$R_{22} = \frac{n a_{Ih}}{a_s} \sin(60 + \theta) \tag{3.7d}$$

From this expression it follows that:

$$R_{21} = \frac{1}{2}(R_{11} - \sqrt{3}R_{12}) \tag{3.8a}$$

$$R_{22} = \frac{1}{2}(\sqrt{3}R_{11} + R_{12}) \tag{3.8b}$$

$$\tan \theta = \frac{R_{12}}{R_{11}} \tag{3.8c}$$

and an immediate consequence is that the R_{ij} values cannot be simultaneously integer because of the irrational $\sqrt{3}$ factor. Thus, a rigid hexagonal lattice on a cubic surface lattice cannot exhibit commensurate registry. However, other symmetries are in principle possible, and this is the case for the (100) cleavage plane of MgO, a rocksalt-type material belonging to the $Fm\bar{3}m$ space group with $a = 4.18$ Å, which has been extensively studied as a substrate for water adsorption. Such a surface is non-polar, exposing the same number of magnesium and oxygen ions, with $E_{ads} = 0.65$ eV/molecule (Engkvist and Stone 1999), so that dissociation should be expected. According to NAPP studies (Newberg et al. 2011) water dissociation already occurs at extremely low RHs (<0.01%) at defect sites leading to coverages of about 0.08 ML of OH. At ~0.1% RH the surface is fully saturated with 1 ML of OH. At lower temperatures water forms $c(4 \times 2)$ patterns (below 180 K) and $p(3 \times 2)$ (between 180 and 210 K) both at coverages above 0.4 ML, as determined by HAS and LEED (Ferry et al. 1996).

Figures 3.20 and 3.21 show DFT-derived structures $p(3 \times 2)$ and $c(4 \times 2)$ as a function of the number of water molecules per unit cell, n and m, respectively (Wlodarczyk et al. 2011). In all structures water undergoes partial dissociation with proton transfer to the MgO surface oxygen atoms. These surface OH groups are denoted as $O_{(s)}H$ and those remaining within the water ML after dissociation are denoted as $O_{(f)}H$ groups.

The two most stable surface structures correspond to $n = 6$ for $p(3 \times 2)$ and $n = 10$ for $c(4 \times 2)$ with two dissociated water molecules per unit cell. The $n = 5$ structure (see Figure 3.20a) contains only one dissociated water molecule per unit cell and the water ML is arranged in stripes separated by voids of about 3 Å in width. The

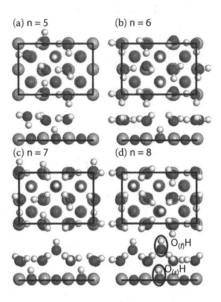

FIGURE 3.20 Side and top views of the most stable $n\mathrm{H_2O}/(3 \times 2)$ structure models on Mg(100). Magnesium, oxygen, and hydrogen atoms are represented by medium gray, dark gray, and white balls, respectively. (Reprinted from R. Wlodarczyk, et al.. *J. Phys. Chem. C* 115:6764–6774, 2011 American Chemical Society. With permission.)

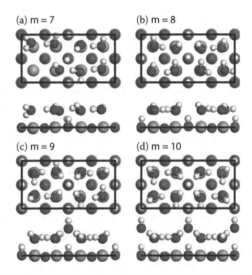

FIGURE 3.21 Side and top views of the most stable $m\mathrm{H_2O}/(4 \times 2)$ structure models on Mg(100). Magnesium, oxygen, and hydrogen atoms are represented by medium gray, dark gray, and white balls, respectively. (Reprinted from R. Wlodarczyk, et al.. *J. Phys. Chem. C* 115:6764–6774, 2011 American Chemical Society. With permission.)

molecular plane of one water molecule is twisted perpendicularly to the surface with a distance between the hydrogen and surface O atoms of only 1.56 Å. All other water molecules are arranged with their molecular plane almost parallel to the surface. For $n = 6$ (see Figure 3.20b) the water ML is virtually flat and two water molecules are dissociated. All O atoms of the water molecules are placed directly above surface Mg sites. The $n = 7$ structure (see Figure 3.20c) also contains two dissociated water molecules per unit cell. A particular feature of this structure is that the $O_{(f)}H$ group oriented perpendicularly to the surface plane is coordinated by hydrogen atoms of four nearby water molecules. This motif is present in all structure models with water coverages higher than one molecule per Mg atom. Finally, for $n = 8$ (see Figure 3.20d) two water molecules per unit cell are dissociated and the water ML contains two $O_{(f)}H$ groups oriented perpendicularly to the surface plane and coordinated by hydrogen atoms of four nearby water molecules. In the case of the $c(4 \times 2)$ symmetry the $m = 7$ structure exhibits two dissociated water molecules per unit cell (see Figure 3.21a). Similarly to the previous $n = 5$ case, the water ML is not completely flat, and one water molecule is twisted with the hydrogen atom pointing toward the surface O atom and a H–O distance of 1.69 Å. Three water molecules per unit cell are dissociated for $m = 8$ with molecular planes oriented parallel to the surface (see Figure 3.21b). The $O_{(f)}H$ groups are tilted up out of the surface. Three dissociated water molecules per unit cell are also found for $m = 9$ (see Figure 3.21c). One $O_{(f)}H$ group is oriented perpendicularly to the surface plane and is coordinated by hydrogen atoms of four nearby water molecules. Finally, for $m = 10$ two water molecules per unit cell are dissociated with both $O_{(f)}H$ groups oriented perpendicularly to the surface plane and coordinated by hydrogen atoms of four surrounding water molecules.

Another important surface with cubic symmetry is Si(100) and the interaction of water with such a surface has been extensively studied from both the experimental and theoretical points of view (Henderson 2002). Under particular cleaning cycles in UHV a 2×1 reconstruction is generated, which is characterized by the formation of asymmetric silicon dimers and dangling sp^3-like bonds (see discussion in Section 1.4.2). The unsaturated character of the surface makes it rather reactive, so that water adsorbs dissociatively on it with (calculated) $E_{ads} = 2.37$ eV, which is to be compared to $E_{ads} = 0.57$ eV from intact adsorption (Cho et al. 2000). Water dissociation is not achieved by a single dangling bond and the contribution of two adjacent dangling bonds is required leading to OH and H fragments at each dangling bond. The fragments can reside at the same dangling bond or on adjacent dangling bonds on different dimers with a preference for the first option.

Rectangular Surface Lattice

In the more general case of a rectangular surface lattice, which indeed includes the previously discussed cubic lattice case, $a_s \neq b_s$, $\alpha_s = 90°$ and $a_{ML} = b_{ML} = na_{lh}$, $\alpha_{lh} = 60°$. Then expression (3.3) transforms into:

$$R_{11} = \frac{na_{lh}}{a_s} \cos\theta \tag{3.9a}$$

$$R_{12} = \frac{na_{lh}}{a_s} \frac{a_s}{b_s} \sin\theta \tag{3.9b}$$

$$R_{21} = \frac{na_{\text{Ih}}}{a_s} \sin(30 - \theta) \tag{3.9c}$$

$$R_{22} = \frac{na_{\text{Ih}}}{a_s} \frac{a_s}{b_s} \sin(60 + \theta) \tag{3.9d}$$

and it follows that

$$R_{21} = \frac{1}{2}(R_{11} - \sqrt{3}\frac{b_s}{a_s}R_{12}) \tag{3.10a}$$

$$R_{22} = \frac{1}{2}(\sqrt{3}\frac{a_s}{b_s}R_{11} + R_{12}) \tag{3.10b}$$

$$\tan\theta = \frac{b_s}{a_s}\frac{R_{12}}{R_{11}} \tag{3.10c}$$

and

$$R_{11}R_{22} - R_{12}R_{21} = \left[\frac{na_{\text{Ih}}}{a_s}\right]^2 \frac{a_s}{b_s}\frac{\sqrt{3}}{2} \tag{3.11}$$

that is, the determinant of the R_{ij}-matrix is independent of θ. When $a_s = b_s$ then expressions (3.7) and (3.8) are obtained. Here, the irrational $\sqrt{3}$ factor can be removed if $a_s = \sqrt{3}b_s$, or alternatively $b_s = \sqrt{3}a_s$, opening the door to commensurate registry. In the $a_s = \sqrt{3}b_s$ case:

$$R_{21} = \frac{1}{2}(R_{11} - R_{12}) \tag{3.12a}$$

$$R_{22} = \frac{1}{2}(3R_{11} + R_{12}) \tag{3.12b}$$

$$\tan\theta = \frac{1}{\sqrt{3}}\frac{R_{12}}{R_{11}} \tag{3.12c}$$

Browsing through available inorganic crystal structure databases for different combinations of integer R_{ij} values fulfilling (3.12) and the conditions $c = \sqrt{3}a$ or $c = \sqrt{3}b$ does not provide any suitable material for commensurate registry with a rigid hexagonal ice Ih lattice. Note that the bulk lattice parameters instead of the surface counterparts have been used for simplicity. However, one might with luck or inspiration find examples for particular orientations that might fulfill the required conditions, but the most extensively studied surface with rectangular symmetry is rutile $TiO_2(110)$, the thermodynamically most stable surface of TiO_2, with a rectangular surface ($a_s = 6.496$ Å and $b_s = 2.959$ Å) which is not suited for commensurate registry with the basal plane of ice Ih because $a_s/b_s = 2.19 > \sqrt{3}$ Å (Diebold 2003). Titanium dioxide is used in many applications because of its photocatalytic activity (discussed in Chapter 5), biocompatibility (used, e.g., as food additive under the term E171), and used as well in cosmetics, paints, pharmaceuticals, sunscreens, solar cells, and a long et cetera, hence its importance. The three most common polymorphs of titania are rutile [tetragonal, $P4_2/mnm$, $a = b = 4.594$ Å, $c = 2.959$

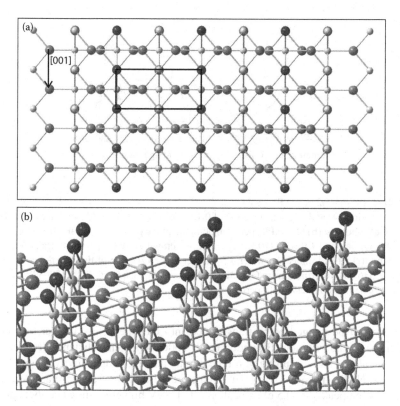

FIGURE 3.22 (a) Top and (b) perspective views of the crystal structure of the $TiO_2(110)$ surface. $P42/mnm$, $a = 4.5937$ Å, $c = 2.9587$ Å (Gonschorek 1982). Oxygen atoms are represented by medium and dark gray balls (bridge oxygens in dark gray) and titanium atoms by light gray balls. The rectangular surface lattice, $a_s = 6.496$ Å and $b_s = 2.959$ Å, is shown.

Å (Gonschorek 1982)] with an electronic bandgap of 3.0 eV, anatase (tetragonal, $I4_1/amd$, $a = b = 3.782$ Å, $c = 9.502$ Å) with an electronic bandgap of 3.2 eV and brookite (rhombohedral, $Pbca$, $a = 5.436$ Å, $b = 9.166$ Å, $c = 5.135$ Å). The rutile $TiO_2(110)$ 1×1 surface undergoes a certain degree of relaxation essentially in the direction perpendicular to the surface for symmetry reasons (Charlton et al. 1997). Two projections are shown in Figure 3.22 highlighting its corrugated and anisotropic character given by the rows formation. Along the [001] direction we observe rows of oxygen atoms, known as the bridging oxygen atoms, that protrude from the surface plane. Such bridging oxygens are twofold coordinated in contrast to the bulklike threefold coordinated oxygen atoms on the surface plane. Bridging oxygens are coordinated with sixfold titanium atoms, whereas terraces contain fivefold titanium atoms, with one dangling bond perpendicular to the surface.

The accepted picture is that oxygen vacancies are able to dissociate the water molecules generating surface hydroxides, which in turn anchor water molecules forming strong $OH–H_2O$ complexes, which act as nucleation centers for further water adsorption. This has been experimentally proven by STM, where individual oxygen

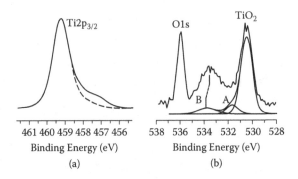

FIGURE 3.23 (a) Ti2$p_{3/2}$ XPS peak of a clean TiO$_2$(110) surface at 420 K before water exposure (solid line) and after introduction of 0.1 mtorr of water vapor (dashed line) measured with 630 eV photons. (b) O1s XPS spectrum obtained in the presence of 17 mtorr water at 298 K (bottom curve), and 1 torr at 270 K (top curve) taken with 690 eV photons. Peaks A and B correspond to OH and molecular water, respectively. The O1s line of the water–gas phase is located at 536 eV. (Reprinted from G. Ketteler, et al.. *J. Phys. Chem. C* 111:8278–8282, 2007. American Chemical Society. With permission.)

vacancies can be imaged being transformed into OH species as a water molecule dissociates in the vacancy (Bikondoa et al. 2006). Oxygen deficiency (reduction) provides color centers resulting in n-type doping and high conductivity. Figure 3.23 shows the NAPP lines of TiO$_2$(110) when exposed to water vapor. Apart from the lattice oxygen peak at 530.5 eV, a peak at 1.1–1.6 eV higher binding energy appears that is characteristic of hydroxyl groups at bridging sites. A second peak appears at 2.4–3.5 eV higher binding energy (peak B), and has been attributed to either hydroxyl or to molecularly adsorbed water on Ti^{+4} sites between bridging O rows (Ketteler et al. 2007). However, the generally accepted view that water adsorbs nondissociatively on defect-free TiO$_2$(110) surfaces has been challenged. Pseudo-dissociation has been claimed, where rapidly switching through proton transfer between dissociated and molecular geometries that are in molecular equilibrium coexist involving OH species in both atop and bridging sites. Pseudo-dissociated water molecules on titanium atoms are mobile along the [001] direction across the bridging oxygen rows, according to STM results (Du et al. 2010), and the previously mentioned coexistence of molecular and dissociated water has been deduced from photoelectron diffraction experiments (Duncan, Allegretti, and Woodruff 2012).

The anisotropic nature of the (110) surface facilitates the formation of 1D chains. This is illustrated with STM images taken at 50 K in Figure 3.24 (Lee et al. 2013). At low water coverages (0.04 ML) water monomers, oxygen vacancies (V$_o$), and bridging hydroxyls (OH$_b$) are observed as well as incipient short water chains along a Ti row (Figure 3.24a). At higher coverages (0.13 ML), longer chains are formed as shown in Figure 3.24b. Upon annealing the samples at 190 K longer chains are obtained due to surface diffusion of the water molecules. Figure 3.24c provides an example where in addition it is observed that oxygen vacancies V$_o$ have been converted to bridging hydroxyls OH$_b$. The diffusion and dissociation of water compete during the annealing process. According to DFT calculations from the same authors the diffusion barrier

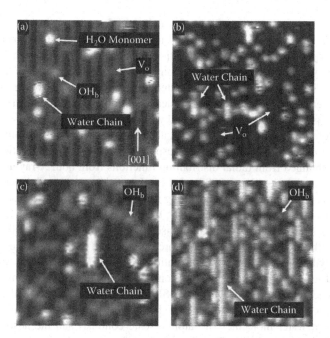

FIGURE 3.24 STM images of water molecules on $TiO_2(110)$ taken at a bias voltage and tunneling current of 1.5 V and 10 pA, respectively. (a) $T = 50$ K and 0.04 ML (9×9 nm^2), (b) $T = 50$ K and 0.13 ML (12×12 nm^2), (c) $T = 50$ K after annealing the sample from (b) to $T = 190$ K for 5 min (9×9 nm^2), and (d) $T = 50$ K at an initial coverage of 0.5 ML after annealing to $T = 190$ K for 5 min (14×14 nm^2). H_2O monomers, bridging hydroxyls (OH_b), oxygen vacancies (V_o), and water chains are indicated. (Reprinted from J. Lee, et al. *J. Phys. Chem. Lett.* 4:53–57, 2013 American Chemical Society. With permission.)

of a water monomer along a Ti row is 0.47 eV whereas the barrier for dissociation of water at V_o sites is 0.50 eV. At an initial coverage of 0.5 ML, ~7 nm long water chains are observed after thermal annealing at 190 K and then cooling back to 50 K for imaging, as shown in Figure 3.24d. Note that no 2D layers are observed, which is due to the presence of the bridging oxygen rows, which are prone to build H-bonds with the water molecules.

3.3 SUBSTRATE-INDUCED STRUCTURING OF WATER MULTILAYERS

3.3.1 UNPOLARIZED SUBSTRATES

We saw in Section 1.3 that a flat surface induces in its close proximity a layered density distribution of water under the action of a Lennard–Jones potential or within the simpler hard wall approximation in the case $D_W \gg D_w$. Such a layering or structuring effect has been observed experimentally by means of high-resolution

X-ray reflectivity studies using SR performed on high-grade muscovite mica (001) surfaces (Cheng et al. 2001). The derived interfacial water structure consists of an oscillatory oxygen density profile as a function of the oxygen–oxygen distance (X-rays predominantly scatter from the oxygen atoms in water) with a narrow first layer and substantially damped subsequent layers, extending to about 10 Å above the surface. An adsorbed layer located at 1.3 Å above the mean position of the relaxed surface oxygen is identified as both H_3O^+ ions exchanged with surface K^+ ions and adsorbed H_2O. The first hydration layer is located at 2.5 Å above the surface, which is comparable to the $\simeq 2.8$ Å oxygen–oxygen distance corresponding to the first coordination shell of bulk water obtained from radial distribution functions from both X-ray and neutron diffraction experiments (Soper 2007), that is taken as the mean molecular diameter. Such results are supported by 3D scanning force microscopy (SFM) experiments in the frequency-modulation detection mode, where the water distribution at the same interface has been visualized with atomic-scale resolution (Fukuma et al. 2010). In such mode, the tip–sample interaction force is detected as a resonance frequency shift of the vibrating cantilever. In conventional (2D) SFM, the vertical tip position is regulated to keep the frequency shift constant while laterally scanning the tip to obtain a 2D height image. However, in 3D-SFM the tip is also scanned in the direction perpendicular to the substrate surface. The hydration layer reveals a uniform lateral distribution of water molecules (no atomic-scale contrast) and the adsorbed layer exhibits atomic-scale contrast. The coexistence of water molecules having ordered (adsorbed) and disordered (hydration) distributions is of great interest because it reconciles the two opposing ideas of icelike and liquidlike water molecules at the mica/water interface.

Apart from layering, other kinds of ordering can be induced by substrates. In the case of Pt(111) surfaces, water films grown between 120 and 137 K can become ferroelectric, meaning that a net dipole is obtained (Su et al. 1998). The presence of polar ordering of water molecules in the films is demonstrated by the strong enhancement of OH stretch resonances with film thickness as demonstrated by means of SFG spectroscopy. A similar example, although limited to the bilayer regime, was discussed above (see Figure 3.15) for water on BaF_2(111) surfaces. It is noteworthy to mention here that metastable cubic ice has been observed on Pt(111) when grown between 120 and 150 K (Thürmer and Bartelt 2008). Cubic ice emerges from screw dislocations in the crystalline ice film caused by the mismatch between the atomic step of the substrate surface and the ice–bilayer separation.

3.3.2 Polarized Substrates: Electrofreezing

In this part we consider the effect of electric fields at and close to the surfaces of given materials to the ordering of water. When freezing of supercooled water is enhanced by the action of local, both external and internal, electric fields, the term *electrofreezing* is used (Pruppacher 1973). Electrofreezing is closely related to electrowetting, where the contact angle can be externally controlled by electric fields, a point briefly discussed in Chapter 4 (see Figure 4.17). The water density profile up to about 1 nm from a metallic surface depends on the polarity of such a surface. X-ray scattering experiments performed on Ag(111) single crystals have shown that, on average, water dipoles are pointing outward or toward the surface when the charge is positive or negative, respectively (Toney et al. 1994). This is the expected result from straightforward electrostatic considerations. In addition, the density of this first layer is larger than that corresponding to bulk water. When liquid water is confined between two parallel metallic surfaces at distances much larger than the mean water diameter to avoid the effects due to confinement (think of an ideal capacitor filled with water) then the electric field between both electrodes not only is able to orient the water molecules but they may induce crystallization, that is, the growth of ice. This problem has been studied by means of MD computer simulations using two parallel Pt(100) surfaces as electrodes under the application of external electric fields, which induce opposed surface charge densities (Xia and Berkowitz 1995). The outcome of such calculations is that water restructures itself and eventually crystallizes into domains of strained cubic ice in order to adapt to the new environment imposed by the homogeneous external field. This way the H-bonded network can remain intact and the orientational polarization of water can be aligned along the field. The transition of liquid water to crystalline water occurs for surface charge densities around 30 $\mu C\ cm^{-1}$. This has been experimentally achieved by Choi et al. (2005) using an electrochemical STM with a gold tip and an Au(111) substrate. The interfacial water undergoes a sudden, reversible phase transition to ice in electric fields of $10^6\ Vm^{-1}$ at RT.

Zwitterionic amino acids, with positively charged ammonium and negatively charged carboxylic groups, provide rather large molecular dipoles (about 15 D) and the electric fields generated by such dipoles are strong enough to align water molecules through dipole–dipole interactions thus facilitating the formation of ice on the surfaces of amino acid crystals, in particular in defects. Polar surfaces induce freezing of water at higher temperatures than the corresponding apolar surfaces, with a difference of 3 to 5°C (Gavish et al. 1992). For example, the freezing point of water on hydrophobic L-valine crystal surfaces is −5.6°C, and for the racemic DL-valine crystals is −9.9°C. However, for L-alanine the freezing point is −7.5°C and for DL-alanine it is −2.6°C because in this case the surface of L-alanine is nonpolar whereas for DL-alanine it is polar. This point is addressed in Chapter 4.

Charged surfaces can also be achieved in a controlled way by using pyroelectric materials, because surface polarization can be externally controlled with temperature. This has been shown with $LiTaO_3$ crystals and $SrTiO_3$ thin films (Ehre et al. 2010). Positively charged surfaces promote ice nucleation, whereas the same surfaces when negatively charged reduce the freezing temperature. Droplets of water cooled down on a negatively charged $LiTaO_3$ surface and remaining liquid at −11°C freeze immediately when this surface is heated to −8°C, as a result of the replacement of the negative surface charge by a positive one. Using a different material, in this case thin films of $Pb(Zr_{0.20}Ti_{0.80})O_3$ grown on metallic (001) Nb-doped $SrTiO_3$ substrates, Segura (2012) has shown that interfacial ambient water already becomes ordered at positive temperatures (about 10–15°C) on previously poled regions due to the preferential alignment of substrate dipoles. Selected regions of the surface were polarized with a metallic AFM tip in contact mode and the written patterns were measured with KPFM in noncontact. When the surface is covered by at least one ML of water, the polarized regions are not detected by KPFM at RT (left image of Figure 3.25). However, when the temperature is decreased, the KPFM signal increases due to the

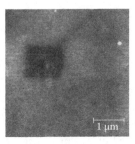

FIGURE 3.25 Evolution of the KPFM signal with temperature on a poled region (dark) of a Pb(Zr$_{0.20}$Ti$_{0.80}$)O$_3$ film grown on a metallic (001) Nb-doped SrTiO$_3$ substrate surface taken at 20°C (left), 15°C (middle), and 10°C (right). The increasing contrast is due to the progressive accumulation of structured water on the polarized region.

combined contribution from the polarized region and the aligned water molecules (middle and right of Figure 3.25). Thus, electrofreezing can be achieved by a careful selection of the electrostatic nature of surfaces.

Water not only is able to order due to the presence of electric fields, but it is a very efficient tool to remove excess charge. This is illustrated next with graphene sheets deposited on SiO$_2$/Si wafers (Verdaguer et al. 2009). Graphene sheets can be polarized again with a metallic AFM tip. Because the sheets are isolated on silicon oxide, the accumulated charge will vary with time as a function of RH decaying exponentially with time constants on the order of tens of minutes. Figure 3.26a shows the time evolution of KPFM images of a few-layer graphene film charged at +8 V and at 30% RH. The images show how the positive charge spreads over the silicon substrate and the initially confined graphene film discharges. Graphene discharges through the water film adsorbed on the SiO$_2$ surface. The time evolution of the mean contact potential difference measured on the graphene film at different RH conditions exhibits an exponentiallike decay (Figure 3.26b).

3.4 CONFINED WATER

Here the 1D and 2D confinement of water at the nanometer scale is discussed highlighting opposing phenomena such as the promotion or hindrance of crystallization. A short example of 3D nanoconfinement was given in Section 2.1.2 when describing reversed micelles. Following the philosophy of this book, the essentials and fundamentals are introduced, and those readers eager to learn more on the subject are recommended to read Brovchenko and Oleinikova (2008).

3.4.1 2D CONFINEMENT

As illustrated in Figure 1.9, when the distance between two parallel surfaces D_W is comparable to $D_w \simeq 0.28$ nm then water becomes spatially confined and the structural isotropy characteristic of bulk water is lost. Israelachvili and Pashley (1983) showed that the force between two curved mica surfaces as a function of D_W, when

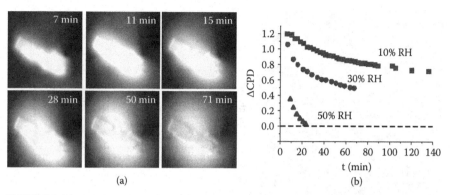

FIGURE 3.26 (a) Time evolution of KPFM images of a graphene sheet charged at +8 V and at 30% RH. (b) Time evolution of the contact potential difference (CPD) after charging the graphene sheet at +8 V for 10%, 30%, and 50% RH. (Reprinted from A. Verdaguer, et al.. *Appl. Phys. Lett.* 94:233105, 2009, American Institute of Physics. With permission.)

they approach along the direction perpendicular to both surfaces, shows a repulsive oscillatory behavior below 2 nm with a period of $\simeq 0.25$ nm due to the ordered layering of water molecules. When shear (lateral) forces are applied, by displacing two surfaces parallel to each other in a shear force microscope, both the amplitude and phase of the response show a stepwise behavior with a periodicity of about 0.25 nm (Antognozzi, Humphris, and Miles 2001). Using surfaces with contrasting affinities to water (hydrophilic and hydrophobic) leads to interesting differences. As expected, hydration forces are less pronounced for hydrophobic surfaces such as graphite (Li et al. 2007). For subnanometer hydrophilic confinement, the lateral force measurements show orders of magnitude increase of the viscosity with respect to bulk water whereas no viscosity increase is observed for hydrophobic surfaces. This is in contrast to experiments using two curved mica surfaces, where the viscosity of water remains within a factor of three of its bulk value when confined below 3.5 nm (Raviv, Laurat, and Klein 2001). When one of the surfaces is replaced by the tip of an AFM one can study the viscoelastic response of confined water. Using small oscillation amplitudes ($\sim 0.5 - 1.0$ Å) and small approaching speeds it has been observed that the elastic (solidlike) and viscous (liquidlike) responses oscillate with molecular layering as the tip–surface gap is reduced, corresponding to short-range ordering of the water in the tip–surface gap. The mechanical response changes dramatically when the approach speed is increased. Above a certain threshold rate, the liquid behaves solidlike with low viscosity and high elasticity when the gap is commensurate with molecular size, retaining a liquidlike, high-viscosity state when the gap is incommensurate with the molecular size (Khan et al. 2010).

The layering effect has been experimentally shown by means of AFM measurements in contact mode using a protective graphene sheet as one of the confining surfaces. Xu, Cao, and Heath (2010) transferred graphene sheets on mica surfaces in ambient conditions and showed that the trapped water layers grow in a layer-by-layer (Frank–van der Merwe) mode, forming atomically flat structures with heights

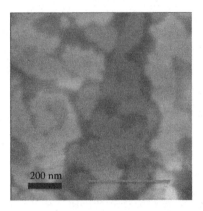
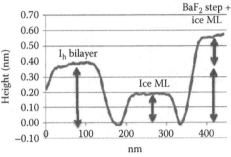

FIGURE 3.27 (Left) Contact mode AFM image of few-layer graphene deposited on freshly cleaved $BaF_2(111)$ surfaces at 50% RH. A cross-section, corresponding to the line shown in the figure, is shown at the right side. (Reprinted from A. Verdaguer, et al. *J. Chem. Phys.* 138:121101, 2013, American Institute of Physics. With permission.)

of 0.37 nm, the distance between adjacent bilayers in ice Ih (see Figure 1.12). The same layering effect has been observed in $BaF_2(111)$ surfaces covered with graphene flakes (Verdaguer et al. 2013). This is a further interesting example because such a surface has a relatively small surface lattice mismatch with the basal plane of ice Ih (see Table 3.2), conferring a certain degree of order to the adsorbed water layers, and in addition, the step height is 0.36 nm (see Figure 3.14), also in unison with the interbilayer distance in ice Ih. Figure 3.27 shows an AFM image taken in contact mode of water confined between graphene and a $BaF_2(111)$ surface prepared at RH $\sim 50\%$. Water layers cover the terraces limited by triangular steps. The cross-section shows the layered structure with ~ 0.4 nm steps, indicative of the presence of ice Ih. Here, the $BaF_2(111)$ steps are clearly seen and can be used as internal height references because of the well-known 0.36 nm value. In addition, ~ 0.2 nm structures are observed, which certainly cannot be ascribed to ice Ih, after an inspection of Figure 1.12.

Water molecules on $BaF_2(111)$ surfaces exposed to the atmosphere an extremely mobile (Foster, Trevethan, and Shluger 2009; Cardellach et al. 2010) and become trapped at defects such as vacancies and steps. The effect of the graphene is to limit such mobility on the surface efficiently allowing for 2D nucleation and growth, emulating the effect of low temperatures in a vacuum. On the other hand, graphene hinders any displacement out of the surface. Thus, the combined effect of both surfaces eliminates one degree of freedom (perpendicular to the interface), forces the saturation of hydrogen bonds (graphene) and triggers nucleation (substrate). The ~ 0.2 nm high features are of the order of the mean molecular diameter, suggesting that the molecules lie flat on the interface. Such a configuration is compatible with a cross-linked structure in which all water molecules are at the same vertical position (e.g., no bilayer is formed; Nutt and Stone 2002). In this case the interaction with the substrate is strong enough to avoid the formation of ice Ih bilayers probably caused by a local lower density.

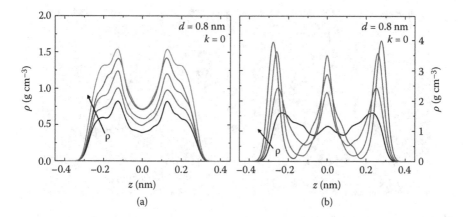

FIGURE 3.28 Density profiles for water confined between hydrophobic walls with $k = 0$ and $d = 0.8$ nm ($d \equiv D_W$ as in the text and in Figure 1.9). The wall surfaces are located at $z = \pm 0.4$ nm. (a) The underlying structure is bilayer when $\rho \leq 0.8$ g cm^{-3}, and (b) trilayer when $\rho \geq 0.9$ g cm^{-3}. On plot (a) the average densities increase from 0.4 g cm^{-3} (bottom) to 0.8 g cm^{-3} (top) in 0.1 steps. On plot (b) the average densities increase from 0.9 g cm^{-3} to 1.2 g cm^{-3}. Arrows indicate the direction of increasing density. (Reproduced from T.G. Lombardo, N. Giovambattista, and P.G. Debenedetti, *Faraday Discuss.* 141:359–376, 2009. With permission of The Royal Society of Chemistry.)

Computer simulations also show the layering induced upon confinement (Zangi and Mark 2003). Lombardo, Giovambattista, and Debenedetti (2009) have studied the behavior of glassy water confined between two hydrophobic surfaces of silica. In the calculations, which use the SPC/E model, it is assumed that the walls have no charge and that H atoms at the wall have no interaction, which is parametrized by a factor k, a normalized magnitude of a surface dipole moment that quantifies the polarity of the surface ($k = 0$ in the present example). The density profiles for $k = 0$ and $D_W = 0.8$ nm are shown in Figure 3.28 for densities $0.4 \leq \rho \leq 1.2$ g cm^{-3}. For $\rho = 0.8$ g cm^{-3} two symmetric bilayers are formed (see Figure 3.28a) but at $\rho \geq 0.9$ g cm^{-3} a third layer of water molecules develops as depicted from Figure 3.28b. Examining configurations of water molecules that belong to a single peak in the density profile provides valuable structural information. Figure 3.29 shows a characteristic arrangement of water molecules that constitute one of the two symmetric outer peaks (e.g., the O atom of the water molecule is located at $|z| > 0.18$ nm) of the density profile for $\rho = 1.2$ g cm^{-3}. This ML of water molecules is approximately 0.2 nm thick and self-assembles into a hexagonal lattice. In the perfect lattice, water molecules can adopt two configurations: (i) with a single H atom pointing away from the wall and (ii) with both H atoms in a plane nearly parallel to the surface. These configurations alternate, and water molecules form six-membered rings, giving rise to a hexagonal lattice. Thus, each water molecule participates in four H-bonds: three of these point roughly parallel to the wall and correspond to H-bonds with water molecules in the same layer, and one points normal to the substrate surface and corresponds to a H-bond with a water molecule in the adjacent layer. A closer

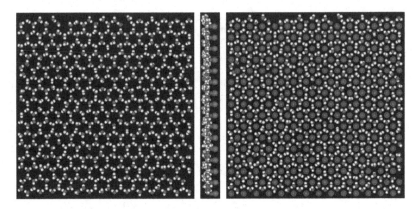

FIGURE 3.29 (left) Inherent structure configuration showing water molecules within 0.22 nm of a hydrophobic wall with $k = 0$, $\rho = 1.2$ g cm^{-3}, and $d = 0.8$ nm ($d \equiv D_W$ as in the text and in Figure 1.9). (middle) Profile view of the same layer with surface wall atoms included and (right) with surface Si and O wall atoms included. Si, O, and H atoms are represented by gray, dark gray, and white spheres, respectively. (Reproduced from T.G. Lombardo, N. Giovambattista, and P.G. Debenedetti, *Faraday Discuss.* 141:359–376, 2009. With permission of The Royal Society of Chemistry.)

inspection of the ML in Figure 3.29 reveals that the center of each hexagonal ring of water molecules is occupied by an oxygen atom on the surface of the wall. This strict correlation between the substrate and water suggests that the lattice is templated by the wall. At high density, water molecules are pushed into the small spaces between the top layer of Si and O atoms on the substrate surface. These water molecules self-assemble into a hexagonal lattice in order to participate in four H-bonds. Thus, it is the atomic structure of the wall that imposes the hexagonal lattice on the water. Without the hexagonal template provided by the wall, no long-range order exists for water molecules in the middle of the film.

Note that the here-discussed ice Ih bilayer structure is formed at RT, so that the term "RT ice" is fully justified. We know from Chapter 1 that temperature is not the only relevant parameter and that hydrostatic pressure also has to be considered, as in any phase diagram. By applying external pressures in dedicated pressure cells at RT above 2 GPa, ice VII is formed and well above such a value, at 65.8 GPa, another polymorph (ice X) is obtained (see Table 1.4). However, large local pressures of few GPa can be obtained in SFMs. In such instruments nN forces are applied on nanometer-sized areas, hence the large applied pressures that can even plastically deform solid surfaces (Fraxedas et al. 2002). Thus, in principle, ice could be obtained at RT under the confinement induced by a tip and a substrate and at least two research groups have succeeded. Ice has been imaged in contact mode on HOPG cleaved surfaces in air (Teschke 2010). Hexagonal lattices with periods of 0.24 and 0.45 nm have been measured depending on experimental conditions of scanning speed and RH. The first one is in good agreement with the known lattice constant of graphite, 0.246 nm (Trucano and Chen 1975), and the second one corresponds to the lattice constant of ice Ih (see Figure 1.12). The registry of this heteroepitaxial system would correspond to the $\theta = 30°$ case, as discussed in Section 3.2.2. Before such experiments Jinesh and Frenken (2008), using a high-resolution friction force microscope, demonstrated that water rapidly transforms into crystalline ice at RT. At ultralow scan speeds and modest RHs the tip exhibits stick–slip motion with a period of 0.38 nm, clearly closer to ice Ih than to HOPG.

3.4.2 1D CONFINEMENT

Here we explore the case of confinement induced by nanochannels, where the diameter of such channels is larger than the mean water molecular diameter (otherwise the molecules could not enter) but negligible when compared to their lengths. Let us remember here that the critical radius of growth r^* is about ~ 1 nm according to (2.9) at 273 K, so that nanochannels with diameters of such order can trigger or frustrate confined crystallization. We can differentiate two cases: (i) when the nanochannel is defined by an individual entity as in the cases of open (uncapped) carbon nanotubes (CNTs), and (ii) when the nanochannels are distributed in arrays, either ordered or disordered, in materials such as zeolites and nanoporous silica.

We start first with CNTs and in particular with single-walled CNTs (SWNTs). Their pore diameters can be controlled to a great extent, they can reach really long axial dimensions (hundreds of microns), and they exhibit the interesting property of having hydrophobic walls (Iijima 1991). The first question that arises is if water can enter easily inside the tube for really small diameters and under which conditions, and the answer comes from both theory and experiment. MD simulations have predicted that CNTs immersed in water become filled building 1D chains of water molecules (Hummer, Rasaiah, and Noworyta 2001) and experiments using Raman spectroscopy have shown that filling is achieved for diameters down to 0.548 nm, so that the radial constraint permits only one water molecule to fit in (Cambré et al. 2010).

According to the referred MD simulations, the water molecules entering the nanotube lose on average two out of four H-bonds and only a fraction of the lost energy (~ 10 kcal mol^{-1}) can be recovered through vdW interactions with the carbon atoms of the nanotube (~ 4 kcal mol^{-1}). It is interesting that a pulselike transmission of water through the nanotubes is expected and a reduction in the attraction between the tube wall and water dramatically affects pore hydration, leading to sharp dry–wet transitions between empty (dry) and filled (wet) states on a ns timescale. The referred simulations use short uncapped, single-walled nanotubes 1.34 nm long with a diameter of 0.81 nm solvated in a water reservoir with an elapsed time of 66 ns. Despite its strongly hydrophobic character, the initially empty central channel of the nanotube is rapidly filled by water from the surrounding reservoir, and remains occupied by about five water molecules during the entire 66 ns forming a H-bonded chain. A high molecular flow is predicted (about 20 water molecules per ns). The same conclusions have been achieved through a conceptually simple Ising model based on a coarse-grained lattice on a cubic grid (Maibaum and Chandler 2003). The tube is modeled by an ensemble of parallelepipedic cells building an internal open channel. The free surfaces can be exposed either to vapor (dry) or liquid (wet). The appearance of a density fluctuation at the tube mouth is necessary to facilitate the emptying of a filled tube, which is associated with a large entropic barrier. The filling of an empty tube, however, is associated with a large energetic barrier. Such a hydrophobic–hydrophilic transition has been experimentally observed for low-defect 1.4-nm diameter SWNTs when decreasing temperature from 22.1 to 8.0°C, with a transition temperature of about 18.4°C (Wang et al. 2008). For SWNTs with diameters above 1.6 nm, an anomalous behavior has been reported (Kyakuno et al. 2011). In this case the open SWNTs become filled at RT and emptied below a characteristic temperature (around 250 K̇) based on X-ray diffraction, nuclear magnetic resonance (NMR), and electrical resistance measurements.

Such concepts have triggered the use of CNTs in nanofluidics, in particular in water filtering. Holt et al. (2006) have developed a microelectromechanical system (MEMS)–compatible fabrication process (Figure 3.30a) for sub-2–nm nanotube pore membranes. The process uses catalytic chemical vapor deposition (CVD) to grow a dense, vertically aligned array of double-walled CNTs (DWNTs) on the surface of a silicon chip (Figure 3.30b), followed by conformal encapsulation of the nanotubes (1.6 nm diameter) by a hard, low-pressure chemical vapor–deposited silicon nitride (Si_3N_4) matrix (Figure 3.30c). This process produces gap-free membranes over the length scale of the whole chip. The excess silicon nitride is removed from both sides of the membrane by ion milling, and the ends of the nanotubes are opened up with reactive ion etching. The membranes remain impermeable until the very last etching step. The transport occurs exclusively through the inner pores of the CNTs spanning the membrane. Water flux is rather high, evaluated to 10–40 water molecules nm^{-2} ns^{-1}.

The 1D arrangement caused by the reduced space and the hydrophobic character of the walls opens the door to new ordering possibilities, 1D ice, also termed ordered ice nanotubes, not included in the phase diagram of (bulk) water shown in Figure 1.16a. In a MD prediction of the spontaneous ice formation in CNTs it was shown that the confined water freezes into square, pentagonal, hexagonal, and heptagonal ice nanotubes depending on the diameter of CNTs or the applied pressure (Koga et al. 2001). The phase behavior of water inside nanotube diameters in the 0.9–1.7 nm range and in the 100–300 K temperature range under a fixed 0.1 MPa pressure has been explored with MD simulations finding nine ordered 1D phases (Takaiwa et al. 2008). Among them, the square ice nanotube has the highest melting point, 290 K, for a 1.08 nm diameter. Water remains faithful to its versatility even under nanoconfinement.

Coming to the second case of materials exhibiting arrays of nanochannels or pores of radius r_{pore}, in particular mesoporous silica and zeolites with r_{pore} typically below 10–12 nm, it has been experimentally verified that water in the inner region of the pores crystallizes at a temperature T_m^{pore} below the bulk freezing point, T_m^{bulk}, and that the dependence of $\Delta T_m = T_m^{bulk} - T_m^{pore}$ on r_{pore} follows the expression (Faivre, Bellet, and Dolino 1999):

$$\Delta T_m \simeq \frac{k_{GT}}{r_{pore}} \qquad (3.13)$$

The k_{GT} coefficient is defined as $k_{GT} = 2T_m^{bulk} \Delta \gamma V_m / \Delta H_m$, where $\Delta \gamma = \gamma_{ws} - \gamma_{wl}$, representing the balance between the involved pore wall/ice and pore wall/water interfacial energies, and V_m and ΔH_m stand for the molar volume of the liquid and the molar enthalpy of melting, respectively [see (2.4) and (2.6) for comparison]. When $\Delta \gamma > 0$ (pore surface has higher affinity to the liquid phase) then $k_{GT} > 0$ and, as a

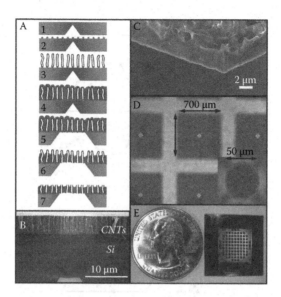

FIGURE 3.30 (a) Schematic of the fabrication process: (step 1) microscale pit formation (by KOH etching), (step 2) catalyst deposition/annealing, (step 3) nanotube growth, (step 4) gap filling with low-pressure chemical vapordeposited Si_3N_4, (step 5) membrane area definition (by XeF_2 isotropic Si etching), (step 6) silicon nitride etch to expose nanotubes and remove catalyst nanoparticles (by Ar ion milling), (step 7) nanotube uncapping (reactive ion etching). (b) SEM cross-section of the as-grown DWNTs. (c) SEM cross-section of the membrane, illustrating the gap filling by silicon nitride. (d) Open membrane areas. The inset shows a close-up of one membrane. (e) Membrane chip containing 89 open 50-μm diameter windows compared to a quarter dollar coin. (Reprinted from J. K. Holt, et al.. *Science* 312:1034–1037, 2006. With permission from AAAS.)

consequence, $T_m^{bulk} > T_m^{pore}$. When $\Delta \gamma = \gamma_{sl}$, then (3.13) reproduces the well-known Gibbs–Thomson equation, hence the GT subscript. For mesoporous silicas MCM-41, $T_m^{pore} \simeq 215$ K for $r_{pore} \simeq 3$ nm, that is, well below the $\sim$235 K limit of homogeneous nucleation of bulk water, so that $\Delta T_m \sim 50$ K (Schreiber, Ketelsen, and Findenegg 2001).

The experimental determination of ΔT_m leads to the empirical expression (Schmidt et al. 1995):

$$\Delta T_m \simeq \frac{k_{GT}}{r_{pore} - t_w} \tag{3.14}$$

which is a modification of (3.13), where t_w represents the effective thickness of a surface layer of nonfreezable water at the pore wall exhibiting values of 0.4 nm for some mesoporous silica materials such as MCM-41. The presence of such a liquidlike structure hinders the formation of ice even at low temperatures. This is shown in Figure 3.31 for the particular case of the aluminophosphate microporous crystal (AlPO4-54) with pore diameters of 1.2 nm.

The figure shows the experimentally derived crystal structures taken at both 293 K (top) and 173 K (bottom). At 293 K we observe that the pores are partially filled with

293 K

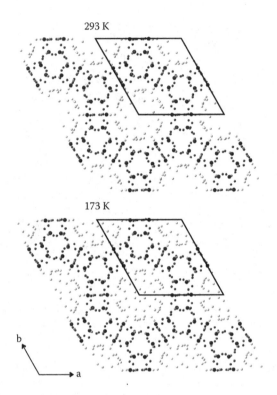

173 K

FIGURE 3.31 Experimental crystal structure of water in the aluminophosphate $AlPO_4$-54 at 293 (top) and 173 K (bottom), as determined by X-ray diffraction (Alabarse et al. 2012). The pore diameter is 1.2 nm. Water oxygens, aluminophosphate oxygens, phosphorous, and aluminum atoms are represented by light gray, dark gray, medium gray, and black balls, respectively.

water molecules that stay closer to the walls (the center is empty). However, upon decreasing the temperature a regular pattern is generated in the inner part of the pore. The hydrophilic pore walls induce an orientational order of water in contact with it, and at 173 K a cylindrical core of glassy water in the pore center is formed. Thus, even at such a low temperature crystallization does not occur in the core region due to the small curvature imposed by the pores that hinders the formation of a tetrahedrally coordinated network.

3.4.3 ELECTROCHEMICAL NANOPATTERNING

An AFM tip close enough to a flat surface in ambient conditions is attracted toward such a surface, provided the cantilever force constant is sufficiently low, due to the formation of a water meniscus between the tip and the surface, which is a further example of confined water. The expression for the capillary force is shown in Appendix B. Thanks to the development of environmental scanning electron microscopes (SEMs) it is possible to image such nanometer-sized water menisci. These instruments are capable of obtaining images with base pressures typically below 10 kPa, thus enabling the study of surfaces in nearly real conditions. The first images were obtained using tungsten tips and coated mica surfaces (Schenk, Füting, and Reichelt 1998). The effect of capillary forces between an AFM tip and a flat surface was previously shown in Figure 2.3, when dealing with force curves on confined water nanodroplets, when discussing the case of BaF$_2$(111) surfaces (see Figure 3.16) and comes later in Section 4.3.2 when dealing with surfaces exhibiting dual hydrophilic–hydrophobic character, but here we are interested in the case when an electric field is applied between a conductive tip and a surface indeed in the presence of water. Under certain conditions water is split and electrochemical action can take place leading, for example, to the well-known local oxidation nanolithography technique. Dagata et al. (1990) were the first to modify a H-terminated silicon surface using an STM tip and Day and Allee (1993) obtained similar results with an AFM. The conductive tips act as a cathode and the water meniscus provides the electrolyte building a nanometer-size electrochemical cell. For silicon the half-cell reactions are:

$$Si + 2h^+ + 2(OH)^- \rightarrow Si(OH)_2 \rightarrow SiO_2 + 2H^+ + 2e^-$$
$$2H^+(aq) + 2e^- \rightarrow H_2$$

The dimensions of the obtained structures (arrays and dots) can be controlled with the shape of the water meniscus (by varying the tip–surface distance) reproducibly obtaining sub-10 nm structures. An illustrative example is shown in Figure 3.32 where arrays of dots, rings, and the first ten lines of Cervantes' *Don Quijote de la Mancha* have been written by local oxidation (García, Martínez, and Martínez 2006). A long list of of electronic and mechanical devices with nanometer-scale features has been obtained with local oxidation nanolithography including data storage memories, conducting nanowires, side-gated field-effect transistors, single electron transistors, and superconducting quantum interference devices to mention a few.

3.5 WHEN IONS COME ON THE SCENE

3.5.1 ION HYDRATION

So far we have intentionally ignored the presence of ions in order to concentrate on the basics of pure water at interfaces. However, the role of ions is extremely important, particularly in biological systems, and is introduced next. Let us start by considering the simplest case of an individual ion with charge ze and radius r_{ion} interacting with an electrical dipole $\mu = ql$ of radius r_μ (q and l represent the dipole charge and separation, respectively). Both objects are separated by a center-to-center distance r and define an angle θ as schematized in Figure 3.33.

From basic electrostatics it can be shown that the interaction energy is:

$$w(r) = -\frac{ze\mu}{4\pi\varepsilon\varepsilon_0 l} \left\{ \frac{1}{\sqrt{(r - \frac{l}{2}\cos\theta)^2 + (\frac{l}{2}\sin\theta)^2}} - \frac{1}{\sqrt{(r + \frac{l}{2}\cos\theta)^2 + (\frac{l}{2}\sin\theta)^2}} \right\}$$
$$(3.15)$$

FIGURE 3.32 Nanopatterns of silicon surfaces obtained by local oxidation with an AFM. (a) Periodic array of 10-nm silicon oxide dots with a lattice spacing of 40 nm. (b) Alternating insulating (bright) and semiconducting rings. (c) First paragraph of *Don Quijote de la Mancha*. Reproduced from R. García, R.V. Martínez, and J. Martínez, *Chem. Soc. Rev.* 35:29–38, 2006. With permission of The Royal Society of Chemistry.

If $z > 0$ (cation) then $|w(r)|$ is maximum when $\theta = 0$, so that (3.15) reduces to:

$$w(r) = -\frac{ze\mu}{4\pi\varepsilon\varepsilon_0}\frac{1}{r^2 - (l/2)^2} \tag{3.16}$$

The same solution would be obtained for the case of anions ($z < 0$) with $\theta = \pi$. Assuming that $r = r_{ion} + r_\mu$, as in the figure, and further simplifying by considering $r \gg l/2$, (3.16) then becomes:

$$w(r) \simeq -\frac{ze\mu}{4\pi\varepsilon\varepsilon_0(r_{ion} + r_\mu)^2} \tag{3.17}$$

From this equation it can be inferred that $|w|$ will achieve larger values for $|z| > 1$ and low r_{ion}. Table 3.3 displays calculated values of the interaction energy using (3.17) for some representative ions taking $\varepsilon = 1$ and $\mu = 1.85$ D and $r_\mu = 0.14$ nm (half of the mean molecular diameter) for the water dipole. We observe that those exhibiting higher energies are Al^{3+}, Mg^{2+}, Ca^{2+}, Li^+, and so on.

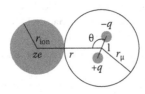

FIGURE 3.33 Scheme of the Coulomb interaction between an ion with charge ze and ionic radius r_{ion} and a dipole $\mu = ql$ of mean radius r_μ separated a distance r and with a relative orientation represented by the angle θ.

TABLE 3.3
Ionic and Hydrated Radii, Interaction Energy, Hydration Number and Lifetime of Selected Ions in Water

Ion	Ionic Radius [nm]	Hydrated Radius [nm]	Interaction Energy [eV]	Hydration Number	Lifetime [s]
Al^{3+}	0.050	0.48	4.6	6	0.1–1
Mg^{2+}	0.065	0.43	2.6	6	10^{-6}
Ca^{2+}	0.099	0.41	1.9	6	10^{-8}
Li^{+}	0.068	0.38	1.3	5–6	5×10^{-9}
Na^{+}	0.095	0.36	1.0	4–5	10^{-9}
K^{+}	0.133	0.33	0.7	3–4	10^{-9}
Cs^{+}	0.169	0.33	0.6	1–2	5×10^{-10}
F^{-}	0.136	0.35	0.7	2	
Cl^{-1}	0.181	0.33	0.5	1	10^{-11}
Br^{-1}	0.195	0.33	0.5	1	10^{-11}
I^{-}	0.216	0.33	0.4	0	10^{-11}
OH^{-}	0.176	0.30	0.5	3	
NO_3^{-}	0.264	0.34	0.3	0	

(Adapted from J. Israelachvili, *Intermolecular & Surface Forces*, 2011. Amsterdam: Elsevier.)

If we now consider the ions in bulk water, there will be a competition between the water dipole orientation and H-bonding among water molecules. Those ions with $|w|$ clearly exceeding the values corresponding to H-bonding (small ions with high charge densities) will tend to structure water around them deserving the term *kosmotropes* (structure makers). As opposed to such ions, *chaotropes* (structure breakers) will consist of large ions with low charge densities because of their limited ability to distort H-bonding (Hribar et al. 2002). A direct consequence is that the hydration number, the average number of water molecules structured around the ion in a first shell, is larger for kosmotropes (4–6) and lower for chaotropes (≤ 2). Cations are generally more solvated than anions because they are smaller due to less electronic repulsion. For anions, the hydration (or residence) lifetime is of the order of 10^{-11} s whereas for cations such characteristic times are about 10^{-9} s in the case of monovalent cations and in the $10^{-8} - 10^{-6}$ s range for divalent cations. This is to be compared to the characteristic H-bonding lifetime, which is of a few ps in bulk water (Luzar and Chandler 1996). The lifetime of the vibrational bending mode (ν_2) is longer for water bound to halogenic anions (~ 1 ps) than for bulk liquid water (~ 380 fs), an indication of the local ion-induced stabilization. The observed bending mode associated with water bound to anions exhibits a redshift with respect to the bulklike band, which increases in the halogenic series Cl^-, Br^-, and I^- (Piatkowski and Bakker 2011).

Ions can be classified according to the criteria established by Hofmeister, and known as the Hofmeister series, based on the particular case of their influence on the solubility of proteins (Hofmeister 1888; Collins and Washabaugh 1985). The resulting

TABLE 3.4
Ionic Surface Tension Increments, in mN m^{-1} M^{-1} Units, of Selected Aqueous Ions at Ambient Conditions

Al^{+3}	Mg^{+2}	Ca^{+2}	Na$^+$	K$^+$	Li$^+$	NH$_4^+$
2.65	2.25	2.10	1.20	1.10	0.95	0.70
OH$^-$	Cl$^-$	F$^-$	HPO$_4^{2-}$	Br$^-$	NO$_3^-$	I$^-$
1.05	0.90	0.80	0.70	0.55	0.15	-0.05

From Y. Marcus, *Langmuir* 29: 2881–2888, 2013. With permission.

ranking for anions and cations is, respectively:

$$SO_4^{2-} > F^- > HPO_4^{2-} > CH_3COO^- > Cl^- > Br^- > NO_3^- > I^- > ClO_4^- > SCN^-$$
$$NH_4^+ > K^+ > Na^+ > Li^+ > Mg^{2+} > Ca^{2+}$$

The Hofmeister series reflects specific ion effects on the long-range structure of water and has to be considered as a reliable empirical rule of thumb rather than an absolute scale. It is in general accepted that anions on the left of the series are kosmotropes and those on the right side are chaotropes whereas for cations it is just the opposite: those on the left are chaotropes and those on the right, kosmotropes. However, this structure maker–breaker concept has been challenged by some authors when studying alkali halide solutions. Smith, Saykally, and Geissler (2007) conclude that the observed changes in the vibrational Raman spectra of liquid water upon addition of potassium halides is a direct consequence of the electric fields that anions exert on adjacent hydrogen atoms rather than due to structural rearrangements in the H-bonding network. Neutron diffraction data on aqueous solutions of NaCl and KCl have been interpreted in terms of water molecules in the hydration shells of potassium cations being more orientationally disordered than those hydrating sodium cations (Mancinelli et al. 2007). Even if both cations are considered as structure-breakers, the sodium cation has a larger effect on water–water correlations.

Concerning γ_{lv}, it increases when ions are added to bulk water and this increase is proportional to the ionic concentration with a proportionality factor that can be expressed in terms of the individual ionic contributions weighted with the corresponding stoichiometric coefficients (Markus 2013). Table 3.4 gives a list of ionic surface tension increments of aqueous ions at ambient conditions for selected common electrolytes. We observe that anions and cations follow the direct and reverse Hofmeister series, respectively, so that the ability of ions to enhance γ_{lv} decreases from left to right in the table (Allen et al. 2009).

The differentiated structuring ability of anions is also observed at the water/air interface. One should expect chaotrope anions to segregate toward such an interface because of their limiting structuring action and indeed this is demonstrated by MD simulations which have shown that the propensity to adsorb at such an interface increases from left to right in the Hofmeister series (Jungwirth and Tobias 2006). Figure 3.34 shows snapshots from the simulations of sodium halide solutions (left) together with the density profiles (right) of the ionic species and of water oxygen atoms

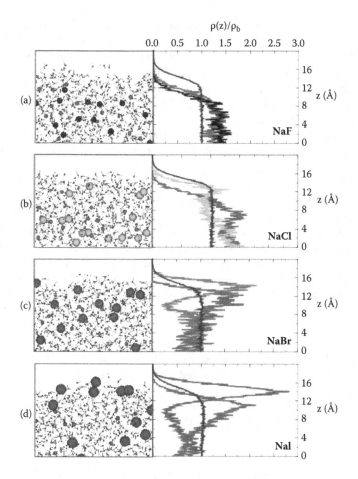

FIGURE 3.34 (Left) snapshots and (right) ion densities from MD simulations of alkali halide solutions/air interfaces: (a) NaF, (b) NaCl, (c) NaBr, and (d) NaI. Halide anions are represented by larger spheres. Ion densities *versus* distance, $\rho(z)$, are plotted from the center of the slabs in the direction normal to the interface and normalized to the bulk water density (ρ_b). See the original color figure for the identification of ion density distributions. (Reprinted from P. Jungwirth, and D.J. Tobias, *J. Phys. Chem. B* 105:10468–10472, 2001, American Chemical Society. With permission.)

(Jungwirth and Tobias 2001). In the NaF solution (Figure 3.34a), both ions are strongly repelled from the surface and bromine and iodine anions are significantly distributed at the air/water interface (see Figures 3.34c and d). In fact for both anions their concentration at the interface is enhanced with respect to the bulk. The interfacial anion concentration is much higher than the cation concentration, whereas just below the surface the cation concentration dominates, resulting in the creation of an electrical double layer. This surface enhancement of bromide relative to chloride is expected to enhance the formation of acid rain as discussed in Section 5.3. Using NAPP the composition of the liquid/vapor interface for deliquesced samples of potassium bromide and potassium iodide have been measured (Ghosal et al. 2005). In both cases, the surface composition of the saturated solution is enhanced in the halide anion compared with the bulk of the solution, an enhancement that is more patent for the iodide anion.

In the case of ions at the ice/vapor interface MD simulations have shown that when ice grows from a supercooled ionic solution, ions are rejected by the growing ice (with cubic structure) and confined to the QLL (Carignano, Shepson, and Szleifer 2007). This is in fact a way of purifying water upon freezing. In fact, when a solution is cooled below its eutectic point pure ice and salt crystallize separately. This property has been proposed for the purification of water at an industrial level (van der Ham et al. 1998).

What happens when ions are located on a solid surface? This can be best answered when considering the simpler case of alkali atoms on metallic surfaces. In the absence of water charge is transferred from the atoms into the surface inducing the formation of surface dipoles μ_s and, as a consequence, modifying the work function ϕ of the metal. In the case of Na^+, K^+, Cs^+, and so on, ϕ decreases for increasing coverage Θ down to a minimum value and then increases. Such behavior is not exclusive of alkali atoms but characteristic of electron donors (Fraxedas et al. 2011). The change in the work function ($\Delta\phi$) can be correlated with the induced dipole through the Helmholtz equation:

$$\Delta\phi \simeq -\frac{\mu_s}{\varepsilon_0 A} \tag{3.18}$$

where A stands for the surface area per molecule. According to Langmuir's depolarization model, the strength of the induced dipoles due to charge transfer decreases with the number of dipoles (Langmuir 1932). If we assume a linear dependence, $\mu_s \simeq \mu_s(0) - \alpha\Theta$, where $\mu_s(0)$ represents the induced dipole at very low coverages and α to a linear coefficient, then, through (3.18) $\Delta\phi$ depends quadratically on Θ:

$$\Delta\phi \propto -\mu_s(0)\Theta + \alpha\Theta^2 \tag{3.19}$$

because $A \propto 1/\Theta$. For alkali ions, $\mu_s(0) \sim 10$ D, a rather high value.

When water is coadsorbed on the surface the dipolar interactions determine the spatial distribution of the water molecules, in the case where they remain intact, which is true for low coverages. As a result of the dipolar interaction the orientation of the water molecules changes with respect to the orientation found on the ion-free substrates. For increasing coverages there is a competition between water–alkali electrostatic interactions, water–water H-bonding and water–metal interactions. Water coadsorbed with a sufficiently low coverage of potassium (<0.3 ML) does not dissociate at 85 K, as has been demonstrated by HREELS, with an estimated number of water molecules affected per potassium atom of about 3 (Chakarov, Österlund, and Kasemo 1995). At larger potassium coverages the coadsorption results in water dissociation. ϕ is indeed affected by the presence of water. Its slope as a function of water coverage is positive for low coverage but changes to negative at a certain water coverage (Bonzel, Pirug, and Ritke 1991). This behavior can be explained in a two-step model. The adsorption of water starts at sites in the vicinity of the alkali atom where water molecules form hydrationlike complexes, which in turn results in an increase in the work function. Clustering of water due to H-bonding interactions does not occur until all the hydration sites around the alkali atoms have been occupied. Once these sites are filled, adsorption takes place on sites that are not influenced by the alkali atoms, leading to the decrease in ϕ.

3.5.2 ELECTRICAL DOUBLE LAYER

We come now to the case of a flat surface with charges (e.g., positive) distributed on it, which define a given surface charge density, submerged in a solution containing electrolytes. The electrostatic field generated by the interfacial charges will distort the otherwise homogeneous distribution of ions inducing a gradient of concentration. As a result of the positive charge, a negative charge distribution close to the interface will be generated, together building an electrical (Stern) double layer (Stern 1924). The Stern layer can be subdivided into an inner and an outer (Helmholtz) layer (Helmholtz 1853). The inner layer sets the location of the negative charges closest to the surface (positive) charges, both planes forming a simple capacitor, and the outer layer defines the border with the diffuse (Gouy–Chapman) layer (Gouy 1910; Chapman 1913), introduced below. An illustrative scheme is given in the top part of Figure 3.35. A detailed description of the charge distribution close to the water/solid interfaces as well as the involved forces can be found in Israelachvili (2011). Here only a glimpse of fundamental concepts is given.

Let us consider a solution containing different types of electrolytes ($i = 1, \ldots, N$) of valency z_i in the simple 1D case arbitrarily set along the x-axis (see Figure 3.35). The spatial charge density $\rho_{ch}(x)$ can be expressed in terms of the individual ionic concentrations $n_i(x)$ through the expression:

$$\rho_{ch}(x) = \sum_{i=1}^{N} z_i e n_i(x) \tag{3.20}$$

where $n_i(x)$ can be expressed in terms of the Boltzmann distribution:

$$n_i(x) = n_i(\infty)e^{-z_i e \varphi(x)/k_B T} \tag{3.21}$$

where $\varphi(x)$ stands for the electrostatic potential and $n_i(\infty)$ represents the bulk ionic concentrations (away from the surface). $\varphi(x)$ can be determined using the Poisson–Boltzmann equation:

$$\frac{d^2\varphi(x)}{dx^2} = -\frac{\rho_{ch}(x)}{\varepsilon\varepsilon_0} = -\frac{1}{\varepsilon\varepsilon_0}\sum_i z_i e n_i(\infty)e^{-z_i e \varphi(x)/k_B T} \tag{3.22}$$

In the limit of small electrostatic potentials, $e^{-z_i e \varphi(x)/k_B T} \simeq 1 - z_i e \varphi(x)/k_B T$, so that (3.22) reduces to:

$$\frac{d^2\varphi(x)}{dx^2} \simeq \kappa_D^2 \varphi(x) \tag{3.23}$$

given that $\sum_{i=1}^{N} z_i e n_i(\infty) = 0$, where

$$\kappa_D^2 = \frac{e^2}{\varepsilon\varepsilon_0 k_B T} \sum_{i=1}^{N} z_i^2 n_i(\infty) \tag{3.24}$$

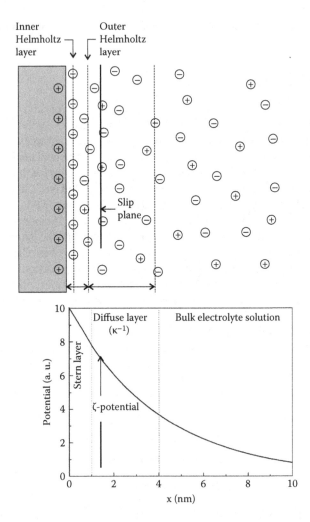

FIGURE 3.35 Spatial charge distribution of an aqueous electrolyte solution induced by a fixed (positive) surface charge distribution on a solid surface (top) and calculated electrostatic potential as a function of the distance to the interface using (3.23) for a Debye length of 4 nm (bottom). The positions of the inner and outer Helmholtz layers and of the diffuse layer are indicated by discontinuous lines. The slip plane (continuous line) and the ζ-potential have been arbitrarily located close to the outer Helmholtz layer position. A linear variation of the electrostatic potential has been taken within the Stern layer.

which defines an exponential decay for the electrostatic potential, $\varphi(x) \simeq \varphi(0)$ $e^{-\kappa_D x}$. Note that $\varphi(\infty) = 0$.

The inverse of κ_D corresponds to a characteristic decay length, known as the Debye length, that defines the width of the diffuse layer (see Figure 3.35). Such a layer can be regarded as the transition between the more localized (rigid) Stern–Helmholtz layer and the bulk region, where the charges are mobile and under negligible influence from the charges at the interface. As an example, for an aqueous solution of NaCl, κ_D^{-1} varies, for example, from about 30 nm at 10^{-4} M to about 1 nm at 0.1 M. In the lower part of Figure 3.35, $\varphi(x)$ is represented for the particular case of $\kappa_D^{-1} = 4$ nm from $x = 1$. The decay in the Stern region is taken as linear as in a simple capacitor with a constant internal electric field.

The spatial charge density can be calculated combining (3.20), (3.21), and (3.23):

$$\rho_{ch}(x) \simeq -\varepsilon\varepsilon_0\kappa_D^2\varphi(x) \simeq -\varepsilon\varepsilon_0\kappa_D^2\varphi(0)e^{-\kappa_D x} \qquad (3.25)$$

For the particular case here of a positive interfacial charge, $\rho_{ch}(x) < 0$ and $\rho_{ch}(\infty) = 0$.

There is still one important spatial location to be considered that defines the ζ-potential (zeta potential). It is well known from electrokinetic measurements that when an aqueous solution moves tangentially to a charged surface a thin water layer remains immobilized, which is known as the stagnant layer. The plane separating the stagnant layer and the mobile part of the fluid is known as the slip plane (Lyklema, Rovillard, and De Coninck 1998). In general, it is assumed that this plane is located very close to the outer Helmholtz plane. This means that, in practice, the ζ-potential lies between the diffuse-layer potentials at the outer Helmholtz layer and at the Debye length ($\varphi(1) \geq \zeta \geq \varphi(4)$ in Figure 3.35). One has to bear in mind that all planes referred to here are nothing other than practical abstractions. The outer Helmholtz layer is interpreted as a sharp boundary between the diffuse and the nondiffuse parts of the electric double layer and the slip plane as a sharp boundary between the hydrodynamically mobile and immobile fluids. In reality, none of these transitions is sharp. However, liquid motion may be hindered in the region where ions experience strong interactions with the surface (Delgado et al. 2005). Experimental determination of the ζ-potential at the interface between air bubbles and deionized water gives values of about -65 mV (Graciaa et al. 1995). This is further proof that the air/water interface is negatively charged due to the preferential accumulation of OH^- or depletion of H_3O^+ ions at these interfaces, as previously discussed in Section 2.1.1.

If we now consider two identically charged surfaces, such as the one from Figure 3.35, approaching each other along the x-direction, they will undergo the opposite effects of electrostatic repulsion due to the corresponding double layers and attraction induced by vdW forces. The balance will depend, among other factors, on the surface charge, electrolyte type and concentration, surface shape, and the like. As an interfacial phenomenon it will become more important for systems with a large surface-to-volume ratio such as particles and colloids in the micronanometer range. Colloidal stability is satisfactorily described by the DLVO theory, that incorporates both contrasting interactions, and the acronym includes the names of their developers, namely Derjaguin and Landau (1941), Verwey and Overbeek (1948).

3.5.3 DISSOLUTION OF AN IONIC SURFACE

We end this chapter by analyzing the case of an ionic solid surface, NaCl(001), with water on top of it showing how the hydration of ions controls the surface charge state (Verdaguer et al. 2008). Figure 3.36a shows the NAPP spectra taken at a base water vapor of 2 torr of both Cl2p (top) and Na2s (bottom) core levels measured at 5%

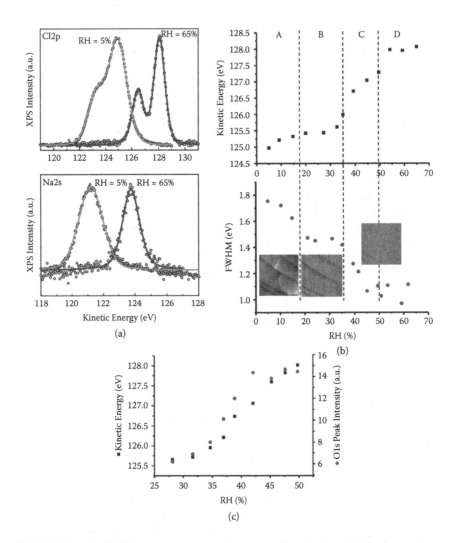

FIGURE 3.36 (a) NAPP spectra corresponding to the Cl2p (top) and Na2s (bottom) core levels measured at 5 and 65% RH, respectively, at 2 torr water vapor pressure. $K \sim 125$ eV for both core levels, ensuring a similar probing depth for both elements and high surface sensitivity. The experimental data (open circles) are compared to least-square fits to Voigt functions (lines). (b) Evolution of K (top curve) and of FWHM (bottom curve) of the Cl2p line excited with 335 eV photons *versus* RH. Different regions (A, B, C, D) are defined according to the observed evolution of KE and FWHM with RH. KPFM images of the surfaces are shown. (c) K (left) and surface oxygen content (right) versus RH. (Reprinted from A. Verdaguer, et al. *J. Phys. Chem. C* 112:16898–16901, 2008. American Chemical Society. With permission.)

and 65% RH, respectively. In both cases the full width at half maximum (FWHM) decreases and K increases for increasing RH values, a characteristic behavior of surface discharge as observed with photoemission. The RH dependence of both K (top) and FWHM (bottom) for the Cl$2p$ line is summarized in Figure 3.36b divided in regions labeled A, B, C, and D. Region A, from the lowest RH to ~20%, exhibits a small increase in K of about 0.3 eV reaching an intermediate constant value at 30–35% RH (region B). Above 35% K increases up to 50% (region C) and above RH ~55% (region D) the K value remains unchanged. The total energy shift is 3 eV. For the Na$2s$ line, the total shift is smaller (2.5 eV). The FWHM curve shows a similar dependence on RH. Region A exhibits a monotonous decrease of FWHM whereas in region B the FWHM remains approximately constant, until decreasing strongly starting at about 35% RH, eventually reaching a value of 1 eV. An increase in the RH from this point on does not lead to a further decrease in the FWHM value. The onset of strong discharging at 35% RH has been related to the modification of the surface step structure due to ion mobility on this surface (Dai, Hu, and Salmeron 1997). In addition, IR studies have shown that this change is associated with a sudden increase of water coverage from sub-ML coverage to ~3 MLs (Foster and Ewing 2000). Above about 50% RH the surface seems to be effectively discharged, given that the K of the core levels remains essentially constant. In Figure 3.36c the K values (left scale) of the Cl$2p$ peak together with the area of the O$1s$ peak from adsorbed water are plotted. From 32% to 42% both curves are roughly coincident, indicating that surface discharging through ionic mobility is correlated with the formation of molecularly thin water layers. Thus, discharging becomes more efficient once surface ions become mobile due to solvation after water adsorption.

3.6 SUMMARY

- On close-packed surfaces of transition metals water monomers lie nearly flat on atop positions, with the H–O–H plane above the surface plane, due to preferential bonding of molecular $1b_1$ states to d metal states. The electronic structure of the monomer/metal system can be rationalized in terms of frontier orbitals, as introduced by R. Hoffmann.
- Water dimers on surfaces exhibit, in general, buckled geometries. The addition of more water molecules one by one leads to the formation of trimers, tetramers, pentamers, hexamers, and so on. Cyclic hexamers constitute the smallest pieces of ice and can be thought of as LEGO-like building blocks. Ice bilayers can be regarded as buckled monolayers.
- Interfacial registry, either commensurate or incommensurate, can be classified using matrix formalisms in a systematic way. This strictly geometrical method allows the identification of potential surfaces that may stabilize the growth of ice multilayers, although disregarding chemical aspects.
- Water monolayers can become hydrophobic. On some surfaces water seems to induce a protective wetting hindering layer-by-layer growth because of unfavorable H-bonding architectures.
- Wetting of metals at near-ambient conditions is controlled by the presence of hydroxyl groups on the surface that stabilize water molecules via H-bonding.

- Hexagonal ice can be obtained at room temperature upon nanometer-scale confinement. However, cylindrical pores of hydrophilic materials are able to reduce the melting temperature of bulk water.
- The solid phase diagram of water is enlarged by the action of carbon nanotubes that induce varied one-dimensional ordering depending on their internal tube diameters.
- The behavior of anions at the air/water interface can be understood in terms of the Hofmeister series. Chaotrope anions accumulate at such an interface building an electrical double layer whereas kosmotrope anions are excluded from such an interface.

REFERENCES

1. Alabarse, F.G., Haines, J., Cambon, O., Levelut, C., Bourgogne, D., Haidoux, A. et al.. 2012. Freezing of water at the nanoscale. *Phys. Rev. Lett.* 109:035701.
2. Allen, H.C., Casillas-Ituarte, N.N., Sierra-Hernández, M.R., Chen, X., and Tang, C.Y. 2009. Shedding light on water structure at air–aqueous interfaces: Ions, lipids, and hydration. *Phys. Chem. Chem. Phys.* 11:5538–5549.
3. Andersson, S., Nyberg, C., and Tengstål, C.G. 1984. Adsorption of water monomers on Cu(100) and Pd(100) at low temperatures. *Chem. Phys. Lett.* 104:305–310.
4. Antognozzi, M., Humphris, A.D.L., and Miles, M.J. 2001. Observation of molecular layering in a confined water film and study of the layers viscoelastic properties. *Appl. Phys. Lett.* 78:300–302.
5. Árnadóttir, L., Stuve, E.M., and Jónsson, H. 2010. Adsorption of water monomer and clusters on platinum (111) terrace and related steps and kinks. I. Configurations, energies and hydrogen bonding. *Surf. Sci.* 604:1978–1986.
6. Beaglehole, D. and Christenson, H.K. 1992. Vapor adsorption on mica and silicon: Entropy effects, layering, and surface forces. *J. Phys. Chem.* 96:3395–3403.
7. Bikondoa, O., Pang, C.L., Ithnin, R., Muryn, C.A., Onishi, H., and Thornton, G. 2006. Direct visualization of defect-mediated dissociation of water on $TiO_2(110)$. *Nature Mater.* 5:189–192.
8. Bluhm, H., Inoue, T., and Salmeron, M. 2000. Formation of dipole–oriented water films on mica substrates at ambient conditions. *Surf. Sci.* 462:L599–L602.
9. Bonzel, H.P., Pirug, G., and Ritke, C. 1991. Adsorption of H_2O on alkali-metal-covered Pt(111) and Ru(001): a systematic comparison. *Langmuir* 7:3006–3011.
10. Bradshaw, A.M. and Richardson, N.V. 1996. Symmetry, selection rules and nomenclature in surface spectroscopies (IUPAC Recommendations 1996). *Pure & Appl. Chem.* 68:457–467.
11. Brovchenko, I. and Oleinikova, A. 2008. *Interfacial and Confined Water*. Amsterdam: Elsevier.
12. Bührer, W., Nicklow, R.M., and Brüesch, P. 1978. Lattice dynamics of β-(silver iodide) by neutron scattering. *Phys. Rev. B* 17:3362–3370.
13. Burley, G. 1963. Structure of silver iodide. *J. Chem. Phys.* 38:2807–2812.
14. Cambré, S., Schoeters, B., Luyckx, S., Goovaerts, E., and Wenseleers, W. 2010. Experimental observation of single-file water filling of thin single–wall carbon nanotubes down to chiral index (5,3). *Phys. Rev. Lett.* 104:207401.

15. Cantrell, W. and Ewing, G.E. 2001. Thin film water on muscovite mica. *J. Phys. Chem. B* 105:5434–5439.

16. Cardellach, M., Verdaguer, A., and Fraxedas, J. 2011. Defect–induced wetting on $BaF_2(111)$ and $CaF_2(111)$ at ambient conditions. *Surf. Sci.* 605:1929–1933.

17. Cardellach, M., Verdaguer, A., Santiso, J., and Fraxedas, J. 2010. Two–dimensional wetting: The role of atomic steps on the nucleation of thin water films on $BaF_2(111)$ at ambient conditions. *J. Chem. Phys.* 132:234708.

18. Carignano, M.A., Shepson, P.B., and Szleifer, I. 2007. Ions at the ice/vapor interface. *Chem. Phys. Lett.* 436:99–103.

19. Carrasco, J., Michaelides, A., and Scheffler, M. 2009. Insight from first principles into the nature of the bonding between water molecules and $4d$ metal surfaces. *J. Chem. Phys.* 130:184707.

20. Chakarov, D.V., Österlund, L., and Kasemo, B. 1995. Water adsorption and coadsorption with potassium on graphite (0001). *Langmuir* 11:1201–1214.

21. Chapman, D.L. 1913. A contribution to the theory of electrocapillarity. *Phil. Mag.* 25:475–481.

22. Charlton, G., Howes, P.B., Nicklin, C.L., Steadman, P., Taylor, J.S.G., Muryn, C.A. et al. 1997. Relaxation of $TiO_2(110)$-(1×1) using surface X-ray diffraction. *Phys. Rev. Lett.* 78:495–498.

23. Chen, H.–T., Musaev, D.G., and Lin, M.C. 2007. Adsorption and dissociation of H_2O on a W(111) surface: A computational study. *J. Phys. Chem. C* 111:17333–17339.

24. Cheng, L., Fenter, P., Nagy, K.L., Schlegel, M.L., and Sturchio, N.C. 2001. Molecular-scale density oscillations in water adjacent to a mica surface. *Phys. Rev. Lett.* 87:156103.

25. Cho, J.-H., Kim, K.S., Lee, S.H., and Kang, M.H. 2000. Dissociative adsorption of water on the Si(001) surface: A first-principles study. *Phys. Rev. B* 61:4503–4506.

26. Choi, E.-M., Yoon, Y.-H., Lee, S., and Kang, H. 2005. Freezing transition of interfacial water at room temperature under electric fields. *Phys. Rev. Lett.* 95:085701.

27. Clay, C., Haq, S., and Hodgson, A. 2004. Intact and dissociative adsorption of water on Ru(0001). *Chem. Phys. Lett.* 388:89–93.

28. Collins, K.D. and Washabaugh, M.W. 1985. The Hofmeister effect and the behaviour of water at interfaces. *Quart. Rev. Biophys.* 18:323–422.

29. Dagata, J.A., Schneir, J., Harary, H.H., Evans, C.J., Postek, M.T., and Bennett, J. 1990. Modification of hydrogen-passivated silicon by a scanning tunneling microscope operating in air. *Appl. Phys. Lett.* 56:2001–2003.

30. Dai, Q., Hu, J., and Salmeron, M. 1997. Adsorption of water on NaCl (100) surfaces: Role of atomic steps. *J. Phys. Chem. B* 101:1994–1998.

31. Day, H.C. and Allee, D.R. 1993. Selective area oxidation of silicon with a scanning force microscope. *Appl. Phys. Lett.* 62:2691–2693.

32. Delgado, A.V., González-Caballero, F., Hunter, R.J., Koopal, L.K., and Lyklema, J. 2005. Measurement and interpretation of electrokinetic phenomena. *Pure Appl. Chem.* 77:1753–1805.

33. Derjaguin, B.V. and Landau, L. 1941. Theory of the stability of strongly charged lyophobic sols and of the adhesion of strongly charged particles in solution of electrolytes. *Acta Physicochimica URSS* 14:633–662.

34. Diebold, U. 2003. The surface science of titanium dioxide. *Surf. Sci. Rep.* 48:53–229.

35. Du, Y., Deskins, N.A., Zhang, Z., Dohnalek, Z., Dupuis, M., and Lyubinetsky, I. 2010. Water interactions with terminal hydroxyls on $TiO_2(110)$. *J. Phys. Chem. C* 114:17080–17084.

36. Duncan, D.A., Allegretti, F., and Woodruff, D.P. 2012. Water does partially dissociate on the perfect $TiO_2(110)$ surface: A quantitative structure determination. *Phys. Rev. B* 86:045411.

37. Ehre, D., Lavert, E., Lahav, M., and Lubomirsky, I. 2010. Water freezes differently on positively and negatively charged surfaces of pyroelectric materials. *Science* 327:672–675.

38. Engkvist, O. and Stone, A.J. 1999. Adsorption of water on the MgO(001) surface. *Surf. Sci.* 437:239–248.

39. Faivre, C., Bellet, D., and Dolino, G. 1999. Phase transitions of fluids confined in porous silicon: A differential calorimetry investigation. *Eur. Phys. J. B* 7:19–36.

40. Feibelman, P.J. 2002. Partial dissociation of water on Ru(0001). *Science* 295:99–102.

41. Feibelman, P.J. 2003. Comment on Vibrational recognition of hydrogen–bonded water networks on a metal surface. *Phys. Rev. Lett.* 91:059601.

42. Fernández–Torrente, I., Monturet, S., Franke, K.J., Fraxedas, J., Lorente, N., and Pascual, J.I. 2007. Long–range repulsive interaction between molecules on a metal surface induced by charge transfer. *Phys. Rev. Lett.* 99:176103.

43. Ferry, D., Glebov, A., Senz, V., Suzanne, J., Toennies, J.P., and Weiss, H. 1996. Observation of the second ordered phase of water on the MgO(100) surface: Low energy electron diffraction and helium atom scattering studies. *J. Chem. Phys.* 105:1697–1701.

44. Firment, L.E. and Somorjai, G.A. 1976. The surface structures of vapor–grown ice and naphthalene crystals studied by low–energy electron diffraction. *Surf. Sci.* 55:413–426.

45. Fomin, E., Tatarkhanov, M., Mitsui, T., Rose, M., Ogletree, D.F., and Salmeron, M. 2006. Vibrationally assisted diffusion of H_2O and D_2O on Pd(111). *Surf. Sci.* 600:542–546.

46. Foster, A.S., Trevethan, T., and Shluger, A.L. 2009. Structure and diffusion of intrinsic defects, adsorbed hydrogen, and water molecules at the surface of alkali-earth fluorides calculated using density functional theory. *Phys. Rev. B* 80:115421.

47. Foster, M.C. and Ewing, G.E. 2000. Adsorption of water on the NaCl(001) surface. II. An infrared study at ambient temperatures. *J. Chem. Phys.* 112:6817–6826.

48. Fraxedas, J. 2006. *Molecular Organic Materials*. Cambridge, UK: Cambridge University Press.

49. Fraxedas, J., García-Gil, S., Monturet, S., Lorente, N., Fernández-Torrente, I., Franke, K.J. et al.. 2011. Modulation of surface charge transfer through competing long-range repulsive versus short-range attractive interactions. *J. Phys. Chem. C* 115:18640–18648.

50. Fraxedas, J., Garcia-Manyes, S., Gorostiza, P., and Sanz, F. 2002. Nanoindentation: Toward the sensing of atomic interactions. *Proc. Natl. Acad. Sci. USA* 99:5228–5232.

51. Fukuma, T., Ueda, Y., Yoshioka, S., and Asakawa, H. 2010. Atomic-scale distribution of water molecules at the mica-water interface visualized by three-dimensional scanning force microscopy. *Phys. Rev. Lett.* 104:016101.

52. García, R., Martínez, R.V., and Martínez, J. 2006. Nano-chemistry and scanning probe nanolithographies. *Chem. Soc. Rev.* 35:29–38.

53. Gavish, M., Wang, J.-L., Eisenstein, M., Lahav, M., and Leiserowitz, L. 1992. The role of crystal polarity in alpha-amino acid crystals for induced nucleation of ice. *Science* 256:815–818.

54. Ghosal, S., Hemminger, J.C., Bluhm, H., Mun, B.S., Hebenstreit, E.L.D., Ketteler, G. et al. 2005. Electron spectroscopies of aqueous solution interfaces reveals surface enhancement of halides. *Science* 307:563–566.

55. Giri, A.K. and Mitra, G.B. 1985. Extrapolated values of lattice constants of some cubic metals at absolute zero. *J. Phys. D Appl. Phys.* 18:L75–L78.

56. Gladys, M.J., El Zein, A.A., Mikkelsen, A., Andersen, J.N., and Held, G. 2008. Chemical composition and reactivity of water on clean and oxygen-covered Pd(111). *Surf. Sci.* 602:3540–3549.

57. Glebov, A., Graham, A.P., Menzel, A., and Toennies J.P. 1997. Orientational ordering of two-dimensional ice on Pt(111). *J. Chem. Phys.* 106:9382–9385.

58. Gonschorek, W. 1982. X-Ray charge density study of rutile (TiO$_2$). *Z. Kristallogr.* 160:187–203.

59. Gouy, M. 1910. Sur la constitution de la charge électrique à la surface d'un électrolyte. *J. Phys. Theor. Appl.* 9:457–468.

60. Graciaa, A., Morel, G., Saulner, P., Lachaise, J., and Schechter, R.S. 1995. The ζ-potential of gas bubbles. *J. Colloid and Interf. Sci.* 172:131–136.

61. Gupta, V., Hampton, M.A., Nguyen, A.V., and Miller, J.D. 2010. Crystal lattice imaging of the silica and alumina faces of kaolinite using atomic force microscopy. *J. Colloid Int. Sci.* 352:75–80.

62. Hallett, J. 1961. The growth of ice crystals on freshly cleaved covelite surfaces. *Phil. Mag.* 6:1073–1087.

63. Haq, S., Harnett, J., and Hodgson, A. 2002. Growth of thin crystalline ice films on Pt(111). *Surf. Sci.* 505:171–182.

64. Held, G. and Menzel, D. 1995. Structural isotope effect in water bilayers adsorbed on Ru(001). *Phys. Rev. Lett.* 74:4221–4224.

65. Helmholtz, H. 1853. Über einige Gesetze der Vertheilung elektrischer Ströme in körperlichen Leitern mit Anwendung auf die thierisch-elektrischen Versuche. *Ann. Physik* 89:211–233.

66. Henderson, M.A. 2002. The interaction of water with solid surfaces: Fundamental aspects revisited. *Surf. Sci. Rep.* 46:1–308.

67. Hodgson, A. and Haq, S. 2009. Water adsorption and the wetting of metal surfaces. *Surf. Sci. Rep.* 64:381–451.

68. Hoffmann, R. 1988. A chemical and theoretical way to look at bonding on surfaces. *Rev. Mod. Phys.* 60:601–628.

69. Hofmeister, F. 1888. Zur Lehre von der Wirkung der Salze. *Arch. Exp. Path. Pharmakol.* XVII:247–260.

70. Holt, J.K., Park, H.G., Wang, Y., Stadermann, M., Artyukhin, B., Grigoropoulos, C.P. et al.. 2006. Fast mass transport through sub-2 nanometer carbon nanotubes. *Science* 312:1034–1037.

71. Hooks, D.E., Fritz, T., and Ward, M.D. 2001. Epitaxy and molecular organization on solid substrates. *Adv. Mater.* 13:227–241.

72. Hribar, B., Southall, N.T., Vlachy, V., and Dill, K.A. 2002. How ions affect the structure of water. *J. Am. Chem. Soc.* 124:12302–12311.

73. Hu, J., Xiao, X.–D., and Salmeron, M. 1995. Scanning polarization force microscopy: A technique for imaging liquids and weakly adsorbed layers. *Appl. Phys. Lett.* 67:476–478.

74. Hu, J., Xiao, X.–D., Ogletree, D.F., and Salmeron, M. 1995. Imaging the condensation and evaporation of molecularly thin films of water with nanometer resolution. *Science* 268:267–269.

75. Hu, X.L. and Michaelides, A. 2007. Ice formation on kaolinite: Lattice match or amphoterism. *Surf. Sci.* 601:5378–5381.

76. Hummer, G., Rasaiah, J.C., and Noworyta, J.P. 2001. Water conduction through the hydrophobic channel of a carbon nanotube. *Nature* 414:188–190.

77. Iijima, S. 1991. Helical microtubules of graphitic carbon. *Nature* 354:56–58.

78. Israelachvili. J. 2011. *Intermolecular & Surface Forces*. Amsterdam: Elsevier.
79. Israelachvili, J.N. and Pashley, R.M. 1983. Molecular layering of water at surfaces and origin of repulsive hydration forces. *Nature* 306:249–250.
80. Jacobi, K., Bedürftig, K., Wang, Y., and Ertl, G. 2001. From monomers to ice–new vibrational characteristics of H_2O adsorbed on Pt(111). *Surf. Sci.* 472:9–20.
81. Jinesh, K.B. and Frenken, J.W.M. 2008. Experimental evidence for ice formation at room temperature. *Phys. Rev. Lett.* 101:036101.
82. Jungwirth, P. and Tobias, D.J. 2001. Molecular structure of salt solutions: A new view of the interface with implications for heterogeneous atmospheric chemistry. *J. Phys. Chem. B* 105:10468–10472.
83. Jungwirth, P. and Tobias, D.J. 2006. Specific ion effects at the air/water interface. *Chem. Rev.* 106:1259–1281.
84. Ketteler, G., Yamamoto, S., Bluhm, H., Andersson, K., Starr, D.E., Ogletree, D.F. et al. 2007. The nature of water nucleation sites on $TiO_2(110)$ surfaces revealed by ambient pressure X-ray photoelectron spectroscopy. *J. Phys. Chem. C* 111:8278–8282.
85. Khan, S.H., Matei, G., Patil, S., and Hoffmann, P.M. 2010. Dynamic solidification in nanoconfined water films. *Phys. Rev. Lett.* 105:106101.
86. Kimmel, G.A., Petrik, N.G., Dohnálek, Z., and Kay, B.D. 2005. Crystalline ice growth on Pt(111): Observation of a hydrophobic water monolayer. *Phys. Rev. Lett.* 95:166102.
87. Koga, K., Gao, G.T., Tanaka, H., and Zeng, X.C. 2001. Formation of ordered ice nanotubes inside carbon nanotubes. *Nature* 412:802–805.
88. Kyakuno, H., Matsuda, K., Yahiro, H., Inami, Y., Fukuoka, T., Miyata, Y. et al. 2011. Confined water inside single-walled carbon nanotubes: Global phase diagram and effect of finite length. *J. Chem. Phys.* 134:244501.
89. Langenbach, E., Spitzer, A., and Lüth, H. 1984. The adsorption of water on Pt(111) studied by IR–reflection and UV–photoemission spectroscopy. *Surf. Sci.* 147:179–190.
90. Langmuir, I. 1932. Vapor pressures, evaporation, condensation and adsorption. *J. Am. Chem. Soc.* 54:2798–2832.
91. Lee, J., Sorescu, D.C., Deng, X., and Jordan, K.D. 2013. Water chain formation on $TiO_2(110)$. *J. Phys. Chem. Lett.* 4:53–57.
92. Lehmann, A., Fahsold, G., König, G., and Rieder, K.H. 1996. He-scattering studies of the $BaF_2(111)$ surface. *Surf. Sci.* 369:289–299.
93. Li, T.-D., Gao, J., Szozskiewicz, R., Landman, U., and Riedo, E. 2007. Structured and viscous water in subnanometer gaps. *Phys. Rev. B* 75:115415.
94. Lombardo, T.G., Giovambattista, N., and Debenedetti, P.G. 2009. Structural and mechanical properties of glassy water in nanoscale confinement. *Faraday Discuss.* 141:359–376.
95. Luzar, A. and Chandler, D. 1996. Hydrogen-bond kinetics in liquid water. *Nature* 379:55–57.
96. Lyklema, J., Rovillard, S., and De Coninck, J. 1998. Electrokinetics: The properties of the stagnant layer unraveled. *Langmuir* 14:5659–5663.
97. Maibaum, L. and Chandler, D. 2003. A coarse-grained model of water confined in a hydrophobic tube. *J. Phys. Chem. B* 107:1189–1193.
98. Mancinelli, R., Botti, A., Bruni, F., Ricci, M.A., and Soper, A.K. 2007. Hydration of sodium, potassium, and chloride ions in solution and the concept of structure maker/breaker. *J. Phys. Chem. B* 109:13570–13577.
99. Marcus, Y. 2013. Individual ionic surface tension increments in aqueous solutions. *Langmuir* 29:2881–2888.
100. Marx, D. and Wiechert, H. 1996. Ordering and phase transitions in adsorbed monolayers of diatomic molecules. *Adv. Chem. Phys.* 95:213–394.

101. Mehlhorn, M. and Morgenstern, M. 2007. Faceting during the transformation of amorphous to crystalline ice. *Phys. Rev. Lett.* 99:246101.
102. Meng, S., Wang, E.G., and Gao S. 2004. Water adsorption on metal surfaces: A general picture from density functional theory studies. *Phys. Rev. B* 69:195404.
103. Michaelides, A. 2007. Simulating ice nucleation, one molecule at a time, with the DFT microscope. *Fadaray Discuss.* 136:287–297.
104. Michaelides, A. and Morgenstern, K. 2007. Ice nanoclusters at hydrophobic metal surfaces. *Nature Mater.* 6:597–601.
105. Michaelides, A., Ranea, V.A., de Andrés, P.L., and King D.A. 2003. General model for water monomer adsorption on close-packed transition and noble metal surfaces. *Phys. Rev. Lett.* 90:216102.
106. Miranda, P.B., Xu, L., Shen, Y.R., and Salmeron, M. 1998. Icelike water monolayer adsorbed on mica at room temperature. *Phys. Rev. Lett.* 81:5876–5879.
107. Mitsui, T., Rose, M.K., Fomin, E., Ogletree, D.F., and Salmeron, M. 2002. Water diffusion and clustering on Pd(111). *Science* 297:1850–1852.
108. Morgenstern, K. and Rieder, K.-H. 2002. Formation of the cyclic ice hexamer via excitation of vibrational molecular modes by the scanning tunneling microscope. *J. Chem. Phys.* 116:5746–5752.
109. Motobayashi, K, Matsumoto, C., Kim, Y., and Kawai, M. 2008. Vibrational study of water dimers on Pt(111) using a scanning tunneling microscope. *Surf. Sci.* 602:3136–3139.
110. Mugarza, A., Shimizu, T.K., Ogletree, D.F., and Salmeron, M. 2009. Chemical reactions of water molecules on Ru(0001) induced by selective excitation of vibrational modes. *Surf. Sci.* 603:2030–2036.
111. Nakamura, M. and Ito, M. 2004. Ring hexamer like cluster molecules of water formed on a Ni(111) surface. *Chem. Phys. Lett.* 384:256–261.
112. Nauta, K. and Miller, R.E. 2000. Formation of cyclic water hexamer in liquid helium: The smallest piece of ice. *Science* 287:293–295.
113. Neder, R.B., Burghammer, M., Grasl, T.H., Schulz, H., Bram, A., and Fiedler, S. 1999. Refinement of the kaolinite structure from single–crystal synchrotron data. *Clays Clay Minerals* 47:487–494.
114. Newberg, J.T., Starr, D.E., Yamamoto, S., Kaya, S., Kendelewicz, T., Mysak, E.R. et al. 2011. Autocatalytic surface hydroxylation of MgO(100) terrace sites observed under ambient conditions. *J. Phys. Chem. C* 115:12864–12872.
115. Nutt, D.R. and Stone, A.J. 2002. Adsorption of water on the $BaF_2(111)$ surface. *J. Chem. Phys.* 117:800–807.
116. Odelius, M., Bernasconi, M., and Parrinello, M. 1997. Two dimensional ice adsorbed on mica surface. *Phys. Rev. Lett.* 78:2855–2858.
117. Ogasawara, H., Brena, B., Nordlund, D., Nyberg, M., Pelmenschikov, A., Pettersson, L.G.M., and Nilsson, A. 2002. Structure and bonding of water on Pt(111). *Phys. Rev. Lett.* 89:276102.
118. Ogasawara, H., Yoshinobu, J., and Kawai, M. 1999. Clustering behavior of water (D_2O) on Pt(111). *J. Chem. Phys.* 111:7003–7009.
119. Pápai, I. 1995. Theoretical study of the $Cu(H_2O)$ and $Cu(NH_3)$ complexes and their photolysis products. *J. Chem. Phys.* 103:1860–1870.
120. Pascual J.I., Lorente N., Song Z., Conrad, H., and Rust, H.-P. 2003. Selectivity in vibrationally mediated single-molecule chemistry. *Nature* 423:525–528.
121. Piatkowski, L. and Bakker, H.J. 2011. Vibrational dynamics of the bending mode of water interacting with ions. *J. Chem. Phys.* 135:214509.

122. Pruppacher, H.R. 1973. Electrofreezing of supercooled water. *Pure Appl. Geophys.* 104:623–634.
123. Ranea, V.A., Michaelides, A., Ramírez, R., de Andrés, P.L., Vergés, J.A., and King, D.A. 2004. Water dimer diffusion on Pd(111) assisted by an H-bond donor-acceptor tunneling exchange. *Phys. Rev. Lett.* 92:136104.
124. Raviv, U., Laurat, P., and Klein, J. 2001. Fluidity of water confined to subnanometer films. *Nature* 413:51–54.
125. Reischl, B., Köfinger, J., and Dellago, C. 2009. The statistics of electric field fluctuations in liquid water. *Mol. Phys.* 107:495–502.
126. Repp, J., Meyer, G., Stojkovic, S.M., Gourdon, A., and Joachim, C. 2005. Molecules on insulating films: Scanning-tunneling microscopy imaging of individual molecular orbitals. *Phys. Rev. Lett.* 94:026803.
127. Richardson, S.M. and Richardson, Jr., J.W. 1982. Crystal structure of a pink muscovite from Archer's post, Kenya: Implications for reverse pleochroism in dioctahedral micas. *Am. Miner.* 67:69–75.
128. Röttger, K., Endriss, A., Ihringer, J., Doyle, S., and Kuhs, W.F. 1994. Lattice constants and thermal expansion of H_2O and D_2O ice Ih between 10 and 265 K. *Acta Cryst.* B50:644–648.
129. Santos, S., Verdaguer, A., Souier, T., Thomson, N.H., and Chiesa, M. 2011. Measuring the true height of water films on surfaces. *Nanotechnology* 22:465705.
130. Schenk, M., Füting, M., and Reichelt, R. 1998. Direct visualization of the dynamic behavior of a water meniscus by scanning electron microscopy. *J. Appl. Phys.* 84:4880–4884.
131. Schmidt, R., Hansen, E.W., Stöcker, M., Akporiaye, D., and Ellestad, O.H. 1995. Pore size determination of MCM-41 mesoporous materials by means of ^1H NMR spectroscopy, N_2 adsorption and HREM. A preliminary study. *J. Am. Chem. Soc.* 117:4049–4056.
132. Schreiber, A., Ketelsen, I., and Findenegg, G.H. 2001. Melting and freezing of water in ordered mesoporous silica materials. *Phys. Chem. Chem. Phys.* 3:1185–1195.
133. Segura, J.J. 2012. *Dipole-induced water adsorption on surfaces.* PhD Thesis, Autonomous University of Barcelona.
134. Shevkunov, S.V. 2007. Nucleation of water vapor in microcracks on the surface of β-AgI aerosol particles: 1. The structure of nuclei. *Colloid J.* 69:360–377.
135. Shevkunov, S.V. 2009. Numerical simulation of water vapor nucleation on electrically neutral nanoparticles. *J. Exp. Theor. Phys.* 108:447–468.
136. Smith, J.D., Saykally, R.J., and Geissler, P.L. 2007. The effects of dissolved halide anions on hydrogen bonding in liquid water. *J. Am. Chem. Soc.* 129:13847–13856.
137. Soper, A.K. 2007. Joint structure refinement of X-ray and neutron diffraction data on disordered materials: application to liquid water. *J. Phys. Condens. Matter* 19:335206.
138. Stern, O. 1924. Zur Theorie der Elektrolythischen Doppelschicht. *Z. Elektrochem.* 30:508–516.
139. Stipe, B.C., Rezaei, M.A., and Ho, W. 1998. Single-molecule vibrational spectroscopy and microscopy. *Science* 280:1732–1735.
140. Su, X., Lianos, L., Shen, Y.R., and Somorjai, G.A. 1998. Surface-induced ferroelectric ice on Pt(111). *Phys. Rev. Lett.* 80:1533–1536.
141. Takaiwa, D., Hatano, I., Koga, K., and Tanaka, H. 2008. Phase diagram of water in carbon nanotubes. *Proc. Natl. Acad. Sci. USA* 105:39–43.
142. Tang, Q.L. and Chen, Z.X. 2007. Density functional slab model studies of water adsorption on flat and stepped Cu surfaces. *Surf. Sci.* 601:954964.

143. Teschke, O. 2010. Imaging ice-like structures formed on HOPG at room temperature. *Langmuir* 26:16986–16990.

144. Thiam, M.M., Kondo, T., Horimoto, N., Kato, H.S., and Kawai, M. 2005. Initial growth of the water layer on (1×1)-oxygen-covered Ru(0001) in comparison with that on bare Ru(0001). *J. Phys. Chem. B* 109:16024–16029.

145. Thiel, P.A. and Madey, T.E. 1987. The interaction of water with solid surfaces: Fundamental aspects. *Surf. Sci. Rep.* 7:211–385.

146. Thiel, P.A., Hoffmann, F.M., and Weinberg, W.H. 1981. Monolayer and multilayer adsorption of water on Ru(001). *J. Chem. Phys.* 75:5556–5572.

147. Thissen, P., Grundmeier, G., Wippermann, S., and Schmidt, W.G. 2009. Water adsorption on the α-Al$_2$O$_3$(0001) surface. *Phys. Rev. B* 80:245403.

148. Thürmer, K. and Bartelt, N.C. 2008. Growth of multilayer ice films and the formation of cubic ice imaged with STM. *Phys. Rev. B* 77:195425.

149. Toney, M.F., Howard, J.N., Richer, J., Borges, G.L., Gordon, J.G., Melroy, O.W. et al. 1994. Voltage-dependent ordering of water molecules at an electrode-electrolyte interface. *Nature* 368:444–446.

150. Trucano, P. and Chen, R. 1975. Structure of graphite by neutron diffraction. *Nature* 258:136–137.

151. Tunega, D., Gerzabek, M.H., and Lischka, H. 2004. *Ab initio* molecular dynamics study of a monomolecular layer on octahedral and tetrahedral kaolinite surfaces. *J. Phys. Chem. B* 108:5930–5936.

152. van der Ham, F., Witkamp, G.J., de Graauw, J., and van Rosmalen, G.M. 1998. Eutectic freeze crystallization: application to process streams and waste water purification. *Chem. Eng. Proc.* 37:207–213.

153. Verdaguer, A., Cardellach, M., and Fraxedas, J. 2008. Thin water films grown at ambient conditions on BaF$_2$(111) studied by scanning polarization force microscopy. *J. Chem. Phys.* 129:174705.

154. Verdaguer, A., Cardellach, M., Segura, J.J., Sacha, G.M., Moser, J., Zdrojek, M. et al. 2009. Charging and discharging of graphene in ambient conditions studied with scanning probe microscopy. *Appl. Phys. Lett.* 94:233105.

155. Verdaguer, A., Sacha, G.M., Bluhm, H., and Salmeron, M. 2006. Molecular structure of water at interfaces: Wetting at the nanometer scale. *Chem. Rev.* 106:1478–1510.

156. Verdaguer, A., Segura, J.J., Fraxedas, J., Bluhm, H., and Salmeron, M. 2008. Correlation between charge state of insulating NaCl surfaces and ionic mobility induced by water adsorption: A combined ambient pressure X-ray photoelectron spectroscopy and scanning force microscopy study. *J. Phys. Chem. C* 112:16898–16901.

157. Verdaguer, A., Segura, J.J., López-Mir, L., Sauthier, G., and Fraxedas, J. 2013. Growing room temperature ice with graphene. *J. Chem. Phys.* 138:121101.

158. Verwey, E.J.W. and Overbeek, J.T.G. 1948. *Theory of the Stability of Lyophobic Colloids*. Amsterdam: Elsevier.

159. Vogt, J. 2007. Helium atom scattering study of the interaction of water with the BaF$_2$(111) surface. *J. Chem. Phys.* 126:244710.

160. Vonnegut, B. 1947. The nucleation of ice formation by silver iodide. *J. Appl. Phys.* 18:593–595.

161. Wang, H.-J., Xi, X.-K., Kleinhammes, A., and Wu, Y. 2008. Temperature-induced hydrophobic-hydrophilic transition observed by water adsorption. *Science* 322:80–83.

162. Wlodarczyk, R., Sierka, M., Kwapień, K., Sauer, J., Carrasco, E., Aumer, A. et al. 2011. Structures of the ordered water monolayer on MgO(001). *J. Phys. Chem. C* 115:6764–6774.

163. Wood, E.A. 1964. Vocabulary of surface crystallography. *J. Appl. Phys.* 35:1306–1312.

164. Wu, Y., Mayer, J.T., Garfunkel, E., and Madey, T.E. 1994. X-ray photoelectron spectroscopy study of water adsorption on $BaF_2(111)$ and $CaF_2(111)$ surfaces. *Langmuir* 10:1482–1487.

165. Xia, X. and Berkowitz, M.L. 1995. Electric-field induced restructuring of water at a platinum-water interface: A molecular dynamics computer simulation. *Phys. Rev. Lett.* 74:3193–3196.

166. Xu, K., Cao, P., and Heath, J.R. 2010. Graphene visualizes the first water adlayers on mica at ambient conditions. *Science* 329:1188–1191.

167. Yamamoto, S., Beniya, A., Mukai, K., Yamashita, Y., and Yoshinobu, J. 2005. Water adsorption on Rh(111) at 20 K: From monomer to bulk amorphous ice. *J. Phys. Chem. B* 109:5816–5823.

168. Zangi, R. and Mark, A. E. 2003. Monolayer ice. *Phys. Rev. Lett.* 91:025502.

4 Hydrophobicity and Hydrophilicity

Humida, la pell lliscava
com si la cobrís una capa d' oli
A. Sánchez Piñol, La pell freda

In this chapter we discuss the affinity of water to surfaces, with special emphasis on the two extreme cases of high and low affinity. Such sympathy or antipathy was described in the previous chapter, but limited to the formation of water monolayers. Here, we consider the case of bulk water in contact with such surfaces. *Hydrophilic* surfaces become *wetted* (covered) by water and *hydrophobic* surfaces exhibit quite the opposite behavior, a phenomenon known as *dewetting*. Mother Nature provides us with several examples of hydrophobic surfaces, such as the well-known case of lotus leaves, and there is an enormous interest in controlling the water-repellent character of surfaces, with applications in different domains, which can be achieved through an engineered manipulation of such surfaces with artificial coatings. In addition to the intrinsic chemical affinity the structuration of surfaces leads to increased hydrophobicity, which leads to the term *superhydrophobicity* with almost perfect repellence. This is one of the many relevant activities in nanotechnology. We also explore the curious case of objects (nanoparticles, materials, surfaces, etc.) exhibiting both high and low affinity to water, deserving the adjective *amphiphilic*.

4.1 WETTING AND CONTACT ANGLE

In Chapter 2 we introduced Young's equation (2.12) for the liquid/solid/vapor interfaces for the particular case of liquid water on ice. Here we rewrite this equation, which is valid for perfectly flat surfaces, in the form:

$$\cos \theta_c = \frac{\gamma_{sv} - \gamma_{sl}}{\gamma_{lv}} \tag{4.1}$$

where $\gamma_{sv} > 0$, $\gamma_{sl} > 0$, and $\gamma_{lv} > 0$, making patent that the (cosines of) contact angle corresponds to the ratio between the difference of interfacial energies involving the solid phase and the vapor/liquid interface (de Gennes 1985). Interestingly, T. Young (1805) did not derive his formula. He wrote that "for each combination of a solid and a fluid, there is an appropriate angle of contact between the surfaces of the fluid, exposed to the air, and to the solid. This angle, for glass and water, and in all cases where a solid is perfectly wetted by a fluid, is evanescent: for glass and mercury, it is about 140°, in common temperatures and when the mercury is moderately clean," and envisioned a discrete nature of liquids held by some cohesion forces: "but whenever

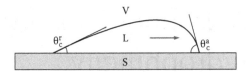

FIGURE 4.1 Definition of advancing (θ_c^a) and receding (θ_c^r) contact angles for a liquid (L) droplet moving on a solid (S) surface in the direction indicated by the arrow. V stands for the vapor phase. See Figure 2.8 for comparison.

there is a curved or angular surface, it may be found by collecting the actions of the different particles, that the cohesion must necessarily prevail over the repulsion, and must urge the superficial parts inwards with a force proportionate to the curvature, and thus produce the effect of a uniform tension of the surface." Note that his work was published in 1805, well before J. W. Gibbs and J. D. van der Waals, as well as the rest of founders of modern thermodynamics, were born. Pure genius.

Thus, the contact angle θ_c characterizes the extent of contact of a droplet with a solid surface. The higher θ_c is, the lower the contact area between the liquid droplet and the solid is, the lower the adhesion and the lower the resistance for the droplet to move. The contact angle hysteresis is the difference between the advancing (a) and receding (r) contact angles, θ_c^a and θ_c^r, respectively, which correspond to the maximum and minimum contact angles at the front and back of the advancing droplet, respectively. Both parameters are sketched in Figure 4.1. Note that $\theta_c^a > \theta_c^r$ and that the droplet will flow more easily when their difference (hysteresis) attains lower values ($\theta_c^a \simeq \theta_c^r$). From (4.1) we observe that $\theta_c < \pi/2$ when $\gamma_{sv} > \gamma_{sl}$, so that wetting is favored by high-energy surfaces, that for larger γ_{sv} values corresponds to surfaces of materials built from strong interactions (e.g., covalent). When $\gamma_{sv} < \gamma_{sl}$ then $\theta_c > \pi/2$ leading to partial wetting that corresponds to low surface energies, that is, materials built from weak interactions (van de Waals, H-bonding, etc.). In the extreme case of $\theta_c = 0$ (perfect wetting or superhydrophilicity), then $\gamma_{sv} = \gamma_{sl} + \gamma_{lv}$ and for $\theta_c = \pi$ (perfect dewetting or superhydrophobicity), then $\gamma_{sv} = \gamma_{sl} - \gamma_{lv}$. Below we show spectacular examples of $\theta_c \simeq \pi$. The reported lowest surface free energy of any solid is 6.7 mJ m^2 and corresponds to the surface of regularly aligned closest hexagonal packed $-CF_3$ groups (Nishino et al. 1999). In this case $\theta_c = 119°$. This value is far from $\theta_c = \pi$ and this is due to the large surface tension of water. From (4.1) we see that the higher γ_{lv} is, the lower $\cos\theta_c$ is. Thus superhydrophobicity, with an arbitrarily set lower limit of the contact angle of 150°, cannot be achieved solely by reducing molecular interactions, a point that is discussed later in detail.

4.2   HYDROPHOBICITY AT DIFFERENT LENGTH SCALES

Objects exhibiting different length scales, spanning from the 0D case of small single molecules to the 2D case of large flat surfaces, that share the property of interacting weakly with water (only through vdW interactions), do not challenge the Bernal–Fowler ice rules in contact with bulk water. In fact, such objects tend to structure the involved water/object interfaces, because water molecules tend to saturate their four

H-bonds in view of the impossibility to create chemical bonding with the intruding objects (Lum, Chandler, and Weeks 1999). If we consider the simplest case of methane, a true tetrahedral molecule with poor affinity to water as discussed in Section 1.3, water ends up by surrounding such a molecule by building well-defined structures called clathrate hydrates through H-bonding, that are discussed next. In this way methane becomes sequestered, which is seen as a general strategy to store gases in order to eliminate them from the atmosphere or to use natural reservoirs as combustion fuel. In the case of extended flat surfaces, with dimensions much larger than the mean water diameter, the induced H-bond network at the interface is unable to surround such surfaces and, as a result, a depletion layer is formed: a (pseudo)gap is generated at the water/surface interface. The profile of such an interface is similar to that of the liquid/vapor interface, as pointed out by Stillinger (1973) and mentioned in Section 2.2. Another interesting phenomenon is known as hydrophobic interaction. When two parallel hydrophobic surfaces submerged in liquid water approach below a critical distance, the water between both surfaces is expelled forming a void or cavity and both surfaces experience an attraction due to the pressure exerted by the liquid. This point is discussed in Section 6.2 within the framework of protein hydration. We start next with the 0D case of clathrate hydrates and see that, far from being satisfied with sequestration, water builds complex 3D crystalline structures in honor of the distinguished guest molecules, structures not found in their absence. Such ices cannot be strictly considered as ice polymorphs because of the additional ordered impurity network.

4.2.1 CLATHRATE HYDRATES

According to Frank and Evans (1945) water surrounds hydrophobic solutes acquiring a well-ordered (clathrate) structure that melts as temperature is increased. This model is often referred to as the *iceberg model* and is consistent with the experimentally observed negative excess entropy and large heat capacity. The hydration of methane at RT gives an entropy decrease of about $-65 \, J\,K^{-1}\,mol^{-1}$ and a heat capacity increase of about $240 \, J\,K^{-1}\,mol^{-1}$, respectively (Crovetto, Fernández-Prini, and Japas 1982). Clathrate hydrates represent a fascinating example of crystalline host–guest solids, where water is the building block of the H–bond based host structure which includes small guest molecules encapsulated in cages (Sloan and Koh 2007). In general, no chemical bond is formed between water molecules and the trapped molecules thus interacting through vdW interactions. We saw in Section 1.4.2 that ice II, a high-pressure proton-ordered polymorph, exhibits a columnar structure with cavity diameters inside the columns of about 3 Å, which can allocate guest atoms thus forming host-guest hydrates (see Figure 1.16b). The formation of clathrate hydrates is more stringent, because the guest molecules become encapsulated by water. The solvating water molecules show slower orientational dynamics than molecules in liquid water. This has been studied with fs 2D mid-infrared and polarization-resolved pump-probe spectroscopy of the O–H stretch vibration of HDO in solutions of amphiphilic solutes in isotopically diluted water (Bakulin et al. 2011). The addition of the amphiphilic solutes leads to a strong slowing down of both the spectral and the orientational dynamics.

Large quantities of methane clathrates are found on the deep sea floor as well as in permafrost regions. They are indeed regarded as potential energy resources although the extraction is not trivial. Hydrocarbon clathrates can form inside gas pipelines often resulting in plug formation, causing serious problems in the extraction process. Uncontrolled release of methane from the decomposition of such deposits may have serious consequences inasmuch as methane would efficiently contribute to the greenhouse effect, being more efficient than carbon dioxide. It is obvious that this has to be considered quite seriously. On the other hand, carbon dioxide might be partially removed from the atmosphere if clathrates could be deposited in the deep sea floor.

Clathrate structures are determined mainly by the size of the gas molecules. Among the many different crystal structures that clathrate hydrates can build only three structures are known to occur in natural environments, namely structures I, II, and H (Buffett 2000). Both structures I and II are cubic, with space groups $Pm\bar{3}n$ and $Fd\bar{3}m$, respectively. Structure I is the most common form of clathrate in natural settings where methane is the main hydrate-forming gas. Structures II and H (hexagonal, $P6/mmm$) are found for larger molecules. Figure 4.2 shows an example of a methane-based crystal structure of type I (Kirchner et al. 2004). The unit cell consists of 46 water molecules forming two types of cages, small and large. The small cages or cavities are dodecahedra with 20 water molecules arranged to form 12 pentagonal faces. Usually the nomenclature 5^{12} is used, highlighting the 12 pentagonal faces. The large cavities are tetradecahedra containing 24 water molecules, forming 12 pentagonal and 2 hexagonal faces ($5^{12}6^2$). The small cavities are located at the center and the four corners of the unit cell to form a bcc structure. Six additional water molecules inside the unit cell link the small cavities to form the large cavities. Each unit cell contains two small and six large cavities. Methane has a molecular diameter of 4.3 Å, whereas the free diameter of the cage is 4.4 Å. This results in free rotation of the hydrocarbon. The unit cell of structure II consists of 136 water molecules, also forming two types of cages, small and large. The small cages (16) again have the shape of a pentagonal dodecahedron (5^{12}) but the large ones (8) are hexadecahedra ($5^{12}6^4$). Structure H is formed by three types of cages, two small and a large one, requiring the cooperation of two guest gases (large and small) to be stable. Typical large molecules are butane and hydrocarbons.

4.2.2 EXTENDED SURFACES

If we replace the specific methane molecule by a generic solute of radius r, then we should expect a different situation because it becomes impossible for the adjacent water molecules to maintain a complete H-bonding network. As a result, water tends to move away from the large solute forming an interface around it similar to that between liquid and vapor (Stillinger 1973). In the case of methane, the volume within the liquid is so small that water adapts easily around the invasive molecule without breaking H-bonds. Chandler (2005) has shown that the solvation free energy changes with solute size with a critical radius of about 1 nm below which the solvation free energy grows linearly with solute volume and above which it grows linearly with surface area. This 1 nm sets a fuzzy frontier line between small and large solutes, a quantity that might seem too small but that has to be compared with the mean radius of the water molecules (about 0.3 nm), thus giving an idea of the length

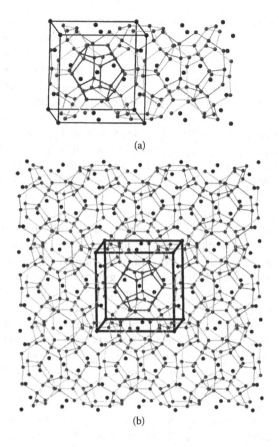

(a)

(b)

FIGURE 4.2 Crystal structure of methane hydrate. Space group $Pm\bar{3}n$, $a = 11.877(3)$ Å at 123 K (Kirchner et al. 2004). Hydrogen atoms are not shown for clarity. (a) Two unit cells highlighting the dodecahedron with thicker dark grey oxygen–oxygen bonds and the adjacent tetradecahedra. (b) Planar view showing the cubic distribution.

range where H-bonds can adapt to deformation. MD simulations show that the water density profile assumes the sigmoidal shape of a flat liquid/vapor interface, modeled in Section 2.1.1 by (2.1) (Mittal and Hummer 2008). With increasing solute size, the interfaces become wider and recede from the solute. This broadening of the solute/water interface is at least in part due to the perturbations imposed by the capillary waves to an intrinsic interfacial density profile. For a spherical droplet, and also for a bubble, the mean square displacement of the surface is a function of the radius of the droplet r_d, specifically proportional to its logarithm (Henderson and Lekner 1978). Thus, the position of the Gibbs dividing surface, z_G, locates away from the solute surface creating a depleted region.

Such a depletion zone or gap has been studied, both from the experimental and theoretical points of view, for extended flat surfaces ($r \rightarrow \infty$). X-ray reflectivity measurements using synchrotron radiation (Jensen et al. 2003; Poynor et al. 2006; Mezger et al. 2010), as well as neutron reflectivity experiments (Maccarini et al. 2007) are interpreted in terms of the formation of a depletion region with a width in the 1–3 Å range. Such a value is of the order of the measured roughness, also determined by X-ray reflectivity (Braslau et al. 1985), so that the obtained gap value is not exempt from uncertainty. In addition, one has to take the roughness of the supporting hydrophobic surface used in the experiments into account (Mezger et al. 2010), a point addressed below. Using very hydrophobic smooth surfaces, it has been shown that the depletion region width increases with contact angle, with a value of $\sim$7 Å at $\sim$120° (Chattopadhyay et al. 2010), thus the election and preparation of the surface turn out to be critical to overcome the intrinsic interfacial fluctuations. As $\sim$120° is the highest contact angle that can be achieved on a flat surface, as previously discussed (Nishino et al. 1999), then the $\sim$7 Å value sets an upper limit to the width of the depletion layer.

The most commonly used hydrophobic surfaces are flat surfaces coated with self-assembled monolayers (SAMs), because of the relative simplicity of the method and the choice of a host of candidate molecules. SAMs are ordered molecular assemblies formed by the adsorption of an active surfactant on a solid inorganic surface. The molecules exhibit two well-differentiated end groups: head and tail. The adsorbate interacts with the surface through its head group, forming strong covalent bonds (typically sulfur/gold, carbon/silicon), thus defining robust interfaces. The tail chemical function can be selected (methyl, carboxylic acid, amides, etc.) and thus the interaction with water can be chemically controlled. In SAMs the packing and ordering are essentially determined by the contribution of the chemisorptive interaction with the surface and of both intra- and interchain interactions (vdW, steric, electrostatic, etc.; Ulman 1996; Schreiber 2004). The molecules can be deposited using both dry (e.g., vacuum sublimation) or wet (e.g., immersion) routes. The molecules used for the experiments mentioned above are octadecyltriethoxysiloxane (Poynor et al. 2006), perfluorodecyltrimethoxysilane (Mezger et al. 2010), and fluoroalkylsilanes (Chattopadhyay et al. 2010). The contact angle essentially depends on the tail chemical function, packing structure of the molecules, defects, grain boundaries, and homogeneity of the coating. Because the contact angle measurements are based on micrometer-sized water droplets, it is clear that nonhomogeneous films will give lower contact angles because water will interact with hydrophilic (noncovered) regions. This is an important point when classifying the degree of hydrophobicity/hydrophilicity of surfaces, that was already mentioned in Chapter 3 when discussing the accumulation of water at step edges on $BaF_2(111)$ and HOPG.

Let us now discuss the effect of atomic-scale surface topography on the hydrophobic effect and study the revealing example of L-alanine [(S)-2-aminopropanoic acid], one of the smallest amino acids, which has a nonreactive hydrophobic methyl group as side chain. Amino acids are the building blocks of proteins and are small organic molecules containing both a positively charged ammonium group (NH_3^+) and a negatively charged carboxylic group (COO^-), linked to a common carbon atom and to an organic substituent known as a radical. This charged configuration is known as the zwitterionic form of the amino acid and is the dominant one in aqueous solution over a wide range of pH values, as well as in the crystalline state. The degree of hydrophilicity/hydrophobicity of amino acids is a rather ambiguous parameter (Karplus 1997). The hydrophilic character is ascribed to the charged groups, and the side chain contributes to the global topological hydrophilic/hydrophobic nature.

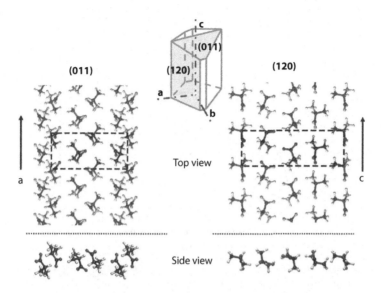

FIGURE 4.3 Crystal habit and RT structure of L-alanine. The crystal structure is orthorhombic, space group $P2_12_12_1$, with four molecules per unit cell and cell parameters $a = 0.6032$, $b = 1.2343$, and $c = 0.5784$ nm (Lehmann, Koetzle, and Hamilton 1972). (Left) Top and side view of the (011) crystal face projected across (top) and along (below) the a-axis. (Right) Top and side view of the (120) crystal face projected across (top) and along (below) the c-axis. Carbon, oxygen, nitrogen, and hydrogen atoms are represented by black, dark grey, medium grey, and white spheres, respectively. (Reprinted from J. J. Segura, et al. *J. Am. Chem. Soc.* 131:17853–17859, 2009, American Chemical Society. With permission.)

L-alanine is considered a hydrophobic amino acid; however, it exhibits a relatively high solubility in water (16.65 g in 100 g of H_2O at 25°C) due to its zwitterionic form.

Figure 4.3 shows a schematic view of the crystal habit of L-alanine indicating the a, b, and c-axes and the (120) and (011) faces, together with their projected RT crystal structures (Lehmann, Koetzle, and Hamilton 1972). Both surfaces display contrasting behaviors when exposed to water: (011) is hydrophilic and (120) is hydrophobic. The hydrophobicity of (120) has been ascribed to the presence of exposed methyl groups on this surface (Gavish et al. 1992), however, an inspection of the crystal structure along the c-axis (bottom right in Figure 4.3) reveals that only one molecule out of every four has its methyl group pointing outwards on this surface, and that carboxylic and amino groups are also clearly exposed. At the (011) surface both polar and nonpolar groups are present (see bottom left in Figure 4.3) and this surface shows a larger rugosity at the molecular scale as compared to (120). Thus, the prediction of the hydrophobic/hydrophilic character solely from the ideal molecular structure may lead to erroneous results.

In Figure 4.4a the MD simulated density profile of water molecules as a function of the height above both (120) and (011) surfaces is plotted, as well as the spatial distribution of this density (Segura et al. 2009). As can be seen in the figure, the density profiles of water on both surfaces at distances smaller than 8 Å are markedly

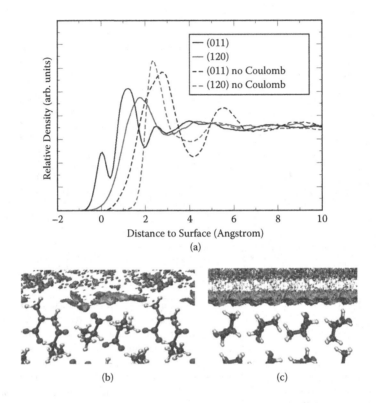

FIGURE 4.4 Distribution of water molecules on L-alanine surfaces. (a) MD calculated probability density of water molecules above surfaces (011) (black) and (120) (grey) of L-alanine. Continuous lines show water probability densities calculated with water–alanine charges switched on and including the dynamics of the alanine molecules, while the dashed lines are the resulting densities obtained without water–alanine electrostatic interactions, and fixing the alanine molecules at perfect crystal positions. (b) (011) Surface viewed along *a* axis and (c) (120) surface viewed along *c* axis. (Reprinted from J.J. Segura, et al. *J. Am. Chem. Soc.* 131:17853–17859, 2009, American Chemical Society. With permission.)

different, with water molecules getting much closer to the outer alanine molecules in the case of the (011) surface than in the (120) (see Figure 4.4b). In fact, in the case of the (011) surface there is a first peak centered at zero, which reveals the presence of water molecules in close contact with the surface. This peak corresponds to H-bond formation between water molecules and surface carboxyl and amino groups. No similar peak is found for the (120) surface, where the distribution has a first peak at a position of ~1.8 Å from the surface. These different patterns confirm the relative hydrophilic/hydrophobic character of the (011) and (120) surfaces, respectively. The results make patent that water molecules move away from extended hydrophobic surfaces forming a depleted density region near such a surface, as discussed above.

One of the nice things about simulations is that they permit us to perform *gedanken* experiments, otherwise impossible to be done in the laboratory. Here, if the electrostatic interactions between the alanine and water molecules are turned off (by setting the partial charges on alanine constituent atoms to zero), the resulting densities are those shown in Figure 4.4a as dashed lines. As can be seen by comparing the distributions obtained with and without electrostatic interactions, these play a crucial role in determining the distribution of water molecules on the (011) surface, but are less important in the case of the (120) surface. Indeed, without electrostatic interactions, the density profile of water on alanine (011) changes dramatically, losing all the structure present when alanine atomic charges are considered. The density takes its first maximum at roughly the same position as in the case of the (120) surface (also without Coulomb interactions). However, in the case of the (120) surface, the exclusion of the electrostatic interactions does not change the form of the water density profile so dramatically, resulting in an approximately rigid shift toward longer distances. Figure 4.4 indicates that corrugated surfaces have a tendency to be more hydrophilic than flat ones (Jensen, Mouritzen, and Peters 2004).

4.2.3 SUPERHYDROPHOBICITY

We saw previously in Section 4.1 that there is a limit in the highest contact angle that can be achieved with a flat, smooth surface exposing chemically inert groups to water of about 120°, in part due to the large surface tension of liquid water. This means that in order to push θ_c toward the π degrees end goal new strategies have to be explored and the most obvious is increasing surface roughness. Roughness increases the solid/liquid contact area, which results in higher contact angles, according to the Wenzel equation (Wenzel 1936):

$$\cos \theta_W = \varrho \cos \theta_c \tag{4.2}$$

where θ_W represents the contact angle for a rough surface (θ_c is defined for a perfectly flat surface) and ϱ is the roughness factor, the ratio between the actual solid/liquid interface and the flat projection. Because $\varrho > 1$, the roughness is able to amplify the affinity to water of the ideal flat surface: a hydrophobic (hydrophilic) surface will become more hydrophobic (hydrophilic).

Let us see now what happens when the surface topography is such that air can be trapped in the cavities of the surface (see Figure 4.5). From the figure it becomes obvious that the solid/liquid contact area has decreased and the liquid/vapor area has increased by roughly the same amount. As a consequence the adhesion decreases, θ_c increases, and the contact angle hysteresis decreases allowing the droplet to roll easily on the solid surface.

Patterned surfaces such as that shown in Figure 4.5 cannot be described with the simplified Wenzel's model. Note that (4.2) becomes inconsistent for those values of ϱ for which $\cos \theta_c > 1$. For patterned surfaces the Cassie–Baxter equation applies (Cassie and Baxter 1944):

$$\cos \theta_{CB} = -1 + \varsigma(\varrho_p \cos \theta_c + 1) \tag{4.3}$$

where θ_{CB} stands for the contact angle for the patterned surface, ς the ratio of the total area of the solid/liquid interface with respect to the total area of solid/liquid and liquid/air interfaces in a plane geometrical area of unity parallel to the rough surface,

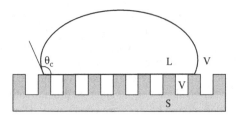

FIGURE 4.5 Scheme of the Cassie–Baxter model, where a liquid (L) droplet sits on a patterned solid (S) surface. The presence of air, represented here by V (vapor), in the grooves reduces the solid/liquid contact area.

and ϱ_p the ratio of the actual wetted area to the projected area. Note that for $\varsigma = 1$ and $\varrho = \varrho_p$ (4.3) reduces to (4.2). From (4.3) it follows:

$$\cos\theta_{CB}^a - \cos\theta_{CB}^r = \varsigma\varrho_p(\cos\theta_f^a - \cos\theta_f^r) \qquad (4.4)$$

If $\varsigma \to 0$, $\theta_{CB} \to \pi$ and $\cos\theta_{CB}^a \to \cos\theta_{CB}^r$. If $\varsigma = 1$ increasing roughness leads to increasing contact angle hysteresis. Hence, superhydrophobicity can only be achieved with patterned surfaces. The fakir's carpetlike patterned surface from Figure 4.5 can be seen as a spacer between the solid and liquid surfaces, efficiently reducing the solid/liquid contact area combining the pillar/valley geometrical dimensions of the pattern and the large water surface tension. Both Wenzel and Cassie–Baxter models have to be regarded as simplified approximations. In real life a mixed state can be expected, where the liquid penetrates to some extent inside the valleys depending on the actual dimensions (Erbil and Cansoy 2009). If the height of the pillars is too small, of the order of the amplitude of capillary waves (few nm), then the liquid could fill the valleys, increasing the liquid/solid area. Thus, such pillars have to be sufficiently high in order to avoid contact.

The combination of the large water surface tension and the superhydrophobicity of the surfaces makes the bouncing of water droplets possible, which exhibit remarkable elasticity (Richard, Clanet, and Quéré 2002). During contact, which lasts about 1–10 ms, millimeter-sized droplets deform differently depending on their kinetic energy. The Weber number, $W_N = \rho v_d^2 r_d / \gamma_{lv}$, where v_d represents the impinging velocity, compares the kinetic energy and surface tension. The greater W_N is the larger the deformation of the droplet. If we consider a water droplet hitting with a velocity v_d a superhydrophobic surface consisting of a regular array of circular pillars with diameter D_p, height H_p, and pitch P_p as shown in Figure 4.6, then the liquid/vapor interface between pillars will deform by a quantity δ_d.

If $\delta_d > H_p$ the droplet will be in contact with the bottom of the cavities between pillars destabilizing the solid/liquid/air interface. To avoid such contact, which will enter the Wenzel regime, the velocity should fulfill the condition (Jung and Bhushan 2008):

$$v_d < \sqrt{\frac{32\gamma_{lv}H_p}{\rho(\sqrt{2}P_p - D_p)^2}} \qquad (4.5)$$

The role of both v_d and pattern parameters in the bouncing of droplets is illustrated in Figure 4.7, which shows snapshots taken with a high-speed camera of 1 mm radius

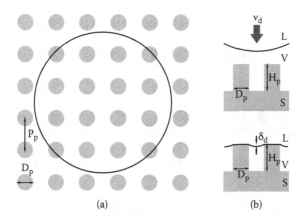

(a)   (b)

FIGURE 4.6 Small water droplet suspended on a superhydrophobic surface consisting of a regular array of circular pillars while hitting at a velocity v_d: (a) top view showing the pitch distance P_p and the diameter D_p of the pillars and (b) cross-section view showing the droplet approaching the surface at v_d (top) and after touch down (bottom) indicating the height H_p of the pillars and the local droplet deformation δ_d invading the valley between pillars. (Adapted from Y.C. Jung and B. Bhushan. *Langmuir* 24: 6262–6269, 2008. With permission.)

water droplets hitting two different micropatterned silicon surfaces based on the scheme from Figure 4.6 (Jung and Bhushan 2008). The geometrical parameters are $P_p = 10\,\mu\text{m}$, $D_p = 5\,\mu\text{m}$, and $H_p = 10\,\mu\text{m}$ and $P_p = 26\,\mu\text{m}$, $D_p = 14\,\mu\text{m}$, and $H_p = 30\,\mu\text{m}$, respectively. From the figure it can be observed that when $v_d = 0.44\text{ m s}^{-1}$ the droplets deform upon hitting and bounce off for both surfaces. The droplets end up by forming a high contact angle after losing their kinetic energy, thus remaining in the Cassie–Baxter regime. For larger impact velocities (0.88 and 0.76 m s^{-1}, respectively) bouncing no longer occurs and the droplets exhibit lower contact angles indicating wetting. We thus observe the transition from a Cassie–Baxter to a Wenzel regime. Such velocities correspond to the critical values discussed in (4.5). According to the equation, v_d decreases with increasing pitch dimensions, in agreement with the observations. From Figure 4.7 we observe that for larger P_p values the critical velocity becomes smaller.

When preparing superhydrophobic surfaces in the laboratory we are in fact emulating Mother Nature, who has an accumulated experience on land plants after more than 400 million years of evolution. Superhydrophobic leaves exhibit a surface hierarchical structure with different levels, also known as orders. Two orders are sufficient, for example, at the micrometer and nanometer range, but some plants exhibit a high degree of complexity with orders as high as six (Koch, Bohn, and Barthlott 2009). The most popular example of a superhydrophobic surface corresponds to the lotus leaves (*Nelumbo nucifera*), with $\theta_c \simeq 160°$ (Barthlott and Neinhuis 1997; Neinhuis and Barthlott 1997). Figure 4.8a shows a picture of a lotus leaf and Figures 4.8b and c SEM images of the upper part of the leaf at two different magnifications, where a hierarchical roughness at both the micrometer and the nanometer scale can be observed, consisting of papillae (Figure 4.8b) with a dense coating of agglomerated wax tubules

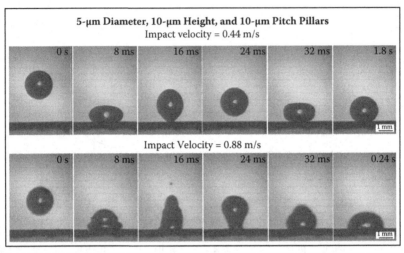

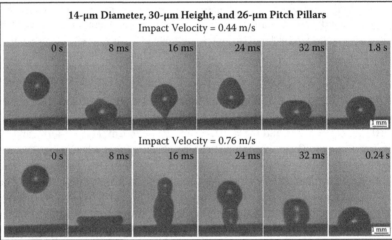

FIGURE 4.7 Snapshots of 1-mm radius droplets hitting with velocities of 0.44, 0.88, and 0.76 m s^{-1} two different micropatterned surfaces: 5-μm diameter, 10-μm height, and 10-μm pitch pillars (upper part) and 14-μm diameter, 30-μm height, and 26-μm pitch pillars (lower part). (Adapted from Y. C. Jung, and B. Bhushan. *Langmuir* 24:6262–6269, 2008, American Chemical Society. With permission.)

(Figure 4.8c). A detail of epidermis cells of the leaf upper side with papillae covered with wax tubules showing the estimated diameter of contact area is given in Figure 4.8d (Ensikat et al. 2011). Because of the low adhesion, the water droplets from rain or vapor condensation can roll easily capturing small particles and contaminants, thus becoming a self-cleaned surface.

Two main strategies for the artificial preparation of superhydrophobic surfaces are at hand: patterning a hydrophobic surface or transforming a patterned surface into hydrophobic. Preparation techniques include lithography, plasma treatment, chemical

FIGURE 4.8 (a) Picture of a lotus leaf. SEM images of the upper leaf side showing a hierarchical surface structure consisting of (b) papillae and (c) wax tubules. (d) A detail of epidermis cells of the leaf upper side with papillae covered with wax tubules showing the estimated diameter of contact area. (Reprinted from H. J. Ensikat, et al. *Beilstein J. Nanotechnol.* 2:152–161, 2011. With permission.)

vapor, and layer-by-layer techniques, aggregation/assembly of particles, and templating to mention some of them. Here we give a few examples. A complete collection can be found in Roach, Shirtcliffe, and Newton (2008) and Yan, Gao, and Barthlott (2011). Onda et al. (1995) prepared fractal surfaces using alkylketene dimer, a kind of wax, and one of the sizing agents for papers, and showed that the contact angle is a function of the fractal dimension. The authors were the first to report contact angles as high as $\theta_c = 174°$, thus close to the 180° frontier, in clear competition with the sacred lotus leaves. However, the comparison is not fair, because plants have to administrate complex biological functions and repellency to water is just one of them.

Microfabricated surfaces, following a top-down approach, provide a higher degree of control of the desired patterns. Bico, Marzolin and Quéré (1999) produced micron-sized patterned surfaces using photolithography. Using spikes on a master surface, they were replicated using an elastomeric mold, which was subsequently used to cast silica features onto a silicon wafer. Following a hydrophobic SAM coating, the spikes gave $\theta_c^a \simeq 170°$. Dual micro-nanoscale hierarchy can be also artificially achieved by two-step lithography. Figure 4.9 shows examples of dual structures prepared by ultraviolet (UV)-assisted capillary molding (Jeong et al. 2009). This method is based on the sequential application of engraved polymer molds followed by a surface treatment. In the example shown in the figure, the microstructures were generated first by using microscaled poly(urethane acrylate) or poly(dimethyl siloxane) molds and then the nanostructures on top of them by using a nanoscaled mold of the first polymer. Partial UV curing was used in the first step to form micropillars, and complete UV curing

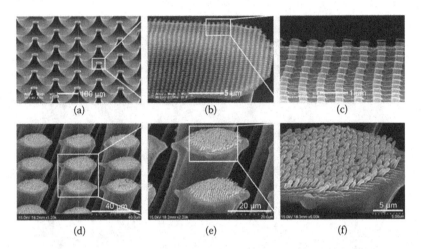

FIGURE 4.9 SEM images of a dual hierarchical structure prepared by UV-assisted capillary force lithography. (a) Regular microstructured pattern built from 30-μm posts of 40-μm spacing and 50-μm height. (b) and (c) Magnified images showing 400-nm dots with spacing of 800 nm and height of 500 nm on the microposts. (d) High-aspect–ratio micropillars with diameters of 20 μm, heights of 100 μm and spacings of 20 μm. (e) and (f) Magnified images showing 400 nm pillars with spacing of 400 nm and height of 2.5 μm on the micropillars. (Reprinted from H. E. Jeong, et al. *J. Colloid Interface Sci.*, 339: 202–207, 2009, With permission from Elsevier.)

was used in the second step to form nanopillars. The fabricated hierarchical surface was chemically treated to enhance hydrophobicity.

Nanoparticles also provide a certain control of roughness because of their shape. If they have nanometer scale dimensions they can be combined with patterned surfaces giving the required hierarchical structure (Yeh, Cho and Chen 2009). Contact angles of 178° have also been reported for structures built from hydrophilic material, thus highlighting the role of surface topography over surface chemistry (Hosono et al. 2005). In this case films of brucite-type cobalt hydroxide coated with lauric acid were prepared exhibiting a needlelike structure with tip diameters of less than 10 nm, in a bottom-up approach.

So far we have dealt with the chemical and geometrical aspects related to water repellency but one can go a step further and make use of the physical properties of the materials that build the patterned surfaces because in this way superhydrophobicity is added to the materials portfolio. A relevant example is ZnO, a wide bandgap semiconductor with outstanding optical, electronic, and catalytic properties (Wöll 2007). ZnO nanorods grown perpendicularly onto silicon wafers expose hexagonal surfaces with dimensions that can be controlled by varying the concentration of precursors. The smaller the diameter is, the smaller the contact angle, thus reducing in a controlled way the ς parameter (Sakai et al. 2009). Figure 4.10 shows SEM images of ZnO nanorods grown by plasma-enhanced CVD (PECVD) yielding static contact angles of ~148° and low water contact angle hysteresis (~12°; Macías-Montero et al. 2012a).

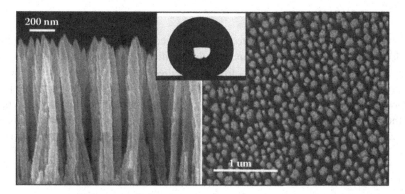

FIGURE 4.10 Cross-section (left) and planar (right) views of SEM micrographs of ZnO nanorod arrays prepared by PECVD. The inset shows a water droplet with a contact angle $\sim$148°. (Reprinted from M. Macías-Montero, et al. *Langmuir* 28:15047–15055, 2012, American Chemical Society. With permission.)

Under UV irradiation the surface of this material becomes photon activated and the water can spread smoothly over the whole internal surface of the nanorod structure, thus becoming superhydrophilic (Feng et al. 2004). Thus, the sympath or antipathy for water can be controlled externally, in this case by irradiation with light, a point discussed below. Photoinduced superhydrophilicity was first reported for TiO_2 in 1997 (Wang et al. 1997). UV light induces the creation of oxygen vacancies favoring water dissociation, building hydrophilic domains because of the presence of hydroxyl groups at the surface, as discussed in Chapter 3.

Superhydrophobic surfaces are nowadays finding varied applications and here only some of them are telegraphically mentioned (Nosonovsky and Bhushan 2009). Apart from the self-cleaning effect noted before, which is useful for glasses, windows, and the like (Blossey 2003), superhydrophilic coatings are also used for antifogging (Chen et al. 2012), preventing the condensation of water in the form of small droplets and often used on transparent glass surfaces of optical instruments (lenses, mirrors, etc.). Superhydrophobic paints, applied to walls, are able to protect them from effects caused by the weather. Self-cleaning textiles are also an interesting issue and can be achieved using TiO_2 nanoparticles or carbon nanotubes on the textile fibers. Superhydrophobic jackets were used by the Swiss sailing team (Alinghi) in the Americas Cup ensuring maximum repellency under the harshest conditions, a clear advantage for the crew. Superhydrophobic coating efficiently protects microdevices making them waterproof (Lee, Kim, and Young 2011). Microfluidics is another relevant domain of application. Here a strict control of the affinity to water is essential in order to permit the flux inside micrometer-sized channels (Vinogradova and Dubov 2012). Air-retaining surfaces are also of great interest because the trapped air leads to a minimization of the water/solid contact area. Underwater superhydrophobicity can prevent the formation of biofilms (algae, bacteria, and other marine organisms) on underwater surfaces of great relevance to vessels.

4.3 AMPHIPHILICITY

Amphiphilicity is the ability shown by certain objects simultaneously to exhibit high and low affinity to water. We explore here differentiated families of such objects. The most popular of them corresponds to amphiphilic molecules, already described in the previous section concerning SAMs, with the hydrophilic and hydrophobic parts separated by molecular spacers (alkyl chains, etc.). Such fundamental property is used for the preparation of LB films as well as the formation of biomembranes. Here we concentrate on other objects, namely particles, patterned and nanostructured surfaces as well as crystalline materials.

4.3.1 JANUS PARTICLES

When half of the surface of a particle is hydrophobic (apolar) and the other half is hydrophilic (polar), the resulting particle is termed *Janus* referring to the mythological Roman god who possessed two faces representing two opposed visions (Granick, Jiang, and Chen 2009). It was P.-G. de Gennes, awarded with the Nobel Prize in Physics in 1991, who drew attention toward such conceptually simple objects in his Nobel Lecture for their potentialities in science and technology (de Gennes 1992). They were first made by Casagrande and Veyssié (1988) and de Gennes considered dense films of such particles, that he called grains, at interfaces involving water because of the interstices between the particles, which should allow for chemical exchange between the two sides of the interface.

Three main strategies are used for the synthesis of Janus particles: masking, phase separation, and self-assembly (Walther and Müller 2008, Lattuada and Hatton 2011), which are schematized in Figure 4.11. Masking involves the protection of one half of the particle followed by chemical modification (see Figure 4.11a). The two commonly used masking techniques are evaporative deposition and suspension at the interface of two immiscible phases. The first one consists of depositing particles on a flat (hard) substrate and then coating them with a metal layer (e.g., gold). The second way permits the fixation of the particles at the interface of water and molten wax, a soft surface, and is usually termed the Pickering emulsion route (Jiang and Granick 2008). Upon cooling the wax solidifies, trapping the particles (e.g., silica nanoparticles), leaving the other half of the particles exposed. Immobilization suppresses rotational diffusion of the particles at the solidified interface. The exposed silica surfaces can then be functionalized through aqueous phase chemistry and then the wax can be dissolved, reconstituting the initial surfaces.

The phase separation technique relies on the segregation of two or more incompatible components in a mixture (see Figure 4.11b). Such incompatibility opens the door to heterogeneous inorganic–inorganic, organic–inorganic and organic–organic particles. Inorganic materials can build heterodimers combining particles with different physical properties. One example, among many others, is the combination of gold nanoparticles, because of their optical properties, with magnetic nanoparticles (Yu et al. 2005). The preparation of such particles is usually undertaken by heteroepitaxial growth of the second one on the first one. The first particle works as a seed, but the unfavorable epitaxial conditions leads to the formation of nanocrystals. Finally, the

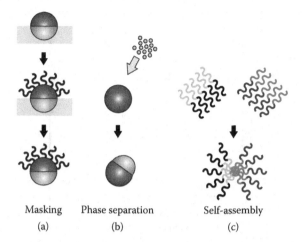

Masking   Phase separation   Self-assembly
(a)   (b)   (c)

FIGURE 4.11 Schematic diagram illustrating examples of the three main strategies for the preparation of Janus nanoparticles: (a) masking, (b) phase separation, and (c) self-assembly. (Adapted from M. Lattuada and T.A. Hatton. *Nano Today* 6: 286–308, 2011. With permission.)

third synthetic route is based on self-assembly (see Figure 4.11c). A pioneering example involves the self-organization of triblock terpolymers into micelles (Erhardt et al. 2001). Janus particles exhibit a variety of complex aggregates. In the case of spherical Janus micelles having hemispheres of polystyrene and poly(methyl methacrylate), aggregation into clusters is observed in various organic solvents. In addition to these amphiphilic particles, dipolar (zwitterionic) Janus particles are also of interest inasmuch as they are water-soluble and build aggregates by dipole–dipole interactions (Hong et al. 2006).

Concerning applications, Janus nanoparticles have been shown to be applicable in textiles (Synytska et al. 2011). They bind with the hydrophilic reactive side of the textile surface, whereas the hydrophobic side is exposed to the environment, thus providing water-repellent behavior. Such particles may also be used in switchable devices for displays, optical probes, biochemical nanosensors, nanomedicine for drug-delivery, and many others and are a matter of intensive research.

4.3.2 JANUS SURFACES

Different surfaces of crystalline materials will, in principle, exhibit distinct affinity to water. We are interested here in those surfaces that show contrasting behavior, hydrophilic and hydrophobic, justifying the term amphiphilic. A first illustrative example is the case of the (111) and (110) surfaces of copper: Cu(111) is hydrophobic and Cu(110) is hydrophilic. As pointed out in Section 3.2 the presence or absence of OH groups on the surface determines the different wetting behavior and this has been clearly shown by means of NAPP experiments (Yamamoto et al. 2007). Cu(111) has a low E_{ads} value, 0.22 eV (Tang and Chen 2007) and as a consequence water does not wet this surface, forming 3D clusters at low temperatures (Mehlhorn and

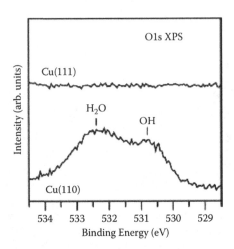

FIGURE 4.12 O $1s$ XPS spectra measured in 1 torr water vapor using 735 eV photons on Cu(110) and Cu(111) surfaces at 295 K (5.0 % RH). (Reprinted from S. Yamamoto, et al. *J. Phys. Chem. C* 111:7848–7850, 2007, American Chemical Society. With permission.)

Morgenstern 2007). Figure 4.12 shows XPS spectra of the O $1s$ line performed on Cu(111) (top) and Cu(110) (bottom) in the presence of 1 Torr of water vapor at 295 K, which corresponds to RH = 5.0%. The featureless spectrum shown for Cu(111) indicates that the surface remains clean and adsorbate-free whereas for Cu(110), with $E_{ads} = 0.39$ eV (Tang and Chen 2007), two broad features are observed at $\simeq 531$ and 532.5 eV binding energy, which are assigned to OH and molecular H_2O, respectively. Thus, although the Cu(111) is hydrophobic, the Cu(110) surface exhibits a hydrophilic character due to the presence of OH groups that efficiently help to adsorb molecular water. This behavior is not exclusive to copper surfaces but is of general applicability; wetting of metals at near-ambient conditions is controlled by the presence of OH groups on the surface that stabilize water molecules via H-bonding. The enhanced stability of the H_2O–OH system over H_2O–H_2O provides a general mechanism for water dissociation, that has been termed autocatalytic, because the activation barrier for water dissociation is lowered (Andersson et al. 2008).

A second example was previously explored in Section 4.2.2, with the (120) and (011) crystal faces of L-alanine as main characters and the origin of their contrasting behavior when exposed to water. Such behavior is not directly related to the presence or absence of OH groups and is discussed in Section 4.4. This system has the additional interest in that water is able to modify the affinity of the hydrophilic (011) surface transforming it into a water-repellent surface. Under ambient conditions water molecules strongly interact with such surfaces promoting solvation and diffusion of L-alanine molecules and leading to the formation of 2D islands with long-range order through Ostwald ripening (Segura et al. 2011a). This is an example of an irreversible self-passivation process induced by water.

Both examples shown here correspond to crystal faces of known materials (inorganic and organic) but the Janus character can be artificially induced. The previously discussed superhydrophobic–superhydrophilic conversion mediated by UV radiation observed in some transition metal oxides can be used to pattern such surfaces with paths that will be able to confine water in a controlled way, a property of huge interest in microfluidics. This has been shown, for example, for TiO_2 surfaces. Water condensation becomes guided by the superhydrophilic stripes due to the extremely large wettability contrast between micrometerwide stripes but not due to changes in topography (no wells or lower channels). Small water droplets ($\sim 1\mu L$) placed on a 500-μm wide stripe spread entirely on the stripe, with a bulge forming in the middle of the stripe and a contact angle of 138° along the stripe (Zhang et al. 2007).

Water-Induced Height Artifacts with AFM

We come now to the revealing case of flat surfaces exhibiting domains with different affinity to water with a well-defined topography (e.g., flat islands on flat surfaces) and next we show that the determination of the surface profile (e.g., height of islands) with AFM in dynamic mode (AM-AFM) is not straightforward. We already saw in Section 3.2.2 that the determination of the height of water patches on well-defined surfaces was not exempt from uncertainty, when measured with AM-AFM in ambient conditions, due to capillary forces (see Appendix B). In the case of surfaces with regions exhibiting different affinity to water, the case we are dealing with in this section, the presence of nanometer-thick water films on both the surface and the tip of the probe, as is usually the case in ambient conditions, can lead to apparent heights markedly different from the real heights due to formation and rupture of water menisci. Depending on the operation parameters (free oscillation amplitude and setpoint), the apparent heights can vary in magnitude and sign, known as contrast inversion (Palacios-Lidón et al. 2010), and, most important, the true heights cannot be obtained with AM-AFM (Santos et al. 2011). When the materials are sufficiently hard, this can be circumvented by using the contact mode, but this is not possible for soft materials. In the case where a thin water layer covers both the tip and the sample surface, the interaction regimes can be divided in three main groups, which are schematized in Figure 4.13a. In the pure nc regime (right in the figure) the oscillation occurs in the absence of any mechanical contact and the water layers remain unperturbed. When the water layers on the surfaces intermittently overlap and/or the capillary neck is formed and ruptured without mechanical contact the resulting regime is termed intermittent contact (ic; center of the figure). Intermittent mechanical contact (mc) is achieved when intermittent contact is made between the tip and the substrate (left in the figure). The case where a hydrated tip interacts with both hydrophilic and hydrophobic surface domains (amphiphilic) while scanning over the sample is shown in Figure 4.13b. We show here the example of the determination of the apparent height of hydrophobic stearic acid SAMs grown on hydrophilic mica surfaces at different RH values (Verdaguer et al. 2012). Because of the expected different response in both regions it may be expected that the measured apparent height will be different from $L - h$, the difference between the true island height (1.6 nm as determined in contact mode) and the height of the water layer (left in Figure 4.13b).

The experimentally derived tip–sample interaction on both mica and stearic SAMs is shown in Figure 4.14. Amplitude and phase (Φ) curves versus tip–sample separation are shown for different free oscillation amplitudes A_0 at RH$\sim$50%. Interaction

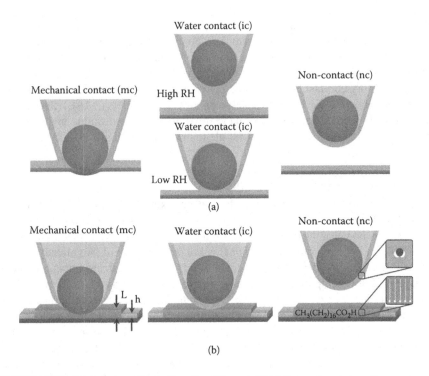

FIGURE 4.13 (a) Schemes illustrating the different AFM tip–surface interaction regimes, noncontact (nc), intermittent contact (ic), and mechanical contact (mc), when both surfaces are covered by a thin water layer of thickness h. Note the formation of a meniscus at high RH. (b) Schemes showing the equivalent regimes but for the case of a hydrated tip on a hydrophobic surface, in particular a stearic acid SAM on mica. (Reprinted from A. Verdaguer, et al. *Phys. Chem. Chem. Phys.* 14:16080–16087, 2012. With permission of the Royal Society of Chemistry.)

regimes, attractive (AR) or repulsive (RR), are determined according to the observed phase lag values. If the phase is set to 90° when far from the surface (no interaction between tip and sample) and when driving at resonance, the perturbed phase will lag above 90° in the AR and below 90° in the RR. AR and RR are represented by $A_0 < A_c$ and $A_0 > A_c$, respectively, whereas transitions from AR to RR by $A_0 \sim A_c$, where A_c stands for the critical free amplitude, which is in general different for each involved tip–surface system (Verdaguer et al. 2012). For $A_0 > A_c$ (Figure 4.14c) the interaction is basically repulsive both on mica and on stearic SAMs. On the contrary for $A_0 < A_c$ (Figure 4.14a) the interaction is purely attractive. However, the most interesting case is that for which $A_0 \sim A_c$. Because A_c will be different from the tip–hydrophobic and tip–hydrophilic domain cases, then the RR can prevail in one part and AR in the other, which necessarily leads to differences in the measured heights. At low RH values and under AR/AR and RR/RR conditions, the islands' heights are 1.1 and 1.3 nm, respectively, far from the nominal 1.6 nm value. At higher RH values the AR/RR imaging conditions become common when $A_0 \simeq A_c$ showing contrast inversion (negative height values) as shown in Figure 4.14d. Thus, when measuring in the conditions discussed here, one should be cautious, otherwise one can get

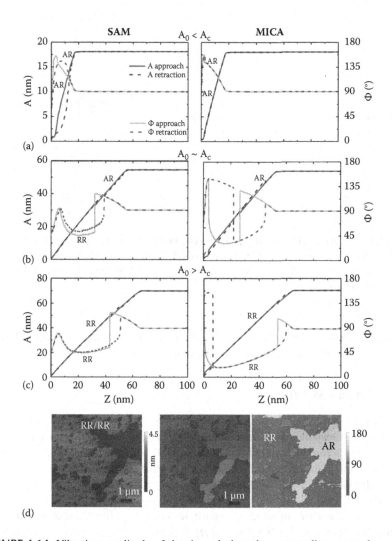

FIGURE 4.14 Vibration amplitude of the tip and phase lag versus distance to the surface on mica and stearic SAMs. Results for three different free amplitudes, A_0, compared to A_c are shown: (a) $A_0 < A_c$, (b) $A_0 \simeq A_c$ and (c) $A_0 > A_c$. In (a) the interaction is mostly attractive ($\Phi > 90°$) whereas in (c) it is mostly repulsive ($\Phi < 90°$). Because A_c is different when measuring on mica and on stearic SAMs due to water neck formation on mica in (b) the most stable interaction regime is repulsive on the stearic SAMs and attractive on mica. (d) AFM images of incomplete stearic SAMs on mica taken in RR/RR and AR/RR conditions at RH~ 70%. Contrast inversion is observed in parts of the AR/RR image that can be easily identified using the phase lag image (right). (Reprinted from A. Verdaguer, et al. *Phys. Chem. Chem. Phys.* 14:16080–16087, 2012. With permission of the Royal Society of Chemistry.)

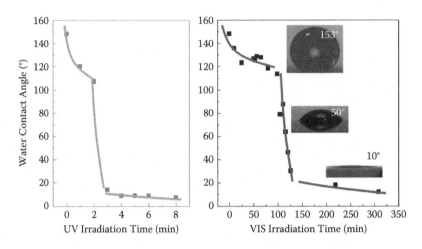

FIGURE 4.15 Changes in the water contact angle induced by irradiation with UV (left) and VIS light (right), respectively, of supported Ag-NPs@ZnO-nanorods. (Reprinted from M. Macías-Montero, et al. *J. Mater. Chem.* 22:1341–1346, 2012. With permission of the Royal Society of Chemistry.)

completely wrong heights, which are usually used to discriminate between different structures, molecules, polymorphs, and the like.

4.3.3 TUNABLE WETTING

There is increasing interest in functional surfaces with controlled wetting properties, which can respond to external stimuli (optical, electrical, mechanical, thermal, etc.). We saw before the example of ZnO nanorod coatings undergoing a superhydrophobic/superhydrophilic transition under UV-irradiation. It turns out that once in the superhydrophilic state, superhydrophobicity can be restored simply by storing the material in the dark for a long period (several days). Figure 4.15 illustrates such a property, in this case for silver nanoparticles embedded in hollow ZnO nanorods (Macías-Montero et al. 2012b). The as-prepared surfaces are superhydrophobic with water contact angles above 150° and turn into the superhydrophilic state with water contact angles below 10° after irradiation under both visible and UV light. Thus, the sympathy or antihpathy for water can be controlled externally, in this case by irradiation with light. However, the long restoration time limits its practical use as a potential active part in a switching device. Reversible superhydrophobicity–superhydrophilicity has been also observed for other nanorod films of transition metal oxides such as TiO_2 (Feng, Zhai, and Jiang 2005) and SnO_2 (Zhu et al. 2006) as well as on nanostructured V_2O_5 films with targeted micro- and nanoscale hierarchical (roselike) structures (Lim et al. 2007).

Reversible switching induced by temperature has been reported by Sun et al. (2004) using a thermoresponsive polymer, poly(N-isopropylacrylamide). When deposited on flat surfaces, θ_c varies from 63.5° at RT to 93.2° at 40°C. However, when deposited on patterned surfaces θ_c reversibly varies from $\sim 0°$ below $T = 29°C$ to $149.5°$ above $40°C$ after many temperature cycles. The same effect can be achieved mechanically. Zhang et al. (2005) described a method to generate reversible wettability by biaxially extending and unloading an elastic polyamide film with triangular netlike structure. Before and after extension $\theta_c \sim 151°$ and $\theta_c \sim 0°$, respectively, thus showing the superwettability transition under pure mechanical stimulus. This effect can be quite simply understood by taking a look at Figure 4.6. Superhydrophobicity relies on the presence of air in the valleys defined by the pillars. If the pitch distance P_p increases and the height H_p decreases under the action of an extension (induced by a pulling force) then the droplet may contact the lower part of the valley. In the particular case of the polyamide film referred to here, the average side of the triangle of the netlike structure goes from about 200 μm before biaxial extension to about 450 μm after extension.

Wettability can be also externally controlled by electrical stimuli. Surfaces strategically functionalized with particular electroactive coatings transform under the application of an electric field exposing either hydrophilic or hydrophobic groups. One example is based on the bending of molecular alkyl chains exhibiting a thiol head group and a carboxylic tail group (e.g., mercaptohexadecanoic acid; Lahann et al. 2003). Submonolayer SAMs of such molecules expose their carboxylic groups under application of a negative voltage (hydrophilic surface), because of electrostatic repulsion. However, when the external polarity is reversed, then the carboxylic groups tend to move toward the polarized surface bending the molecules and exposing the hydrophobic alkyl groups, hence the importance of using submonolayer coverage in order to provide sufficient room for molecular bending. Another chemical strategy relies on functionalization using stable radicals, molecules exhibiting two stable charge states. SAMs of polychlorotriphelylmethyl radicals switch reversibly from the hydrophobic ($\theta \sim 102°$) to the hydrophilic ($\theta \sim 73°$) states when the molecule is reduced from the neutral to the anionic state under electrochemical action (Simão et al. 2011). The switching redox cycles are illustrated in Figure 4.16.

Both of the last examples draw us to the general concept of electrowetting (Mugele and Baret 2005). Electrocapillarity, the basis of electrowetting, was first described by G. Lippmann back in 1875 where he found that the capillary depression of mercury in contact with electrolyte solutions could be varied by applying a voltage between the mercury and electrolyte (Lippmann 1875). If we think of a water droplet between two electrodes (a flat metallic surface and a simple contact or two flat metallic surfaces), then application of large voltages will lead to the electrolytic decomposition of water. Berge (1993) proposed the utilization of a thin insulating layer to avoid direct contact between liquid and metal, which is known as the electrowetting on dielectric (EWOD) method. A scheme is shown in Figure 4.17.

As shown in the figure the contact angle can be varied upon application of an external voltage (U), according to the expression:

$$\cos \theta_c(U) = \cos \theta_c(0) + \frac{\varepsilon_0 \varepsilon_d}{2t_d \gamma_{lv}} U^2 \qquad (4.6)$$

where $\cos \theta_c(0)$ stands for the contact angle in the absence of an applied voltage (our θ_c) and $\varepsilon_d \varepsilon_0$ and t_d correspond to the dielectric constant and thickness of the

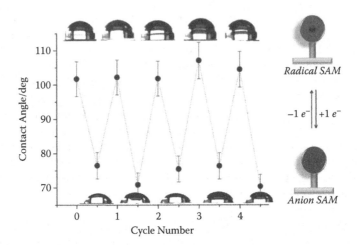

FIGURE 4.16 Contact angle as a function of redox cycles of SAMs of polychlorotriphelyl-methyl radicals. (Reprinted from C. Simão, et al. *Nano Lett.* 11:4382–4385, 2011, American Chemical Society. With permission.)

dielectric material. Under these conditions an electric double layer builds up at the insulator/droplet interface. All parameters of the second term of (4.6) are positive, therefore $\theta_c(U) < \theta_c(0)$. Thus, wetting increases by the application of an external voltage. Using electrowetting it is possible to switch a liquid droplet from the Cassie–Baxter to the Wenzel morphology (Krupenkin et al. 2004). The authors fabricated nanopillars with diameters of 350 nm and heights of 7 μm by dry-etching a Si wafer. Each pillar had a conductive core of Si covered by a thermally grown insulating SiO_2 layer and a hydrophobic coating and the transition was induced at a threshold voltage of 22 eV.

Electrowetting finds applications essentially in lab-on-a-chip systems as well as in flexible lenses, fiber optics, and in display technology (Beni and Hackwood 1981). The configuration allowing droplet transport in a controlled way consists of two parallel flat electrodes confining the liquid. One of the substrates contains the patterned electrodes required for liquid actuation and the other provides electrical contact to the droplet

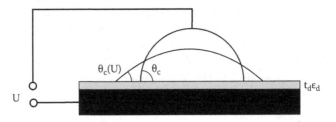

FIGURE 4.17 Scheme of an electrowetting on a dilelectric set-up. When the external voltage $U = 0$, then the contact angle is θ_c, and upon $U \neq 0$, then $\theta_c(U) < \theta_c$. $\varepsilon_d \varepsilon_0$ and t_d stand for the dielectric constant and thickness of the dielectric material, respectively.

independently of its position. Such an electrode is usually a glass substrate coated with a transparent indium–tin–oxide layer which in turn is covered with a thin hydrophobic layer to provide a large contact angle while permitting electrical contact. By applying voltages to the different electrodes in the array droplets of typical volumes of $0.1–1\,\mu\text{L}$ can be controllably moved, transported to containers, combined with other droplets, and so on. Concerning optical applications, liquid lenses typically formed by oil/water interfaces, are flexible. Their curvature and hence their focal length can be tuned by adjusting their shape. This allows for the design of optical systems with variable focal length that can be addressed purely electrically.

4.4   THE ROLE OF DIPOLES

Given the polar character of the water molecule, it should be expected that polar surfaces define their wetting through dipolar interactions. We saw in Section 1.3 that the energy associated with the permanent dipole–permanent dipole interaction (Keesom term) is expressed by (1.27) for fixed dipoles and we find here an illustrative example making use of this equation with both the (120) and (011) surfaces of L-alanine, exhibiting hydrophobic and hydrophilic character, respectively, as discussed above in this chapter. In Figure 4.18 two snapshots obtained from MD simulations of both surfaces exposed to bulk water are shown (Segura et al. 2009). Note first that the alanine dipoles are differently disposed: whereas in the case of the (011) surface such dipoles are arranged in such a way as to form angles of roughly 45° and 135° with the surface normal, in the (120) surface they are contained within the plane of the surface. Such dipolar distribution induces a radically different distribution of the dipoles of the water molecules closer to the surface. Close to the (011) surface, water molecules orient themselves with their dipoles either pointing out or in, depending on whether the dipole of the nearest alanine molecule is pointing out of or into the crystalline surface (see Figure 4.18a). In the case of the (120) surface (see Figure 4.18b), the situation is different: no clear preference for any given orientation of the water dipoles can be discerned, and all orientations can be observed.

The same trends are observed when averaging over time. Figures 4.18c and d show two functions, $\langle P_1(\cos\theta)\rangle$ and $\langle P_2(\cos\theta)\rangle$, where P_1 and P_2 are the first- and second-order Legendre polynomials, defined as $P_1(x) = x$ and $P_2(x) = \frac{1}{2}(3x^2 - 1)$, respectively, and θ is the angle formed by the dipole of a water molecule with the outward pointing surface normal, and the angular brackets indicate an average over water molecules and configurations. P_1 provides information about the average orientation of the water dipoles, whereas P_2 allows us to distinguish between two possible cases leading to the same value of P_1, namely the case of anisotropic orientation of dipoles ($P_1 = 0, P_2 = 0$) and the case of orthogonal orientation to the surface normal ($P_1 = 0, P_2 = -1/2$). From the figure it is apparent that the arrangement and orientation of water molecules is significantly different in both cases. For the (011) surface, we observe that $\langle P_1\rangle$ takes negative values at short distances, consistent with the fact that the nearest water molecules have their dipoles oriented antiparallel to the surface normal. As the distance to the surface is increased there is some oscillation from negative to positive values, reflecting local domains of slight predominance of antiparallel/parallel orientation of the water dipoles to the surface normal, decaying

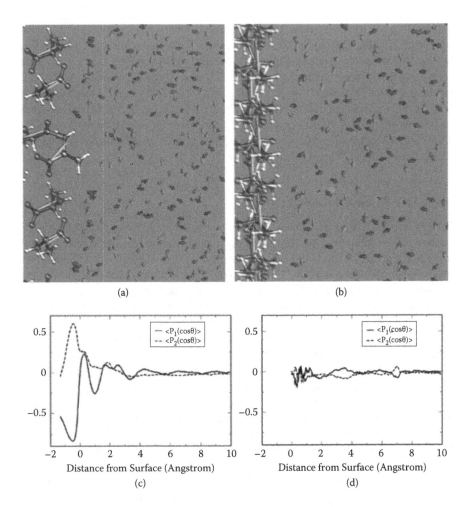

FIGURE 4.18 Distribution of molecular dipoles at L-alanine/water interfaces. Instantaneous configurations of the dipoles of water molecules resulting from MD simulations of the (011) (a) and (120) (b) surfaces exposed to water. Alanine molecular dipoles are shown superimposed on the corresponding molecules. Water molecular dipoles are shown in a grey code, where dark/medium gray and light gray represent nearly orthogonal and parallel distributions with respect to the surface normal, respectively. (c) and (d) display the functions $\langle P_1(\cos\theta) \rangle$ and $\langle P_2(\cos\theta) \rangle$ calculated as a function of distance along the surface normal, for the (011) and (120) surfaces, respectively. (Reprinted from J. J. Segura, et al. *J. Am. Chem. Soc.* 131:17853–17859, 2009, American Chemical Society. With permission.)

to zero farther away from the surface. This decay is indicative of a transition from highly oriented dipolar arrangements close to the surface toward a situation of randomly oriented dipoles as we move into the liquid bulk. In contrast, the values of $\langle P_1 \rangle$ and $\langle P_2 \rangle$ on the (120) surface are always close to zero, regardless of the distance to the surface; clearly, in this case interaction with the surface is not strong enough to induce any favored orientation of the water molecular dipoles.

Thus it can be seen that the orientation of the alanine molecular dipoles is the key to the hydrophobic/hydrophillic nature of the alanine surfaces. According to (1.27) when two dipoles are in a colinear orientation, their interaction energy is twice that corresponding to the antiparallel orientation at the same distance. The most favorable orientation that the water molecules can adopt above the (120) alanine surface is such that their molecular dipole is oriented antiparallel to that of the nearest alanine surface molecule. However, in the case of the (011) surface it is possible for the water molecules to orient themselves such that their dipoles are nearly colinear with that of the nearest alanine surface molecule, leading to a more strongly favorable interaction. This is consistent with the hydrophilic character of the (011) surface and the hydrophobic one of the (120).

This issue has been treated more generally by Giovambattista, Debenedetti, and Rossky (2007). They have performed MD simulations of water in the presence of hydrophobic/hydrophilic walls using a continuous parameter k (introduced in Section 3.4.1), confined to $0 \leq k \leq 1$, such that $k = 1$ represents a hydrophilic surface and $k = 0$ stands for a hydrophobic surface. Using a SPC/E water model (see Table 1.5), they find that the magnitude of the surface dipole is correlated with the contact angle in a one-to-one correspondence. $k = 0.4$ turns out to be a critical value, below which the surfaces are hydrophobic ($\theta_c > 90°$) and above which they are hydrophilic ($\theta_c > 90°$). An important message is that a surface can be hydrophobic in spite of being polar ($k \neq 0$).

Etching and Chirality

The surface dipolar distribution can guide us toward the understanding of etching produced by water. We consider here the revealing example of valine, a nonpolar essential amino acid that has a nonreactive hydrophobic isopropyl group as a side chain building a lamellarlike structure, as shown in Figure 4.19a.

Valine forms bilayers along the crystallographic c-axis, and such bilayers exclusively expose the isopropyl groups, conferring a perfect hydrophobic character (crystals in the form of flakes emerge to the air/water interface during crystallization). Figure 4.19a also represents a surface step, which exposes both carboxylic and ammonium groups (encircled in the figure) with uncompensated dipoles represented by arrows (not to scale). Following the discussion given above, the water molecules will have the tendency to align with such dipoles and even solvate the valine molecules at the step. The result after immersing the crystals in Milli-Q water is the etching of the otherwise hydrophobic surface with the patterns shown in Figures 4.19b, c, and d for enantiomeric L- and D- and racemic DL-valine, respectively (Segura et al. 2011b). Such features consist of regular parallelepipeds, whose sides are steps one bilayer high. The steps are aligned along two well-defined crystallographic directions, namely [010] and [110], with steps along the [010] direction larger than those along the [110] direction. In the case of D-valine (Figure 4.19c), the etching patterns are the specular images of those observed on L-valine crystals. From the images we observe that by visual inspection one can readily identify the chirality. Racemic (001) surfaces show similar patterns but in this case the steps exhibit similar dimensions. Stereoselective etching of α-amino acid crystals has been reported using water dissolutions including additives, a method that has been successfully applied

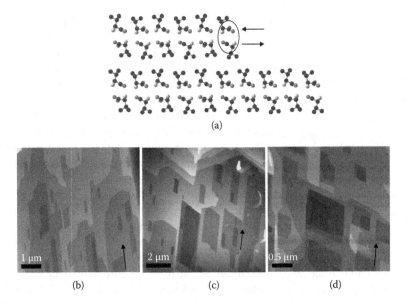

FIGURE 4.19 (a) View along the b-direction of the crystal structure of L-valine. Carbon, nitrogen, and oxygen atoms are represented by black, light grey, and medium grey balls, respectively. Hydrogen atoms are omitted for clarity. The step exposes dipoles nearly parallel to the surface (ab) plane, represented by arrows. Crystallographic data of L-valine: $P2_1$, $a = 0.971$ nm, $b = 0.527$ nm, $c = 1.206$ nm, $\beta = 90.8°$ (Torii and Iitaka 1970). AM-AFM images taken at ambient conditions of water-mediated etched (001) surfaces of: (a) L-valine, (b) D-valine and (c) DL-valine. The arrows indicate the crystallographic b-direction. (Reprinted from J. J. Segura, et al. *Phys. Chem. Chem. Phys.* 13:21446–21450, 2011. With permission of the Royal Society of Chemistry.)

to the direct assignment of their absolute molecular configurations (Shimon, Lahav, and Leiserowitz 1985; Weissbuch, Leiserowitz, and Lahav 2008).

4.5 SUMMARY

- The contact angle of a water droplet on a surface is a microscopic measure of wetting and depends on several factors related to the surface such as its chemical nature, topography at the nano- and microscale, presence of defects (steps, vacancies, etc.), structure of coatings, and so on.
- Apolar solutes can become sequestered by water, which builds a H-bonding network around them for a sufficiently small solute radius (<1 nm). Above this value it becomes impossible for the adjacent water molecules to maintain a complete network, hence water has the tendency to move away from the surface depleting the water bulk density close to it.
- Superhydrophobicity, the nearly absolute water repellency of a surface and quantified by contact angles close to 180°, cannot solely be achieved chemically. By exposing chemical groups exhibiting weak interaction with water

(of van der Waals type) the contact angle is limited to ~120° and bioinspired hierarchical structures in the micro- and nanometer regime are needed.

- Surface affinity to water can be externally tuned in some cases by mechanical action, by irradiating with UV light (sometimes also with visible light), or by varying temperature. Several nano-microstructured surfaces exhibit reversible hydrophobic/hydrophilic transitions.
- The dynamics of AFM cantilevers in ambient conditions are strongly influenced by the presence of water. The measured heights of surface domains with different affinity to water become modulated by capillary forces to the point that it is impossible to obtain the true heights in dynamic mode.
- The orientation of permanent dipoles at surfaces dictates their affinity to water: dipoles oriented perpendicular (parallel) to the geometrical surface induce high (low) affinity. Etching of amino acid surfaces is also governed by dipolar interactions.

REFERENCES

1. Andersson, K., Ketteler, G., Bluhm, H., Yamamoto, S., Ogasawara, H., Pettersson, L.G.M. et al. 2008. Autocatalytic water dissociation on Cu(110) at near ambient conditions. *J. Am. Chem. Soc.* 130:2793–2797.
2. Bakulin, A.A., Pshenichnikov, M.S., Bakker, H.J., and Petersen, C. 2011. Hydrophobic molecules slow down the hydrogen-bond dynamics of water. *J. Phys. Chem. A* 115:1821–1829.
3. Barthlott, W. and Neinhuis, C. 1997. Purity of the sacred lotus, or escape from contamination in biological surfaces. *Planta* 202:1–8.
4. Beni, G. and Hackwood, S. 1981. Electro-wetting displays. *Appl. Phys. Lett.* 38:207–209.
5. Berge, B. 1993. Électrocapillarité et mouillage de films isolants par leau. *C. R. Acad. Sci. Paris* Série II 317:157–163.
6. Bico, J., Marzolin, C., and Quéré, D. 1999. Pearl drops. *Europhys. Lett.* 47:220–226.
7. Blossey, R. 2003. Self-cleaning surfaces–virtual realities. *Nature Mater.* 2:301–306.
8. Braslau, A., Deutsch, M., Pershan, P.S., Weiss, A.H., Als-Nielsen, J., and Bohr, J. 1985. Surface roughness of water measured by X-ray reflectivity. *Phys. Rev. Lett.* 54:114–117.
9. Buffett, B.A. 2000. Clathrates hydrates. *Annu. Rev. Earth Planet Sci.* 28:477–507.
10. Casagrande, C. and Veyssié, M. 1988. Grains Janus: réalisation et premires observations des propriétés interfaciales. *C. R. Acad. Sci. (Paris) II* 306:1423–1425.
11. Cassie, A.B.D. and Baxter, S. 1944. Wettability of porous surfaces. *Trans. Faraday Soc.* 40:546–551.
12. Chandler, D. 2005. Interfaces and the driving force of hydrophobic assembly. *Nature* 437:640–647.
13. Chattopadhyay, S., Uysal, A., Stripe, B., Ha, Y., Marks, T.J., Karapetrova, E.A., and Dutta, P. 2010. How water meets a very hydrophobic surface. *Phys. Rev. Lett.* 105:037803.
14. Chen, Y., Zhang, Y., Shi, L., Li, J., Xin, J., Yang, T., and Guo, Z. 2012. Transparent superhydrophobic/superhydrophilic coatings for self-cleaning and anti-fogging. *Appl. Phys. Lett.* 101:033701.

15. Crovetto, R., Fernández-Prini, R., and Japas, M. L. 1982. Solubilities of inert gases and methane in H_2O and in D_2O in the temperature range of 300 to 600 K. *J. Chem. Phys.* 76:1077–1086.

16. de Gennes, P.G. 1985. Wetting: Statics and dynamics. *Rev. Mod. Phys.* 57:827–863.

17. de Gennes, P.G. 1992. Soft matter. *Rev. Mod. Phys.* 64:645–648.

18. Ensikat, H.J., Ditsche-Kuru, P., Neinhuis, C., and Barthlott, W. 2011. Superhydrophobicity in perfection: The outstanding properties of the lotus leaf. *Beilstein J. Nanotechnol.* 2:152–161.

19. Erbil, H.Y. and Cansoy, C.E. 2009. Range of applicability of the Wenzel and Cassie–Baxter equations for superhydrophobic surfaces. *Langmuir* 25:14135–14145.

20. Erhardt, R., Böker, A., Zettl, H., Kaya, H., Pyckhout-Hintzen, W., Krausch, G. et al. 2001. Janus micelles. *Macromolecules* 34:1069–1075.

21. Feng, X., Feng, L., Jin, M., Zhai, J., Jiang, L., and Zhu, D. 2004. Reversible superhydrophobicity to superhydrophilicity transition of aligned ZnO nanorod films. *J. Am. Chem. Soc.* 126:62–63.

22. Feng, X., Zhai, J., and Jiang, L. 2005. The fabrication and switchable superhydrophobicity of TiO_2 nanorod films. *Angew. Chem. Int. Ed.* 44:5115–5118.

23. Frank, H.S. and Evans, M.W. 1945. Free volume and entropy in condensed systems III. Entropy in binary liquid mixtures; partial molal entropy in dilute solutions; structure and thermodynamics in aqueous electrolytes. *J. Chem. Phys.* 13:507–532.

24. Gavish, M., Wang, J.L., Eisenstein, M., Lahav, M., and Leiserowitz, L. 1992. *Science* 256:815–818.

25. Giovambattista, N., Debenedetti, P.G., and Rossky, P.J. 2007. Effect of surface polarity on water contact angle and interfacial hydration structure. *J. Phys. Chem. B* 111:9581–9587.

26. Granick, S., Jiang, S., and Chen, Q. 2009. Janus particles. *Physics Today* 62:68–69.

27. Henderson, J.R. and Lekner, J. 1978. Surface oscillations and the surface thickness of classical and quantum droplets. *Mol. Phys.* 36:781–789.

28. Hong, L., Cacciuto, A., Luijten, E., and Granick, S. 2006. Clusters of charged Janus spheres. *Nano Lett.* 6:2510–2514.

29. Hosono, E., Fujihara, S., Honma, I., and Zhou, H. 2005. Superhydrophobic perpendicular nanopin film by the bottom-up process. *J. Am. Chem. Soc.* 127:13458–13459.

30. Jensen, M. Ø., Mouritsen, O.G., and Peters, G.H. 2004. The hydrophobic effect: Molecular dynamics simulations of water confined between extended hydrophobic and hydrophilic surfaces. *J. Chem. Phys.* 120:9729–9744.

31. Jensen, T.R., Østergaard Jensen, M., Reitzel, N., Balashev, K., Peters, G.H., Kjaer, K., and Bjørnholm, T. 2003. Water in contact with extended hydrophobic surfaces: Direct evidence of weak dewetting. *Phys. Rev. Lett.* 90:086101.

32. Jeong, H.E., Kwak, M.K., Park, C.I., and Suh, K.Y. 2009. Wettability of nanoengineered dual-roughness surfaces fabricated by UV-assisted capillary force lithography. *J. Colloid Inter. Sci.* 339:202–207.

33. Jiang, S. and Granick, S. 2008. Controlling the geometry (Janus balance) of amphiphilic colloidal particles. *Langmuir* 24:2438–2445.

34. Jung, Y.C. and Bhushan, B. 2008. Dynamic effects of bouncing water droplets on superhydrophobic surfaces. *Langmuir* 24:6262–6269.

35. Karplus, P.A. 1997. Hydrophobicity regained. *Protein Sci.* 6:1302–1307.

36. Kirchner, M.T., Boese, R., Billups, W.E., and Norman, L.R. 2004. Gas hydrate single-crystal structure analyses. *J. Am. Chem. Soc.* 126:9407–9412.

37. Koch, K., Bohn, H.F., and Barthlott, W. 2009. Hierarchical sculptured plant surfaces and superhydrophobicity. *Langmuir* 25:14116–14120.

38. Krupenkin, T.N., Taylor, J.A., Schneider, T.M., and Yang, S. 2004. From rolling ball to complete wetting: the dynamic tuning of liquids on nanostructured surfaces. *Langmuir* 20:3824–3827.
39. Lahann, J., Mitragotri, S., Tran, T., Kaido, H., Sundaram, J., Choi, I.S. et al.. 2003. A reversibly switching surface. *Science* 299:371–374.
40. Lattuada, M. and Hatton, T.A. 2011. Synthesis, properties and applications of Janus particles. *Nano Today* 6:286–308.
41. Lee, S., Kim, W., and Yong, K. 2011. Overcoming the water vulnerability of electronic devices: A highly water-resistant ZnO nanodevice with multifunctionality. *Ad. Mater.* 23:4398–4402.
42. Lehmann, M.S., Koetzle, T.F., and Hamilton, W.C. 1972. Precision neutron diffraction structure determination of protein and nucleic acid components. I. Crystal and molecular structure of the amino acid L-alanine. *J. Am. Chem. Soc.* 94:2657–2660.
43. Lim, H.S., Kwak, D., Lee, D.Y., Lee, S.G., and Cho, K. 2007. UV-driven reversible switching of a roselike vanadium oxide film between superhydrophobicity and super-hydrophilicity. *J. Am. Chem. Soc.* 129:4128–4129.
44. Lippmann, G. 1875. Relations entre les phénomènes électriques et capillaires. *Ann. Chim. Phys.* 5:494-496.
45. Lum, K., Chandler, D., and Weeks, J.D. 1999. Hydrophobicity at small and large length scales. *J. Phys. Chem. B* 103:4570–4577.
46. Maccarini, M., Steitz, R., Himmelhaus, M., Fick, J., Tatur, S., Wolff, M. et al.. 2007. Density depletion at solid-liquid interfaces: A neutron reflectivity study. *Langmuir* 23:598–608.
47. Macías-Montero, M., Borrás, A., Álvarez, R., and Gónzalez-Elipe, A.R. 2012a. Following the wetting of 1D photoactive surfaces. *Langmuir* 28:15047–15055.
48. Macías-Montero, M., Borrás, A., Saghi, Z., Romero-Gómez, P., Sánchez-Valencia, J.R., González, J.C. et al. 2012b. Superhydrophobic supported Ag-NPs@ZnO-nanorods with photoactivity in the visible range. *J. Mater. Chem.* 22:1341–1346.
49. Mehlhorn, M. and Morgenstern, M. 2007. Faceting during the transformation of amorphous to crystalline ice. *Phys. Rev. Lett.* 99:246101.
50. Mezger, M., Sedlmeier, F., Horinek, D., Reichert, H., Pontoni, D., and Dosch, H. 2010. On the origin of the hydrophobic water gap: An X-ray reflectivity and MD simulation study. *J. Am. Chem. Soc.* 132:6735–6741.
51. Mittal, J. and Hummer, G. 2008. Static and dynamic correlations in water at hydrophobic interfaces. *Proc. Natl. Acad. Sci. USA* 105:20130–20135.
52. Mugele, F. and Baret, J.C. 2005. Electrowetting: From basics to applications. *J. Phys. Condens. Matter* 17:R705–R774.
53. Neinhuis, C. and Barthlott, W. 1997. Characterization and distribution of water-repellent, self-cleaning plant surfaces. *Ann. Bot.* 79:667–677.
54. Nishino, T., Meguro, M., Nakamae, K., Matsushita, M., and Ueda, T. 1999. The lowest surface free energy based on -CF$_3$ alignment. *Langmuir* 15:4321–4323.
55. Nosonovsky, M. and Bhushan, B. 2009. Superhydrophobic surfaces and emerging applications: Non-adhesion, energy, green engineering. *Cur. Opinion Colloid Interface Sci.* 14:270–280.
56. Onda, T., Shibuichi, S., Satoh, N., and Tsujii, K. 1995. Super-water-repellent fractal surfaces. *Langmuir* 12:2125–2127.
57. Palacios-Lidón, E., Munuera, C., Ocal, C., and Colchero, J. 2010. Contrast inversion in non-contact dynamic scanning force microscopy: What is high and what is low? *Ultramicroscopy* 110:789–800.
58. Poynor, A., Hong, L., Robinson, I.K., Granick, S., Zhang, Z., and Fenter, P.A. 2006. How water meets a hydrophobic surface. *Phys. Rev. Lett.* 97:266101.

59. Richard, D., Clanet, C., and Quéré, D. 2002. Contact time of a bouncing drop. *Nature* 417:811.

60. Roach, P., Shirtcliffe, N.J., and Newton, M.I. 2008. Progress in superhydrophobic surface development. *Soft Matter* 4:224–240.

61. Sakai, M., Kono, H., Nakajima, A., Zhang, X., Sakai, H., Abe, M., and Fujishima, A. 2009. Sliding of water droplets on the superhydrophobic surface with ZnO nanorods. *Langmuir* 25:14182–14186.

62. Santos, S., Verdaguer, A., Souier, T., Thomson, N.H., and Chiesa, M. 2011. Measuring the true height of water films on surfaces. *Nanotechnology* 22:465705.

63. Schreiber, F. 2004. Self-assembled monolayers: From simple model systems to biofunctionalized interfaces. *J. Phys. Condens. Matter* 16:R881–R900.

64. Segura, J.J., Verdaguer, A., Cobián, M., Hernández, E.R., and Fraxedas, J. 2009. Amphiphillic organic crystals. *J. Am. Chem. Soc.* 131:17853–17859.

65. Segura, J.J., Verdaguer, A., Garzón, L, Barrena, E., Ocal, C., and Fraxedas, J. 2011a. Strong water-mediated friction asymmetry and surface dynamics of zwitterionic solids at ambient conditions: L-alanine as a case study. *J. Chem. Phys.* 134:124705.

66. Segura, J.J., Verdaguer, A., Sacha, G.M., and Fraxedas, J. 2011b. Dipolar origin of water etching of amino acid surfaces. *Phys. Chem. Chem. Phys.* 13:21446–21450.

67. Shimon, L.J.W., Lahav, M., and Leiserowitz, L. 1985. Design of stereoselective etchants for organic crystals. Application for the sorting of enantiomorphs and direct assignment of absolute configuration of chiral molecules. *J. Am. Chem. Soc.* 107:3375–3377.

68. Simão, C., Mas-Torrent, M., Veciana, J., and Rovira, C. 2011. Multichannel molecular switch with a surface-confined electroactive radical exhibiting tunable wetting properties. *Nano Lett.* 11:4382–4385.

69. Sloan, E.D. and Koh, C. 2007. *Clathrate Hydrates of Natural Gases*. Boca Raton, FL: Taylor & Francis.

70. Stillinger, F.H. 1973. Structure in aqueous solutions of nonpolar solutes from the standpoint of scaled-particle theory. *J. Solution Chem.* 2:141–158.

71. Sun, T., Wang, G., Feng, L., Liu, B., Ma, Y., Jiang, L., and Zhu, D. 2004. Reversible switching between superhydrophilicity and superhydrophobicity. *Angew. Chem. Int. Ed.* 43:357–360.

72. Synytska, A., Khanum, R., Ionov, L., Cherif, C., and Bellmann, C. 2011. Water-repellent textile via decorating fibers with amphiphilic Janus particles. *ACS Appl. Mater. Interfaces* 3:1216–1220.

73. Tang, Q.L., and Chen, Z.X. 2007. Density functional slab model studies of water adsorption on flat and stepped Cu surfaces. *Surf. Sci.* 601:954–964.

74. Torii, K. and Iitaka, Y. 1970. The crystal structure of L-valine. *Acta Cryst. B* 26:1317–1326.

75. Ulman, A. 1996. Formation and structure of self-assembled monolayers. *Chem. Rev.* 96:1533–1554.

76. Verdaguer, A., Santos, S., Sauthier, G., Segura, J.J., Chiesa, M., and Fraxedas, J. 2012. Water-mediated height artifacts in dynamic atomic force microscopy. *Phys. Chem. Chem. Phys.* 14:16080–16087.

77. Vinogradova, O.I., and Dubov, A.L. 2012. Superhydrophobic textures for microfluidics. *Mendeleev Commun.* 22:229–236.

78. Walther, A. and Müller, A.H.E. 2008. Janus particles. *Soft Matter* 4:663–668.

79. Wang, R., Hashimoto, K., Fujishima, A., Chikuni, M., Kojima, E., Kitamura, A. et al. 1997. Light-induced amphiphilic surfaces. *Nature* 388:431–432.

80. Weissbuch, I., Leiserowitz, L., and Lahav, M. 2008. Direct assignment of the absolute configuration of molecules from crystal morphology. *Chirality* 20:736–748.

81. Wenzel, R.N. 1936. Resistance of solid surfaces to wetting by water. *Indust. Eng. Chem.* 28:988–994.

82. Wöll, C. 2007. The chemistry and physics of zinc oxide surfaces. *Prog. Surf. Sci.* 82:55–120.

83. Yamamoto, S., Andersson, K., Bluhm, H., Ketteler, G., Starr, D.E., Schiros, T. et al. 2007. Hydroxyl–induced wetting of metals by water at near–ambient conditions. *J. Phys. Chem. C* 111:7848–7850.

84. Yan, Y.Y., Gao, N., and Barthlott, W. 2011. Mimicking natural superhydrophobic surfaces and grasping the wetting process: A review on recent progress in preparing superhydrophobic surfaces. *Adv. Colloid Interface Sci.* 169:80–105.

85. Yeh, K.-Y., Cho, K.-H., and Chen, L.-J. 2009. Preparation of superhydrophobic surfaces of hierarchical structure of hybrid from nanoparticles and regular pillar-like pattern. *Langmuir* 25:14187–14194.

86. Young, T. 1805. An essay on the cohesion of fluids. *Philos. Trans. R. Soc. London* 95:65–87.

87. Yu, H., Chen, M., Rice, P.M., Wang, S.X., White, R.L., and Sun, S. 2005. Dumbbell-like bifunctional Au-Fe$_3$O$_4$ nanoparticles. *Nano Lett.* 5:379–382.

88. Zhang, J., Lu, Z., Huang, W., and Han, Y. 2005. Reversible superhydrophobicity to superhydrophilicity transition by extending and unloading an elastic polyamide film. *Macromol. Rapid Commun.* 26:477–480.

89. Zhang, X., Jin, M., Liu, Z., Tryk, D.A., Nishimoto, S., Murakami, T., and Fujishima, A. 2007. Superhydrophobic TiO$_2$ surfaces: Preparation, photocatalytic wettability conversion, and superhydrophobic-superhydrophilic patterning. *J. Phys. Chem. C* 111:14521–14529.

90. Zhu, W., Feng, X., Feng, L., and Jiang, L. 2006. UV-manipulated wettability between superhydrophobicity and superhydrophilicity on a transparent and conductive SnO$_2$ nanorod film. *Chem. Commun.* 2753–2755.

5 Water on Real Solid Surfaces

We live, I regret to say, in an age of surfaces.
Oscar Wilde, The Importance of Being Earnest

Perhaps a more appropriate title for this chapter would have been *Real Water on Real Solid Surfaces* or, alternatively *Real (Water on Solid Surfaces)* because here we are no longer involved with pure, pristine, contamination-free water but instead with the water we are used to and on which we depend. On the other hand we leave the ideal, homogeneous, flat, defect-free, perfectly ordered surfaces in well-defined conditions and enter the realm of actual, practical (also termed technical) surfaces. The following sections give a glimpse of selected examples where water/solid interfaces play a key role in real working conditions, for example, when they are exposed to ambient conditions. Such selection aims to cover a wide range of important phenomena and among the plethora of existing examples only a few of them are considered. We show how water can be purified and used as a source of energy by the action of photoactive surfaces as well as the detrimental action of corrosion and degradation caused by heterogeneous chemistry induced by water. Intentional condensation high in the troposphere and the adhesive properties of liquid water are also discussed and we finish with a curious case of water at solid/vacuum interfaces.

5.1 WATER PURIFICATION

Clean water is life. As already discussed in Section 1.5, the ever-growing demand for fresh water due to the continuous increase in population and in human activity, both at the industrial and agricultural levels, calls for a strict quality control and reuse of this ubiquitous precious liquid, keeping in mind that the resources of our planet are limited. Water follows a continuous movement above and below the surface of the Earth, the so-called hydrologic or water cycle, involving evaporation, condensation, precipitation, infiltration, runoff, and subsurface flow. A scheme is represented in Figure 5.1. Surface water includes streams, lakes, wetlands, bays, and oceans as well as snow and ice whereas subsurface water primarily includes groundwater that emerges to the surface, for example, as springs and wells. The water cycle referred to may be visualized as a quite simple process, at least as a first approximation, but in reality it can be rather complex because of the host of variables that can contribute differently such as local climate conditions, orography, geology (chemical composition and structure of the solid mantle), proximity of seawater, modifications due to natural phenomena, anthropogenic actions, and many others.

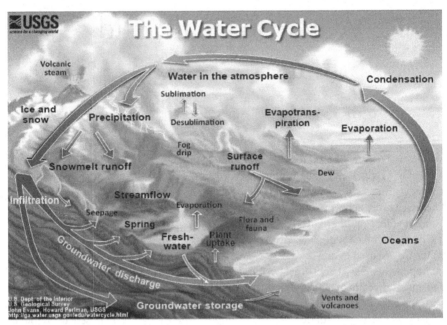

FIGURE 5.1 Scheme of the water cycle. (Courtesy of the US Geological Survey organization (http://www.usgs.gov/). With permission.)

The impact of such natural and human-related processes results in a large list of contaminants that can be found in both surface and groundwaters. Bacteria, viruses, and protozoa belong to the microbial contribution and have different influences in the population and in some cases can even lead to death. Diseases such as typhoid fever, cholera, diarrhea, and dysentery are caused by bacteria. Chemical contamination is achieved by a long list of products, both inorganic and organic. We just mention here toxic metals, nitrates (from agriculture transfered to groundwater aquifers), chloroform, gasoline, pesticides and herbicides from a variety of industrial and agricultural applications, and chlorinated hydrocarbons as well as new types of contaminants such as pharmaceutical products, steroids, hormones, and industrial additives. One has to add radioactive contamination to this portfolio, essentially originated in nuclear plants (e.g., cooling water and accidents). Just remember the devastating effects of the Fukushima nuclear plant in Japan as a result of the 2011 tsunami at the Tohoku coast, which affected hundreds of thousands of residents. Such a long and varied but incomplete list of components may discourage attempts to eliminate them but as we show next there are several strategies that successfully purify water although research is cannot be stopped because we are continuously exposed to different threats. General and updated information, including technical and political aspects, can be found on the UN website http://www.unwater.org/.

The purification of water is usually based in multistage sequential processes, each process targeting different physico(bio)chemical aspects. Here we give only a short summary of this subject and those interested to learn more can browse through the specialized literature (Parsons and Jefferson 2006; Shannon et al. 2008). The conventional technologies for surface water treatment are aggregation, sedimentation, filtration, and disinfection. The eutectic freeze crystallization technique mentioned in Section 3.5.1 could also be used for wastewater purification although it is not indicated for large volumes of water. Aggregation can be achieved by chemical and physical means, leading to coagulation and flocculation, respectively. Coagulation is based on the addition of chemicals to water with the goal of forming sufficiently large particles from smaller ones facilitating their removal. This is a safe and effective strategy that improves the quality of water by reducing the amount of organic compounds, iron and manganese, and suspended particles. Larger particles can also be obtained by physical processes such as flocculation. In this case collisions among particles cause the formation of flocs. After agglomeration the larger aggregates are removed by sedimentation, which uses the gravitation force field. After aggregation/sedimentation the treated water undergoes filtration. This is a physical way of removal of particles by controlling the pore diameter. The most common filters are of dual type, where water flows by gravity through a porous bed of two layers of granular media. Sand filters use a top layer of anthracite coal and a bottom layer of sand. A different kind of filter uses membranes with pores that can be tuned to diameters below the micron range. Ions and small organic molecules cannot be removed with such pore sizes, so other strategies have to be found. One of them was already discussed in Section 3.4.2 with the case of filters based on arrays of CNTs that can discriminate objects below the 1-nm range, the pore size defined by the diameters of the nanotubes (see Figure 3.30). Zeolites are also efficient filters used as selective sorbents in water treatment (Perego et al. 2013). They are porous crystalline aluminosilicates characterized by the presence of regular channels or cages with free dimensions in the 0.3–1.0 nm range (just recall that the mean diameter of the water molecule is about 0.3 nm). The pore diameters can be engineered by chemical synthesis so that particular zeolites can be designed for specific applications (Corma et al. 2004). Another example is reverse osmosis based on the application of a pressure above the osmotic pressure to a solution on one side of a membrane. The combination of pressure and pore size leads to the concentration of solutes on the pressurized side of the membrane and to purified water on the other side. Finally, disinfection is obtained using different chemicals such as chlorine, chlorine-derivatives, and ozone.

But here we are interested in those water treatment processes where surfaces play a key active role (Pan et al. 2010). Perhaps the most important one uses the photocatalytic action of a particular set of semiconductors illuminated with UV and/or visible light that are able to decompose organic molecules, for example, into carbon dioxide and water. Upon irradiation with photon energies above the semiconductor bandgap electron (e^-)-hole (h^+) pairs (excitons) are generated. A few channels are possible for those pairs: (i) recombination with energy dissipation within a few ns, (ii) trapping in metastable surface states, or (iii) reaction with electron donors and electron acceptors adsorbed on the semiconductor surface (Hoffmann et al. 1995). An energy diagram of the processes involved is illustrated in Figure 5.2a. Trapping prevents the recombination of excitons opening the door to redox reactions. Valence band holes are powerful oxidants and conduction band electrons are good reductants that efficiently photodegrade organic molecules. Among the chosen semiconductors with photocatalytic activity the most extensively studied and used is TiO_2 (Fujishima, Zhang, and Tryk 2008; Henderson 2011), already studied in Chapters 3

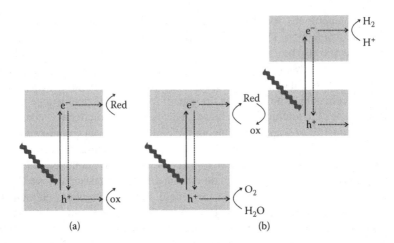

FIGURE 5.2 Schemes of energy diagrams for (a) one-step and (b) two-step photon-induced catalytic processes. A photon generates an electron-hole pair, that may recombine or may diffuse to the surface inducing redox reactions. Water splitting is exemplified in (b).

and 4. TiO$_2$-based materials are nontoxic, sufficiently abundant, chemically inert, photostable, biocompatible, and relatively cheap, properties that make them advantaged candidates from the application point of view in different disciplines (Carp, Huisman, and Reller 2004). However, a clear disadvantage is that titanium oxide, also known as titania, absorbs in the UV region but very poorly in the visible light region, which constitutes 45% of solar energy. In order to extend the absorption range of TiO$_2$ in the visible region the intrinsic bandgap has to be reduced by introducing electronic states in the bulk forbidden energy region. This can be achieved by modifying the surface stoichiometry through doping and/or introducing defects (e.g., vacancies).

The most commonly used dopant is nitrogen (Asahi et al. 2001). N-doping can be achieved, for example, by sputtering, ion beam-assisted deposition, CVD, or laser ablation, among others, using a variety of nitrogen sources such as amines, urea, nitrogen gas, or ammonia. The resulting materials absorb in the visible region and the photocatalytic activity is usually evaluated by measuring decomposition rates of methylene blue as well as other organic compounds such as acetaldehyde. The chemical nature of the nitrogen species introduced at the surface can be best explored with XPS. Figures 5.3a and b show the N1s and Ti2p photoemission lines, respectively, of nanoporous titania films as a function of the nitriding temperature using ammonia (Martínez-Ferrero et al. 2007). The bottom spectrum of Figure 5.3a corresponds to a reference undoped TiO$_2$ sample. A peak at 400.0 eV and a shoulder at ~402 eV can be observed, which correspond to molecularly chemisorbed nitrogen (γ-N$_2$; Asahi et al. 2001). Upon annealing at 500°C a peak at 395.9 eV appears in addition to the 400 eV and 402 eV contribution, which is assigned to atomic nitrogen (β-N) and becomes more prominent with increasing annealing temperature, clearly indicating TiN bonding. At 600°C a peak at 401.9 eV can be seen that is no longer observed at 700°C. Concerning the Ti2p lines shown in Figure 5.3b the bottom spectrum corresponds to the reference sample and exhibits the well-known spin-orbit split lines Ti

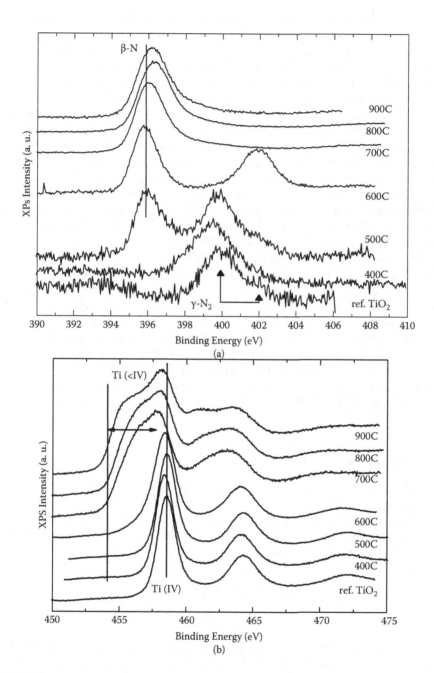

FIGURE 5.3 RT XPS spectra as a function of the annealing temperature of nanostructured titanium oxynitride porous thin films. (a) N1s and (b) Ti2p lines. The spectra have been normalized to their maxima. (Reproduced from E. Martínez-Ferrero, et al. *Adv. Funct. Mater.* 17: 3348–3354, 2007. With permission.)

$2p_{3/2}$ and Ti $2p_{1/2}$ at 458.5 and 464.3 eV, respectively, the energies characteristic of the Ti^{+4} oxidation state. Upon annealing at 600°C both lines become broader on the low-energy side, the asymmetry becoming more obvious for higher annealing temperatures together with a shift of the main peak toward lower energies. This behavior is a signature of the increasing presence of intermediate lower oxidation states.

The generation of oxygen vacancies is also a way to introduce electronic states in the bulk gap, as discussed in Section 3.2.3 (see Figure 3.23 for XPS of TiO$_2$(110) when exposed to water vapor). Substoichiometric TiO$_x$ films, with x ~ 1.8, grown by metal organic CVD, have shown enhanced photocatalytic properties in the visible range as a result of the combination of reduction of the effective gap and surface roughness. Complementary DFT calculations clearly show the generation of electronic states in the forbidden bandgap caused by removal of oxygen atoms (Justícia et al. 2002). The EELS results shown in Figure 5.4 performed on clean and water-covered TiO$_2$(110) surfaces with 0.14 ML of oxygen vacancies confirm such predictions with the observation of valence band-LUMO transitions (Henderson et al. 2003).

The clean surface (Figure 5.4a) exhibits a ~0.9 eV feature associated with localized electrons at defects, the ~3 eV electronic gap and transitions at ~4 (shoulder) and 5.2 eV, respectively. Upon deposition of a thick ice film on the surface (Figure 5.4b) the water bandgap threshold at ~7.2 eV and the HOMO–LUMO transition at ~8.3 eV are observed. Both reference spectra help in the identification of features at lower water coverages. At 1 ML (Figure 5.4c) the defect feature is relatively unaffected by water indicating little charge transfer from Ti^{+3} to adsorbed water and a strong feature at 6.2 eV is observed, which disappears after thermal annealing (see Figures 5.4d and e). This 6.2 eV peak has been assigned to the excitation of valence band electrons to $4a_1$-like states of the adsorbed water, placing the LUMO ~1.2 eV above E_F.

5.2　WATER SPLITTING

Thus far we have seen that the photocatalytic activity of surfaces of given materials is able to degrade molecules and such phenomenology also concerns the water molecule itself that can be split into molecular hydrogen and oxygen. O$_2$ and H$_2$ are produced through valence band holes and conduction band electrons reaching the water/solid interface following the schemes:

$$2h^+ + H_2O \rightarrow \frac{1}{2}O_2 + 2H^+ \tag{5.1a}$$

$$2e^- + 2H^+ \rightarrow H_2 \tag{5.1b}$$

Water is transparent to visible light, so that it cannot be decomposed directly by irradiation unless photons with energies above 6.5 eV are used (see previous section and remember that the separation between the $1b_1$ and $4a_1$ MOs is about 7 eV for the isolated water molecules as depicted from Figure 1.6). Hence a catalyst, in this case a photoactive semiconducting surface, is needed. Fujishima and Honda (1972) were the first to demonstrate experimentally the photolysis of water by electrochemical means using a rutile n-type TiO$_2$(001) surface. A scheme of the cell is shown in Figure 5.5.

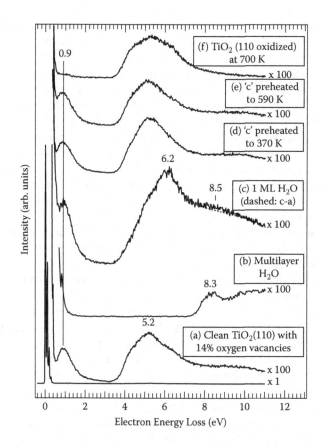

FIGURE 5.4 EELS spectra taken at 120 K of a TiO$_2$(110) surface with 0.14 ML oxygen vacancy sites: (a) clean surface and covered with (b) H$_2$O multilayers, (c) 1 ML H$_2$O, (d) 1 ML H$_2$O after heating to 370 K, (e) 1 ML H$_2$O after preheating to 590 K, and (f) 1 ML H$_2$O after exposure to O$_2$ at 700 K. (Reprinted from M.A. Henderson, et al. *J. Phys. Chem. B*107:534–545, 2003, American Chemical Society. With permission.)

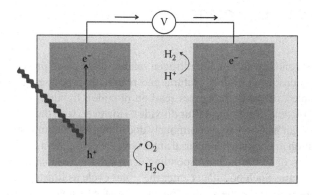

FIGURE 5.5 Scheme of a Fujishima–Honda cell for water splitting using TiO$_2$ (left) and counter (right) electrodes. (Adapted from A. Fujishima and K. Honda, *Science* 238: 37–38, 1972. With permission.)

With such an electrochemical cell, electrical current is generated when the TiO_2 sample is irradiated with photons with energies >3.0 eV. The publication year of this work drives us to an historical note. In 1973 a military conflict in the Middle East engendered a profound worldwide oil crisis due to the embargo from oil producers in the region. This critical situation alerted us to the urgency to seek alternative energy sources to overcome the oil monopoly. The production of hydrogen is of extreme importance because it is a clean source of energy and it can be used in fuel cells to generate electricity (Grätzel 2001). Hydrogen is nowadays mainly produced from fossil fuels such as natural gas by steam reforming where CO_2 is emitted. Thus, it becomes mandatory to produce hydrogen from water using natural sources of energy such as sunlight in order to supply environment-friendly fuel taking advantage of the colossal energy supply from the sun ($\sim 3 \times 10^{24}$ J year^{-1}). This is a step toward a sustainable energy economy (Lewis and Nocera 2006; Edwards, Kuznetzov, and David 2007) and an example of the successful marriage between water and surfaces.

As mentioned above, one disadvantage of TiO_2 is that it absorbs weakly in the visible part of the radiation spectrum, hence the importance of doping in order to reduce its absorption energy gap. Another strategy consists of combining the active semiconducting material with additional materials thus building heterogeneous photocatalysts with a two-step excitation mechanism (Esswein and Nocera 2007; Kudo and Miseki 2009; Maeda and Domen 2010). In this case oxidation takes place with one of the materials and reduction with the other, so that it is the sum of two specialized photocatalysts that guarantees better efficiency, an efficiency that can only be achieved if the electron transfer between materials is efficient. This is illustrated in Figure 5.2b and the process has been in fact inspired by natural photosynthesis in green plants and is called the Z-scheme. Water is split in the photosynthesis process, with the liberation of molecular oxygen and the generation of NADPH (the reduced form of nicotinamide adenine dinucleotide phosphate) used in the fixation of carbon dioxide. High activity and reasonable reaction rates are obtained when oxides such as TiO_2 (anatase), $SrTiO_3$, tantalates, and so on are combined with cocatalysts (Rh, Pt, NiO_x, RuO_2, etc.).

5.3 ATMOSPHERIC AGENTS

ACID RAIN

Water dissolves carbon, sulfur, and nitrogen oxides, three gases present in the atmosphere that are produced both by natural and manmade processes. Natural sources are, for example, metabolic processes (carbon dioxide), lightning strikes (nitrogen oxides), and volcanic eruptions (sulfur dioxide). However, with the industrial revolution the gas budget has increased enormously due to fossil fuel combustion, leading to harmful pollution in populated industrial areas with the consequent public health alert. Under the action of water the corresponding acids are formed that reach the Earth's surface as acid rain. Such pollutants enter the water cycle as illustrated in Figure 5.1 and deleteriously increase the acidic content of surface water, with detrimental effects on living organisms. In addition, acid rain leads to corrosion and degradation of materials. There is a plethora of examples where corrosion is of great relevance at the industrial, infrastructural, and personal levels including bridges, pipelines, water and energy supply, hazardous materials storage, railroads, cars, ships, aircraft, electronics, home appliances, and many others. Think also of the detrimental effects on buildings (degradation of concrete, etc.) as well as on monuments and cultural heritage in general (Van Grieken, Delalieux, and Gysels 1998). Marble is a known example of a material prone to degradation under wet conditions because it is composed of calcite, which is water-soluble. The economic impact is enormous, estimated to be about 3% of the gross national product of the USA.

Gas dissolution in water can be achieved in the absence of solid surfaces but their presence enhances the effect, especially in polluted urban regions. Any surfaces exposed to moist air, and in particular real surfaces (metals, glasses, ceramics, minerals, polymers, etc.) adsorb sufficient water (from nanometer-thick films to droplets) because of their morphology, roughness, chemical composition, defect types and density, and so on, triggering a long list of heterogeneous chemical reactions, indeed including formation of acids (Sumner et al. 2004). However, when interfacial water dissociates it generates hydroxides and oxides. This is in fact a natural way of protection called passivation, which is a barrier to the formation of further reactions. However, under high RH and temperatures and large concentrations of pollutants the oxidation process is accelerated. In addition, corrosion may penetrate the bulk of the material through defects, accelerating the process.

AEROSOLS

In addition to large surfaces, aerosols also play a critical role in atmospheric chemistry. Aerosols are natural or anthropogenic particles suspended in the atmosphere. These particles are either solid (soot, dust, salt, and ice particles) or liquid (water droplets and aqueous marine aerosols), with sizes typically above 100 nm (Finlayson-Pitts and Pitts 2000). Aerosols affect the concentration and size distribution of cloud droplets, which in turn can alter the radiating properties of clouds. The role of aerosols in cloud processes also affects the nature and distribution of rainfall and the subsequent distribution of clouds (we show examples in the next section). Marine aerosols, micron- and submicron-sized water droplets containing sea salts, have a detrimental effect on materials as people living in coastal areas know well (O'Dowd and De Leeuw 2007). Figure 5.6 shows the degradation of the Roman wall of the city of Tarragona, on the Catalonian coast, due to combined effect of the proximity of the Mediterranean sea and of the nearby chemical industry. The accepted mechanism by which marine aerosols enhance acid rain is the surface enhancement of large inorganic anions such as bromine, already discussed in Section 3.5.1 (Finlayson-Pitts 2003).

Coming back to the TiO_2 wonder material, it may help in daytime environmental remediation due to its photocatalytic properties discussed in Section 5.1, provided sufficient active particles are in the atmosphere. The main source of TiO_2 particles suspended in the atmosphere is mineral dust. There are additional sources of airborne particulate matter containing TiO_2 from industrial processes, including the nanotechnology industry, food, and personal care products that could contribute significantly to TiO_2 in the atmosphere. At the ground level, in particular in urban areas, the density and location can be controlled by adding TiO_2 or applying TiO_2 coatings (paints)

FIGURE 5.6 Example of the detrimental effect of marine aerosols on cultural heritage in coastal areas: Roman wall in the city of Tarragona (Catalonia).

to conventional building materials (windows, cement, etc.). CO_2, NO_2, and SO_2 are adsorbed on TiO_2 particles forming carbonates, nitrates, and sulfates with different efficiencies as a function, for example, of RH, with surface hydroxyl groups playing relevant roles (Chen, Nanayakkara, and Grassian 2012). HONO gas is a source of reactive hydroxyl radicals that can be obtained through the TiO_2-mediated reaction:

$$NO_2 + H_2O \rightarrow HONO + HNO_3. \qquad (5.2)$$

CLOUD SEEDING

In Chapter 3 we saw how certain ideal surfaces (of single crystals) were able to nucleate and structure water in the form of ice or other still unknown configurations (solidlike, amorphous, etc.) in controlled conditions depending on external parameters such as temperature or intrinsic parameters such as lattice matching and the nature of surface defects. However, when we come to real surfaces in real ambient conditions, as in tropospheric clouds, things become much more complex and interesting because such surfaces can become principal actors of atmospheric phenomena such as precipitation and can give us a key to intentional weather control (Fletcher 1969). In this case the complexity is due to the large number of variables (morphology, aerosol size, RH, temperature, mass transport, contaminants, etc.) and to the different length scales involved (from nanometers to kilometers). Then, if we are confronted with such complexity we may ask ourselves to what extent can the weather be modified in a controlled way. Let us briefly discuss some considerations on this point.

There are two primary methods employed to stimulate precipitation: hygroscopic seeding, which affects warm clouds (those that have not cooled sufficiently to allow the development of ice) and glaciogenic seeding (inducing the precipitation of ice). Both strategies can be applied from the Earth's surface (ground-based) or from an aircraft (air-based). Here the idea of different length scales emerges quite clearly when one compares the sizes of the clouds with those of the airplanes. The formation of ice in the troposphere at temperatures above $-33°C$ is induced mainly by a foreign body, the so-called ice nucleation agent or sublimation nuclei, in a process known as heterogeneous nucleation (Langmuir 1950; Birstein and Anderson 1955; Cantrell and Heymsfield 2005). A first exploratory trial in order to induce the formation of ice consists of using surfaces mimicking the hexagonal structure of the basal plane of ice Ih. Along this line of thought it was B. Vonnegut (1947) who predicted that surfaces exhibiting lattice constants close to that of ice Ih might induce the nucleation of ice in ambient conditions. The mostly studied compound fulfilling such a condition is β-AgI (see Table 3.2), although its efficiency in cloud seeding is still disputed. We saw in Chapter 3 that the ice Ih bilayer adapts almost perfectly to the basal plane of β-AgI (see Figures 3.12 and 3.13) but that the resulting covered surface was hydrophobic, according to theoretical calculations although nanoparticles would trigger ice nucleation due to defects and electrostatic field (Shevkunov 2009). When silver iodide crystals are submerged in supercooled water in a laboratory environment, then the nucleation statistics shift to higher temperatures by a few degrees C (Heneghan, Wilson, and Haymet 2002), indicating its proclivity to ice nucleation. However, in real ambient conditions the efficiency is disputed and dedicated reports from different associations in the United States diverge when drawing conclusions (see, e.g., the Critical Issues in Weather Modification research report from the National Research Council and the response from the Weather Modification Association in http://www.weathermodification.org/). What seems plausible is that cloud seeding under appropriate atmospheric conditions and when properly conducted has a positive effect on precipitation, but this is of course too ideal for a standardization of the process. There is thus a long way from laboratory-based experiments to field experiments although dedicated test chambers have been developed: cloud expansion (Möhler et al. 2006), continuous flow thermal gradient (Kanji and Abbatt 2009), and so on. One has to bear in mind that the transport and dispersion of the AgI material is done with airplanes, and that there are companies offering such a service, but the success of the missions is subject to many unknowns (nature of the clouds, temperature, cloud displacement, dispersion of material, perturbations induced by the airplane, etc.), therefore the desired effect cannot be guaranteed. In addition, the presence of contamination of anthropogenic origin, as discussed in the previous section, affects the degree of control of the weather.

Apart from silver iodide, other materials are being used such as carbon soot, and mineral dust (desert sand, kaolinite, montmorillonite, etc.), as well as bacteria and biological aerosols such as pollen. We leave the case of protein ice nucleators for Chapter 6. It turns out that dusts coming from the Sahara, Canary Island, and Israel deserts are more effective at forming ice than soot in the $253 > T > 230$ K range (Kanji et al. 2011). In the case of kaolinite, the activation temperature is about $-22°C$ at 23% RH and for montmorillonite is about $-15°C$, thus lower than for kaolinite (Salam et al. 2006). The ice crystals grown on micrometer-sized kaolinite crystals exhibit platelike habits for temperatures ranging from -18 to $-40°C$ at ice supersaturations relevant to the atmosphere (Bailey and Hallett 2002). Finally, it should be mentioned that the understanding of the complex process of nucleation in real conditions cannot be easily achieved without powerful computer modeling.

5.4   CAPILLARY ADHESION

If we consider two clean solid surfaces approaching each other in ambient conditions (imagine that one of them is fixed), capillary water bridges may form when they are sufficiently close, resulting in an attractive (adhesive) force. This was briefly discussed in Appendix B for the particular case of a sphere and a flat surface. Capillary adhesion, which adds to externally applied forces thus playing an indirect role in friction, is particularly important when the dimensions of such objects shrink because then the surface-to-volume ratio increases (in a sphere this ratio is inversely proportional to its radius) thus enhancing the leading character of the surface. In addition, a reduction in shape involves a reduction in weight so that capillary forces can counteract the effect of gravity.

Real surfaces are rough (the real surface is larger than the projected one), with a random distribution of asperities of micro- and nanometer dimensions at the microscopic scale, so that capillarity will be important at the micro- and nanoscale in the presence of water layers contributing more than vdW interactions between asperities. In the laboratory single asperities can be emulated by AFM tips, so that capillarity adhesion and friction can be characterized down to the nN level with such instruments in pseudo-real conditions (Carpick and Salmeron 1997; Jang, Schatz, and Ratner 2004). The term pseudo-real refers to the topographic limitations of AFMs to few micrometers in height.

In the case of small particles (micron- and submicron-sized) the presence of water leads to the well-known sand castle effect (Bocquet et al. 1998). We cannot build sand castles with dry sand because the adherence among particles is too low but neither can we achieve it with completely wet sand, because in this case capillary forces are no longer present (particles are embedded in water). We thus need moist sand, as all kids know. Closely related to this effect is the reduction of capillary forces when working with AFM in ambient conditions. They can either be reduced by achieving very low RH (dry) or by submerging tip and sample in water (wet) when samples have to be in contact with water. Wet granular materials have implications in many diverse fields such as the pharmaceutical, construction, and agricultural industries as well as in a number of geophysical problems.

Adhesion is particularly harmful for microelectromechanical systems (MEMS) and nanoelectromechanical systems (NEMS; Ekinci and Roukes 2005) and is one of the main causes of failure either during fabrication or operation (Zhao, Wang, and Yu 2003; Bhushan 2007). They are particularly susceptible to autoadhesion because the involved separations are small (nano/microrange), the above-mentioned large surface-to-volume ratios, and because they are highly compliant. An example is given in Figure 5.7, where a nanometer-scale polycrystalline silicon (polysilicon) cantilever adheres to the surface due to capillarity during the release of the sacrificial (silicon oxide) layer. For micromachined surfaces of polysilicon it has been shown that the adhesion energies vary from 10^{-3} to 10^2 mJ m^{-2} depending on surface roughness and RH (DelRio et al. 2007). Capillary forces are also important at technological interfaces such as heads on storage discs where they cause stiction, the static friction that has to be overcome to allow relative motion of two stationary objects in contact (see the discussion in Section 3.4.2 on friction force microscopy measurements on HOPG and the formation of ice).

FIGURE 5.7 SEM image of a nanometer-scale polysilicon cantilever adhered to the surface due to capillary forces during the release of the sacrificial layer (silicon oxide). (Courtesy of N. Barniol and F. Pérez-Murano. With permission.)

If we now slide one of the surfaces on the other while applying an external load (again in the presence of water) friction comes on the scene. A quotidian example for drivers involves car wiper blades, which typically work in three operating regimes. The less common is in dry conditions, when the rubber and glass surfaces are in relative motion in the absence of water, and the most common one is when both surfaces are lubricated by liquid water during rain or car washing. Between both regimes, in humid conditions, there is the so-called tacky regime, which is characterized by a surprisingly high friction. In this regime the rubber is pulled toward the glass by capillarity and the contact area becomes larger than for the dry case (Deleau, Mazuyer, and Koenen 2009). The applied pressure in the rubber/glass nominal contact area is very high in wiper blade applications, of the order of MPa, so that the contribution from the capillary bridges must be very large in order to account for the strong increase in friction. The height of the capillary bridges is estimated to be less than 100 nm (Persson 2008).

Capillary forces are the principal mechanism of adhesion in many insects. The leaf beetle (*Hemisphaerota cyanea*) achieves adhesion forces exceeding 100 times its body weight through the parallel action of surface tension across many micron-sized droplet contacts (Eisner and Aneshansley 2000). Flies can walk on a vertical glass window due to capillary bridges formed at the tip of many thin hairlike fibers, which cover the attachment organs of the fly. However, in the case of geckos, those little lizards running on vertical surfaces and watching us upside down on the ceiling, adherence is thought to rely on dry adhesion via vdW interactions. In fact their adhesive capabilities lie in the structure and function of their feet and in the adhesive toe pads on the underside of each digit. These pads have a hierarchical micrometer- and nanometer-size structure (recall the lotus effect in Chapter 4) and consist of a series of modified lamellae, each one covered with uniform arrays of similarly oriented hairlike bristles (setae) formed from β-keratin (Autumn and Peattie 2002).

Inspired by Mother Nature, Vogel and Steen (2010) have developed a switchable electronically controlled capillary adhesion device, where the adhesion can be externally controlled through the application of a bias voltage. The principle of operation is rather simple and a scheme is given in Figure 5.8. During the grabbing state water

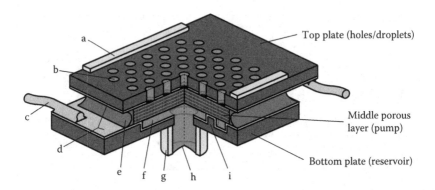

FIGURE 5.8 Scheme of a switchable electronically controlled capillary adhesion device. Main components (not to scale): (a) spacers, (b) holes from which droplets/bridges protrude, (c) wire interconnects to power supply, (d) electrodes, (e) epoxy seal, (f) fluid reservoir, (g) luer connector as reservoir continuation and filling port, (h) reservoir meniscus, and (i) support post. (Reprinted from M. J. Vogel, and P. H. Steen *Proc. Natl. Acad. Sci. USA* 107:3377–3381, 2010. With permission.)

droplets protrude from the holes until a meniscus is formed with a substrate on top (not shown in the image) leading to an adhesive force. Although the force per droplet is small, the total force can be rather large (in the mN range) if a large number of droplets are present.

Here, the presence of a spacer is crucial inasmuch as it will define the bridge length and coalescence of droplets has to be avoided. When the liquid is pumped back, the menisci are broken and adhesion is lost. In the example given here the array of droplets (10 × 10) extends barely above the top plate at 0 V bias. When a 12.5 V is pulse is applied to the pump for 2 s the droplets show up. Grab-and-release is activated by a pump driven by electo-osmosis within a liquid-saturated porous material.

5.5 WATER DESORPTION IN VACUUM SYSTEMS

When UHV is needed because surfaces have to remain truly uncontaminated (atomically clean) for long periods of time (several hours), vacuum vessels are required (on Earth) combining sufficient pumping speed, low outgassing, and absence of leaks. UHV, with base pressures in the low 10^{-10} mbar range, is mandatory for many surface science experiments and in particle accelerators. In Space there is no need for vessels, because the pumping speed is infinite, the main problem being irradiation. Vacuum varies from $\sim 10^{-7}$ mbar in low-earth orbit space to $\sim 10^{-15}$ mbar halfway to the moon. Back on Earth, low outgassing can be achieved by using materials with low vapor pressure at RT, such as metals and alloys (aluminum, copper, stainless steel, etc.). The chamber thickness has to be sufficient in order to be mechanically robust (it works against atmospheric pressure) and to reduce the diffusion of atmospheric hydrogen strongly. There are several strategies to clean such surfaces after mechanization before pumping (using solvents and detergents, baking, electropolishing, coating, etc.) and once in vacuum the rest gas is composed mainly of H_2, H_2O, CO,

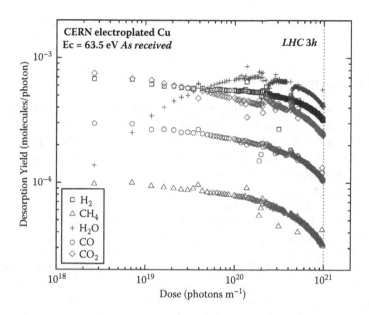

FIGURE 5.9 Desorption yields of H_2, CH_4, H_2O, CO, and CO_2 on accumulated photon dose on a copper-plated stainless steel tubular chamber. (Reprinted from J. Gómez-Goñi. *J. Vac. Sci. Technol. A* 25:1251–1255, 2007, American Vacuum Society. With permission.)

and CO_2. This outgassing is dominated by the desorption of adsorbed gas molecules from the inner surfaces (Redhead, Hobson, and Kornelsen 1993). Externally heating the system above 130°C (bakeout) significantly removes such gases and in conventional systems the base static vacuum is dominated by H_2 in the low 10^{-10} mbar pressure. Thus, interfacial water, which was adsorbed on the surface upon exposure to the atmosphere, can be easily removed but because technical surfaces are covered by oxide layers and hydrogen is present, a reservoir with the required elements is at hand to synthesize water on such real surfaces.

This is exactly what happens when the inner surfaces of vacuum chambers are irradiated with SR photons, as in particle accelerators. Figure 5.9 shows the desorption yields of H_2, CH_4, H_2O, CO, and CO_2 of copper-electroplated stainless steel surfaces exposed to 1×10^{21} photons m^{-1} from the electron–proton accumulator at CERN (Gómez-Goñi 2007). For all gases, except for our eccentric water molecule, the desorption yield decreases as a function of dose, so that the surface becomes cleaner with exposure. The radiation-induced desorption of gas molecules is considered to occur in a two-stage process. Photons produce photoelectrons and gas molecules are desorbed by electron-stimulated desorption from the surfaces (Menzel and Gomer 1964; Redhead 1964; Avouris and Walkup 1989). For water the desorption yield increases initially and then decreases. This is believed to be due to an activated process, hence the delay in desorption, where hydrogen atoms diffuse from the bulk toward the surface and combine with oxygen in the oxide layer forming water, which subsequently desorbs (Gröbner, Mathewson, and Marin 1994). However, the same effect should in principle be expected for H_2, CH_4, CO, O_2, and CO_2. The question thus remains unsolved.

5.6 SUMMARY

- Water can be partially purified through the photocatalytic action of surfaces of certain semiconductors, where molecules become degraded through irradiation with UV and visible light due to the oxidative and reductive power of the generated holes and electrons in the valence and conduction bands, respectively, that reach the surface.
- The ever-growing demand for energy and the quest for environmental sustainability makes the production of hydrogen from water using sunlight, which can be used in fuel cells, a major issue. Such production can be achieved through the photocatalytic activity of surfaces of certain semiconducting materials.
- Interfacial water plays a pivotal role in corrosion and degradation of most of the materials exposed to ambient conditions. Large ions, such as bromine and iodide, have the tendency to accumulate at the water/air interface in marine aerosols, enhancing their reactivity.
- Cloud seeding, a direct way to intentional weather control, is far from being mastered from the fundamental and practical points of view; however, positive effects on precipitation have been shown when properly conducted and performed under appropriate atmospheric conditions.
- Capillarity-induced adhesion is particularly important for objects with small dimensions due to the large surface-to-volume ratio. It can be beneficial for clustering of particles (sand castle effect) but detrimental for micro-nanoelectromechanical systems.
- Water adsorbed on technical surfaces employed for ultrahigh vacuum can be efficiently desorbed by heating the vessels above $130°C$. When irradiated with photons, the behavior of the desorption yield suggests an activated process, where water is formed at the oxide layer through the combination of atomic hydrogen diffusing from the bulk with oxygen from the oxide layer, although the actual mechanisms are still not clear.

REFERENCES

1. Asahi, R., Morikawa, T., Ohwaki, T., Aoki, K., and Taga, Y. 2001. Visible-light photocatalysis in nitrogen-doped titanium oxides. *Science* 293:269–271.
2. Autumn, K. and Peattie, A.M. 2002. Mechanisms of adhesion in geckos. *Integr. Comp. Biol.* 42:1081–1090.
3. Avouris, P. and Walkup, R.E. 1989. Fundamental mechanisms of desorption and fragmentation induced by electronic transitions at surfaces. *Annu. Rev. Phys. Chem.* 40:173–206.
4. Bailey, M. and Hallett, J. 2002. Nucleation effects on the habit of vapour grown ice crystals from -18 to -42°C. *Q. J. R. Meteorol. Soc.* 128:1461–1483.

5. Bhushan, B. 2007. Nanotribology and nanomechanics of MEMS/NEMS and BioMEMS/BioNEMS materials and devices. *Microelec. Eng.* 84:387–412.

6. Birstein, S.J. and Anderson, C.E. 1955. The mechanism of atmospheric ice formation, I: The chemical composition of nucleating agents. *J. Meteor.* 12:68–73.

7. Bocquet, L., Charlaix, E., Ciliberto, S., and Crassous, J. 1998. Moisture-induced ageing in granular media and the kinetics of capillary condensation. *Nature* 396:735–737.

8. Cantrell, W. and Heymsfield, A. 2005. Production of ice in tropospheric clouds. *Bull. Am. Meteorol. Soc.* 86:795–807.

9. Carp, O., Huisman, C.L., and Reller, A. 2004. Photoinduced reactivity of titanium dioxide. *Prog. Solid State Chem.* 32:33–177.

10. Carpick, R.W. and Salmeron, M. 1997. Scratching the surface: Fundamental investigations of tribology with atomic force microscopy. *Chem. Rev.* 97:1163–1194.

11. Chen, H., Nanayakkara, C.E., and Grassian, V.H. 2012. Titanium dioxide photocatalysis in atmospheric chemistry. *Chem. Rev.* 112:5919–5948.

12. Corma, A., Rey, F., Rius, J., Sabater, M.J., and Valencia, S. 2004. Supramolecular self-assembled molecules as organic directing agent for synthesis of zeolites. *Nature* 431:287–290.

13. Deleau, F., Mazuyer, D., and Koenen, A. 2009. Sliding friction at elastomer/glass contact: Influence of the wetting conditions and instability analysis. *Tribol. Int.* 42:149–159.

14. DelRio, F.W., Dunn, M.L., Phinney, L.M., Bourdon, C.J. and de Boer, M.P. 2007. Rough surface adhesion in the presence of capillary condensation. *Appl. Phys. Lett.* 90:163104.

15. Edwards, P.P., Kuznetsov, V.L., and David, W.I.F. 2007. Hydrogen energy. *Phil. Trans. R. Soc. A* 365:1043–1056.

16. Eisner, T. and Aneshansley, D.J. 2000. Defense by foot adhesion in a beetle (Hemisphaerota cyanea). *Proc. Natl. Acad. Sci. USA* 97:6568–6573.

17. Ekinci, K.L. and Roukes, M.L. 2005. Nanoelectromechanical systems. *Rev. Sci. Instrum.* 76:061101.

18. Esswein, A.J. and Nocera, D.G. 2007. Hydrogen production by molecular photocatalysis. *Chem. Rev.* 107:4022–4047.

19. Finlayson-Pitts, B.J. 2003. The tropospheric chemistry of sea salt: A molecular level view of the chemistry of NaCl and NaBr. *Chem. Rev.* 103:4801–4822.

20. Finlayson-Pitts, B.J. and Pitts, J.N. 2000. *Chemistry of the Upper and Lower Atmosphere.* San Diego, CA: Academic Press.

21. Fletcher, N.H. 1969. *The Physics of Rainclouds.* Cambridge, UK: Cambridge University Press.

22. Fujishima, A. and Honda, K. 1972. Electrochemical photolysis of water at a semiconductor electrode. *Science* 238:37–38.

23. Fujishima, A., Zhang, X., and Tryk, D.A. 2008. TiO_2 photocatalysis and related surface phenomena. *Surf. Sci. Rep.* 63:515–582.

24. Gómez-Goñi, J. 2007. Photon stimulated desorption from copper and aluminum chambers. *J. Vac. Sci. Technol. A* 25:1251–1255.

25. Grätzel, M. 2001. Photoelectrochemical cells. *Nature* 414:338–344.

26. Gröbner, O., Mathewson, A.G., and Marin, P.C. 1994. Gas desorption from an oxygen free high conductivity copper vacuum chamber by synchrotron radiation photons. *J. Vac. Sci. Technol. A* 12:846–853.

27. Henderson, M.A. 2011. A surface science perspective on TiO_2 photocatalysis. *Surf. Sci. Rep.* 66:185–297.

28. Henderson, M.A., Epling, W.S., Peden, C.H.F., and Perkins, C.L. 2003. Insights into photoexcited electron scavenging processes on TiO_2 obtained from studies of the reaction of O_2 with OH groups adsorbed at electronic defects on $TiO_2(110)$. *J. Phys. Chem. B* 107:534–545.

29. Heneghan, A.F., Wilson, P.W., and Haymet, A.D.J. 2002. Heterogeneous nucleation of supercooled water, and the effect of an added catalyst. *Proc. Natl. Acad. Sci. USA* 99:9631–9634.
30. Hoffmann, M.R., Martin, S.T., Choi, W., and Bahnemann, D.W. 1995. Environmental applications of semiconductor photocatalysis. *Chem. Rev.* 95:69–96.
31. Jang, J., Schatz, G.C., and Ratner, M.A. 2004. Capillary force in atomic force microscopy. *J. Chem. Phys.* 120:1157–1160.
32. Justícia, I., Ordejón, P., Canto, G., Mozos, J.L., Fraxedas, J., Battiston, G.A. et al. 2002. Designed self-doped titanium oxide thin films for efficient visible-light photocatalysis. *Adv. Mater.* 14:1399–1402.
33. Kanji, Z.A. and Abbatt, J.P.D. 2009. The University of Toronto continuous flow diffusion chamber (UT-CFDC): A simple design for ice nucleation studies. *Aerosol. Sic. Tech.* 43:730–738.
34. Kanji, Z.A., DeMott, P.J., Möhler, O., and Abbatt, J.P.D. 2011. Results from the University of Toronto continuous flow diffusion chamber at ICIS 2007: Instrument inter-comparison and ice onsets for different aerosol types. *Atmos. Chem. Phys.* 11:31–41.
35. Kudo, A. and Miseki, Y. 2009. Heterogeneous photocatalyst materials for water splitting. *Chem. Soc. Rev.* 38:253–278.
36. Langmuir, I. 1950. Control of precipitation from cumulus clouds by various seeding techniques. *Science* 112:35–41.
37. Lewis, N.S. and Nocera, D.G. 2006. Powering the planet: Chemical challenges in solar energy utilization. *Proc. Natl. Acad. Sci. USA* 103:15729–15735.
38. Maeda, K. and Domen, K. 2010. Photocatalytic water splitting: Recent progress and future challenges. *J. Phys. Chem. Lett.* 1:2655–2661.
39. Martínez-Ferrero, E., Sakatani, Y., Boissière, C., Grosso, D., Fuertes, A., Fraxedas, J., and Sanchez, C. 2007. Nanostructured titanium oxynitride porous thin films as efficient visible-active photocatalysts. *Adv. Funct. Mater.* 17:3348–3354.
40. Menzel, D., and Gomer, R. 1964. Desorption from metal surfaces by low-energy electrons. *J. Chem. Phys.* 41:3311–3328.
41. Möhler, O., Field, P.R., Connolly, P., Benz, S., Saathoff, H., Schnaiter, M., Wagner, R. et al.. 2006. Efficiency of the deposition mode ice nucleation on mineral dust particles. *Atmos. Chem. Phys.* 6:3007–3021.
42. O'Dowd, C.D. and De Leeuw, G. 2007. Marine aerosol production: A review of the current knowledge. *Phil. Trans. R. Soc. A* 365:1753–1774.
43. Pan, J.H., Dou, H., Xiong, Z., Xu, C., Ma, J., and Zhao, X.S. 2010. Porous photocatalysts for advanced water purifications. *J. Mater. Chem.* 20:4512–4528.
44. Parsons, S. and Jefferson, B. 2006. *Introduction to Potable Water Treatment Processes*. Oxford, UK: Blackwell.
45. Perego, C., Bagatin, R., Tagliabue, M., and Vignola, R. 2013. Zeolites and related mesoporous materials for multi-talented environmental solutions. *Micropor. Mesopor. Mater.* 166:37–49.
46. Persson, B.N.J. 2008. Capillary adhesion between elastic solids with randomly rough surfaces. *J. Phys. Condens. Matter* 20:315007.
47. Redhead, P.A. 1964. Interaction of slow electrons with chemisorbed oxygen. *Can. J. Phys.* 42:886–905.
48. Redhead, P.A., Hobson, J.P., and Kornelsen, E.V. 1993. *The Physical Basis of Ultrahigh Vacuum*. New York: American Institute of Physics.
49. Salam, A., Lohmann, U., Crenna, B., Lesins, G., Klages, P., Rogers, D. et al.. 2006. Ice nucleation studies of mineral dust particles with a new continuous flow diffusion chamber. *Aerosol Sci. Technol.* 40:134–143.

50. Shannon, M.A., Bohn, P.W., Elimelech, M., Georgiadis, J.G., Mariñas, B. J., and Mayes, A. M. 2008. Science and technology for water purification in the coming years. *Nature* 452:301–310.

51. Shevkunov, S.V. 2009. Numerical simulation of water vapor nucleation on electrically neutral nanoparticles. *J. Exp. Theor. Phys.* 108:447–468.

52. Sumner, A.L., Menke, E.J., Dubowski, Y., Newberg, J.T., Penner, R.M., Hemminger, J.C. et al. 2004. The nature of water on surfaces of laboratory systems and implications for heterogeneous chemistry in the troposphere. *Phys. Chem. Chem. Phys.* 6:604–613.

53. Van Grieken, R., Delalieux, F., and Gysels, K. 1998. Cultural heritage and the environment. *Pure Appl. Chem.* 70:2327–2331.

54. Vogel, M.J. and Steen, P.H. 2010. Capillarity-based switchable adhesion. *Proc. Natl. Acad. Sci. USA* 107:3377–3381.

55. Vonnegut, B. 1947. The nucleation of ice formation by silver iodide. *J. Appl. Phys.* 18:593–595.

56. Zhao, Y.-P., Wang, L.S., and Yu, T.X. 2003. Mechanics of adhesion in MEMS: A review. *J. Adhesion Sci. Technol.* 17:519–546.

6 Water/Biomolecule Interfaces

Alles ist aus dem Wasser entsprungen!
Alles wird durch das Wasser erhalten!
Ozean, gönn' uns dein ewiges Walten!
. . .
Du bist's der das frischeste Leben erhält.
J. W. von Goethe, Faust

We are about to finish the journey that has transported us from the very basics of the isolated water molecule to interfacial water on both ideal and real systems (mainly inorganic). We face in this last chapter the extremely challenging job of understanding water in the most important system, life (as we know it), but restricted to interfaces. In fact this restriction is not limiting at all because biological liquids are characterized by having a rather high concentration of large solutes (biomolecules), so that a large fraction of water molecules resides for a sufficiently long time in the hydration shell: in biosystems interfacial water relegates bulk water to a secondary role. In the simplest picture of the process, the water molecules are static at the interface but this visualization is wrong: hydration is a dynamic process, in the ps range, the timescale of H-bonding. Within the introductory scope of this book, and particularly targeted to the nonexpert in the field, we browse through the sequence of increasing degrees of complexity covering amino acids, peptides, small proteins, nucleic acids, and membranes, concentrating on a few representative examples.

6.1 IS WATER A BIOMOLECULE?

Before giving an answer to such a question (positive, by the way, because without water the cell function would cease to exist) let us briefly summarize what we have learned regarding interfacial water throughout this book. The term *complex simplicity* has been used when referring to the water molecule because of its apparent inoffensive aspect, formed by only three simple atoms, two of them being the simplest ones in the periodic table. However, our knowledge of water is still far from satisfactory and many questions are yet to be answered, especially concerning the condensed phases.

When intermolecular interactions are considered, in particular H-bonding conferring tetrahedral conformation, three scenarios can be highlighted: (i) clusters, (ii) short-range order, and (iii) long-range order. The study of clusters, $(H_2O)_n$, where n stands for the number of water molecules, is extremely illustrative, for one can follow on a molecule-by-molecule basis the molecular distribution, the astonishing cooperative behavior, as well as the characteristic vibrational spectra and enter the condensed phase regime ($n \rightarrow \infty$). An amazing conclusion is that many relevant parameters reach the asymptotic limit for rather small n values, indicating the dominance of the short-range order over the long-range order. The crystalline phases can be rationalized quite simply with the Bernal–Fowler(–Pauling) rules but the large number of polymorphs (16 known to date) is surprising, especially when compared to other molecular systems (to be fair one has to admit that ice has received overwhelming attention over analogous molecular systems). However, in the liquid state the structure of water is still under debate, some authors challenging the dominance of tetrahedral ordering in favor of a more filamentary structure. The short-range order is essentially the same for liquid and solid (here including the amorphous state), the only difference arising from the dynamic–static nature. When considering dynamical processes one has to be cautious when comparing experimental and computational results because the involved timescales might be quite different. The dynamics from theoretical simulations are nowadays limited to the ps–ns range, and each experimental technique has characteristic timescales (from fs to s).

When introducing ideal (flat, rigid, and homogeneous) surfaces we saw that such model structures impose some degree of ordering (layering) to liquid water and that the affinity to water can be opposed, high (hydrophilic) or low (hydrophobic), depending on the chemical nature of the surface, its dipolar distribution, and on the micronanostructure. It is quite remarkable that flat, rigid, and homogeneous surfaces cannot be 100% hydrophobic (the contact angle is limited to about 120°) and that superhydrophobicity can only be achieved through micronanoscale surface patterning: structured roughness at different length scales is mandatory. Ions are also able to modify the local structure of water through hydration and charged surfaces induce a charge distribution within a different reach, a more rigid layering close to the surface (Stern–Helmholtz layers) and a less ordered, dynamic diffuse layer at larger distances (Gouy–Chapman layers). Large means few nm in this context. Water can also be structured by weakly interacting molecules (through vdW interactions) inducing local rearrangements of the H-bonding network leading to molecular sequestration or spectacular host–guest crystalline forms (clathrate hydrates). The H-bond network is depleted for dimensions of objects (solutes) above 1 nm, not that far from the mean molecular diameter of water, which is de facto a critical parameter for interfacial water.

In this last chapter we deal with biomolecules but concentrate on the interfaces they build with water. Biomolecules exhibit surfaces that are nonplanar, nonhomogeneous with hydrophobic and hydrophilic regions, flexible, and dynamic: they are anything but ideal but ideally suited to their biofunctionalities. It is evident that understanding the water/biomolecule interfaces is an extremely difficult task. The surface of biomolecules will modulate the structure of the liquid surrounding them, typically inside cells, which host ions to make things more difficult. But, in addition, water will also modulate the shape, and thus the function, of the biomolecules. This mutual influence is not found for inorganic surfaces. The differentiation between surface bound water and free bulk water (Kuntz and Kauzmann 1974), in line with the mentioned Stern–Helmholtz and Gouy–Chapman layers, appears to be a simple concept but difficulties are encountered when going to a deeper level of knowledge. We explore in the next sections the role of interfacial water with biomolecules and convince ourselves that considering the candidacy of water as a (extremely small) biomolecule is justified

(Ball 2008a). In fact it has been termed the 21st amino acid. Given the extraordinary complexity of biomolecules, we try to concentrate on small biomolecules whenever possible (this is not the case of DNA) that capture the essentials of the phenomena induced by water.

6.2 PROTEINS

6.2.1 WATER-PEPTIDE INTERACTIONS

The primary structure of proteins is given by the chainlike distribution of (20 available) amino acids bound through peptide bonds. In the case of melittin, a small protein found in honeybee venom, the sequence contains 26 amino acids:

GIGAVLKVLTTGLPALISWIKRKRQQ

where G = glycine, I = isoleucine, A = alanine, V = valine, L = leucine, K = lysine, T = threonine, P = proline, S = serine, W = tryptophan, R = arginine, and Q = glutamine. Figure 6.1 shows a view of tetrameric melittin.

This 1D or sequential ordering, where each amino acid has only two neighbors (except for those at the ends), is essential for the specific function of the protein, inasmuch as it is based on the actual amino acid sequence and defines the spatial distribution (folding) of the protein into its final shape, which is known as the secondary structure. Such folding is modulated by the water surrounding the protein building a dynamic interface. The shape of proteins is also crucial for their specific functions because they have to interact with specific molecules with their own amino acid sequence and shape. Water makes such communication possible. The chainlike conformation provides a greater adaptability to proteins, as compared to a more rigid

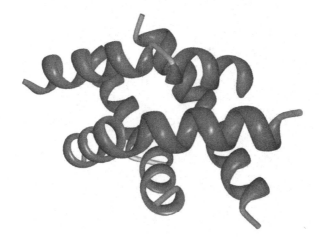

FIGURE 6.1 View of a melittin tetramer. Turn, coil, and helix are represented by light, medium, and dark gray, respectively. From Protein Data Bank ID: 2MLT, T.C. Terwilliger and D. Eisenberg. *J. Biol. Chem.* 257:6010–6015, 1982. The projection has been obtained using J. L. Moreland, et al. *BMC Bioinformatics* 6:21, 2005. (With permission.)

2D distribution. As mentioned in Section 1.5, adaptability is an essential property for life and water contributes decisively to it. Amino acids contain a polar part in the zwitterionic form, with both a positively charged ammonium group (NH_3^+) and a negatively charged carboxylic group (COO^-), linked to a common carbon atom and to a radical. The charged groups confer a hydrophilic character and the radical contributes differently to the affinity to water depending on its chemical nature. In the case of L-alanine the radical is a methyl group, with extremely low affinity to water (only through vdW). The peptide chain of the protein will have a distribution of residues with different local affinity to water, depending on the particular amino acid involved, so that the local hydrophobic/hydrophilic character should play a crucial role in both the secondary and tertiary structures, and thus in the function of the protein in water (Kauzmann 1959).

Before describing protein hydration it is illustrative to have a quick look at short peptides. Peptides are constituted of the same amino acid building blocks as proteins but are shorter. In the case of the L-alanine amino acid referred to, it has been shown through computer simulations that solvated individual molecules (the trivial case with no peptide bonds) exhibit a different spatial conformation as compared to the corresponding structure in the collective water-free crystalline state (Degtyarenko et al. 2008). Figure 6.2 shows a representation of an individual L-alanine molecule in water in its zwitterionic form.

In the gas phase alanine molecules exist only in nonionic form (Blanco et al. 2004). The largest differences between the calculated optimized L-alanine structure in aqueous solution (Degtyarenko et al. 2008) and the experimentally derived crystalline structure (Lehmann, Koetzle, and Hamilton 1972) involve the COO^- group. The C'-O^1 bond length (see Figure 6.2) decreases from 1.290 to 1.240 Å and C'-O^2 from

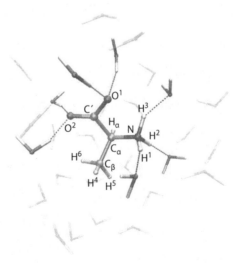

FIGURE 6.2 L-alanine in water. Water molecules within the first hydration shell are shown in solid lines. H-bonding is indicated by dotted lines. (Reprinted from I. Degtyarenko, et al. *J. Comput. Theor. Nanosci.* 5:277–285, 2008. With permission of American Scientific permission conveyed through Copyright Clearance Center, Inc.)

1.273 to 1.257 Å, when comparing calculated and experimental data, respectively, both parameters taken at room temperature. In the crystal, COO^- forms three H-bonds, two of them involving O^2 and only one O^1 atom. However, in aqueous solution, the number of H-bonds varies: COO^- can have from two to six water molecules within its first hydration shell. On average, the O^1 atom is H-bonded with 2.34 water molecules and the O^2 atom with 1.88. A similar difference is also found for NH_3^+. The time-averaged numbers of H-bonds for H^1, H^2, and H^3 atoms are 0.78, 0.92, and 1.0, respectively. The most dramatic difference is related to the $O^1-C'-C_\alpha-H_\alpha$ torsion angle which indicates the COO^- group orientation, varying from 98.7° in the solvated state to $-135.4°$ in the crystalline state.

The first simple peptide formed by alanine is the alanine dipeptide, schematized in Figure 6.3a. This molecule has been extensively used in theoretical studies of backbone conformations in proteins because, in spite of being simple, it contains many of the structural features of the protein backbone: the ϕ and ψ dihedral angles, two amide peptide bonds whose NH and CO groups are capable of participating in H-bonding and a methyl group attached to the α carbon that is considered representative of the side chains in all nonglycine or proline amino acids (Tobias and Brooks 1992). Figures 6.3c and d show the calculated and experimental Ramachandran diagrams in $\phi - \psi$ space, including dipeptide–water interactions, respectively (Cruz, Ramos, and Martínez-Salazar 2011). The calculations are based on MD simulations using the TIP4P water model (see Table 1.5) and the experimental data have been extracted from available protein database libraries selecting structures elucidated by solution NMR. The PPII conformation appears to be preponderant over both the β (sheets) and α_R (helix) structures, in agreement with previous works (Prabhu and Sharp 2006). The location of the conformations in the Ramachandran diagrams in Figure 6.3 indicates the ϕ and ψ ranges for each conformation. In addition, the less favorable Y conformation becomes evident. In Figure 6.3b the Y (left) and α_R (right) conformations are represented. With an increasing number of residues, peptides adopt well-defined secondary structures (e.g., α-helix and β-sheets) in water. Peptide chains with less than 20 residues, especially those rich in alanine content, have been found to show a preference for the PPII conformation. Long alanine-rich chains have a predominantly right-handed α helical structure. In the helix, the peptide backbone is shielded from solvent and the loss of solvent–backbone interactions is partially compensated by the formation of internal H-bonds (Prabhu and Sharp 2006).

6.2.2 HYDROPHOBIC FORCES AND HYDRATION

Protein Folding

The local affinity to water will tend to segregate the peptide chains in hydrophobic and hydrophilic regions in solution with the limitation imposed by the rigidity of the actual backbone structure. In the limit one would expect an inner (hydrophobic) core free of water and a (hydrophilic) surface in contact with liquid water. The simplest approximation consists of modeling a protein as a sphere impenetrable to water (Tanford and Kirkwood 1957). The exclusion of water from the hydrophobic region can be modeled with the so-called dewetting or drying effect. As already mentioned in Section 4.2 when two parallel hydrophobic surfaces submerged in water approach

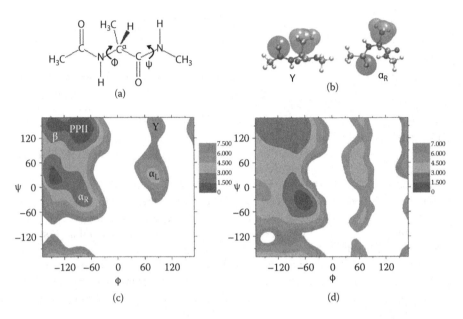

FIGURE 6.3 (a) Scheme of the alanine dipeptide with the flexible backbone dihedral angles, ϕ and ψ. (b) Ball and stick representations of the alanine dipeptide Y (left) and α_R (right) conformations, highlighting both the methyl radical and the carbonyl oxygen atoms. Ramachandran plots including dipeptide–water interactions: (c) calculated and (d) extracted from protein database libraries selecting structures elucidated by solution NMR. Gray scales stand for energy contours in kT units. The dihedral angles are in degrees. (Adapted from V. Cruz, J. Ramos, and J. Martínez-Salazar. *J. Phys. Chem. B* 115:4880–4886, 2011, American Chemical Society. With permission.)

below a critical distance, the water between both surfaces is expelled forming a cavity (vapor phase) and both surfaces experience an attraction due to the pressure exerted by the liquid (Lum, Chandler, and Weeks 1999; Chandler 2005). The attractive character justifies the use of the term hydrophobic interaction or force, but the concept can be somehow misleading inasmuch as the effect is indirect; there is no such hydrophobic force in Nature and the term hydrophobic effect is more indicated (Faraudo 2011). Proteins are thus stabilized via the dewetting mechanism (Levy and Onuchic 2006; Berne, Weeks, and Zhou 2009). Thus, hydrophobicity induces a drying transition in the gap between sufficiently large hydrophobic regions as they approach sufficiently, resulting in the collapse of the protein as well as a depletion of water in the cavities. The dewetting mechanism has been studied preferentially in small proteins, which is understandable due to the complexity of the process and the computational cost. A first illustrative example involves the SH3 protein, which has a hydrophobic core buried within β-sheets in the native structure. According to MD simulations the folding of the SH3 protein is a two-step process, where the fully solvated SH3 protein first undergoes an initial structural collapse to an overall native conformation followed by a second transition in which water molecules are expelled from the hydrophobic core region, resulting in a dry and packed protein (Cheung, García, and Onuchic 2002).

Thus, the structural formation of the protein is achieved before water is expelled from the hydrophobic core. Water molecules play an initial structural role through H-bonding and are gradually expelled afterwards. A similar mechanism has been observed in the folding of BBA5, a 23-residue miniprotein (Rhee et al. 2004). However, buried or internal water molecules can reside for a sufficiently long time, typically on a 0.01–100 μs timescale in solution (Otting, Liepinsh, and Wüthrich 1991; Halle 2004), becoming integrated in the structure of the protein and conferring stability (Takano et al. 1997). Hence, the 21st amino acid label.

The drying transition has also been observed in the collapse of multidomain proteins such as the tetramer of melittin in water (Liu et al. 2005; see Figure 6.1). In the simulations the two dimers of the melittin tetramer are separated to create a nanoscale channel and then are solvated in a water box. A sharp dewetting transition is observed, and essentially all water molecules are expelled from the nanoscale channel after the transition. The critical distance for the melittin tetramer system is approximately 0.5–0.7 nm, which is equivalent to two to three water molecule diameters. The drying is extremely sensitive to single mutations of the three isoleucines to less hydrophobic residues. In fact, such mutations in well-defined locations can switch the channel from dry to wet. However, drying is not observed in the collapse of the two-domain BphC enzyme (Zhou et al. 2004). In this study the two domains were brought together to see if there was a critical distance for drying between them. When either the vdW interaction or both the vdW and the electrostatic interactions between the protein and water were turned off, the region between the domains exhibited a drying transition, but when the full force field was turned on, no complete drying transition was found.

Once folded, the proteins can recover unfolded conformations upon external action either by increasing temperature or by using solvents. An alternative way uses the mechanical action of an AFM and can be performed at the single-molecule level (*in singulo*). The concept is rather straightforward but the operation and interpretation of the experimental results are far from simple. Usually target molecules are adsorbed on a freshly prepared gold surface from a buffered solution in a specific pH. Once adsorbed, the substrate is introduced in the AFM fluid cell. The tip of a cantilever is brought into mechanical contact with the surface so as to permit the attachment of molecules. Then, the cantilever is retracted and its deflection is registered as a function of displacement. If the protein remains attached to both ends (tip and gold surface), then the force-displacement curve exhibits a characteristic discrete sawtooth shape. The peak maxima (several pN force) and peak-to-peak distance (several nm) is a measure of the unfolding of secondary structures. Early works in the literature involved titin (Rief et al. 1997) and tenascin (Oberhauser et al. 1998). Once completely stretched, the proteins can undergo folding again when the applied force is released becoming an ideal benchmark for theoretical calculations (Daggett 2006).

Protein Hydration

The study of protein hydration goes hand-in-hand with the structural determination of proteins, although hydration in crystals is not necessarily the same as in solution. The first experimental atomic structure of a protein was determined for myoglobin and a view of the tertiary structure is illustrated in Figure 6.4a. This was achieved by Kendrew and coworkers (1958) using high-resolution X-ray crystallography giving

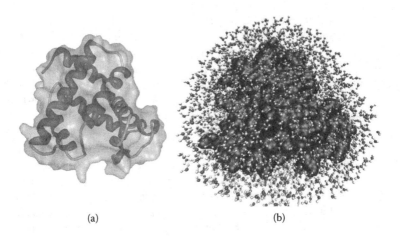

(a)   (b)

FIGURE 6.4 Views of myoglobin. (a) Turn, coil, and helix are represented by light, medium, and dark gray, respectively. Protein Data Bank ID: 1MBN, H.C. Watson. *Prog. Stereochem.* 4:299–333, 1969. The projection has been obtained using J. L. Moreland, et al. *BMC Bioinformatics* 6:21, 2005, and the surface with D. Xu and Y. Zhang, *PLoS ONE* 4:e8140, 2009. (b) Snapshot of hydrated myoglobin. (Reprinted from H. Frauenfelder, et al. *Proc. Natl. Acad. Sci. USA* 106:5129–5134, 2009. With permission.)

birth to structural biology. Since then a large amount of structures have been deciphered and can be freely accessed in the Protein Data Bank (http://www.rcsb.org/pdb/). For this pioneering discovery, J. C. Kendrew was awarded, jointly with M. F. Perutz, the Nobel Prize in Chemistry 1962 for their studies of the structures of globular proteins. Figure 6.4b shows a snapshot of the hydration shell of myoglobin according to computer simulations (Frauenfelder et al. 2009). The whole figure represents a further example of the necessary complementarity between theory and experiment. Because of the considerable size of the protein a large amount of water molecules are involved building a ∼0.5-nm thick shell, that is, two layers mimicking the surface modulated by the local hydrophilic/hydrophobic character, although the distribution is much more complex. As already mentioned above, the diverse protein functions depend on the degree of hydration and the vast majority of proteins cannot function when the hydration shell lies below one layer. Perhaps due to the static representations of structures on paper (and this is valid for any structure), the protein hydration and structure have been regarded as static leading to the misleading term *bound* water as opposed to the *free* bulklike water. The bonding concept was described in an icelike model because it was believed, based on experimental evidence, that the water layer moved rigidly with the protein. The static image will be mitigated in the near future when e-books will be able to display 2D and 3D motion, but at the present time (and this includes this book) we have to carry on with frozen figures in print. This restriction leaves us with a wrong impression, because hydration and protein function are a dynamic process in the ps timescale. Today, thanks to computer simulations and fast GHz hardware we can visualize a more real description, although not completely correct, inasmuch as our timescale of seconds is too large when compared to the referenced ps range. Because of this dynamic nature, the number of configurations is enormous leading to energy landscapes with a large number of local minima and even hierarchically organized: there are energy valleys within energy valleys within energy valleys, in a pseudo-fractal scenario (Ansari et al. 1985).

The first determination of the hydration lifetime was performed by Otting, Liepinsh, and Wüthrich (1991) using the NMR Overhauser effect for oxytocin hormone and bovine pancreatic trypsin inhibitor globular protein. This seminal work established a sub-ns timescale for water residing in the hydration shell. Since then other experimental techniques as well as MD simulations have been used leading to a broad timescale spanning from a few ps to ns (Halle 2004; Zhong, Pal, and Zewail 2011). In spite of the discrepancy in the values a bimodal picture emerges, where water molecules are slowed down when they reside at distances below 1 nm from the protein surface (wherever the origin is), with characteristic times much larger than the (H-bonding) $\sim$1 ps lifetime of bulk water.

A consequence of slowing down water motion is that the average density of the hydration shell is higher than that of bulk water. This has been determined both experimentally by means of X-ray and neutron scattering (Svergun et al. 1998) and theoretically (Merzel and Smith 2002) for lysozyme, a small enzyme that attacks the protective cell walls of bacteria (natural antibiotic). In this case the local increase in water density is about 10%. However, the physical origin of such an increase is somehow tricky. According to Merzel and Smith (2002) about two-thirds of the observed density increase over bulk water arises from a geometrical effect caused by the definition of the surface. On top of this effect, however, a $\sim$5% density increase is caused by perturbation of the average water structure from bulk water. About half of this density increase arises from shortening the average water–water distances, and the other half arises from an increase in the coordination number. Although the nearest-neighbor water molecules are generally farther from the protein atoms than they are from water atoms in bulk water, the higher density of protein atoms in the protein surface constrains the water density on the protein surface to be higher than in any comparable shell in pure water. Functional proteins are forcefully hydrated, therefore it seems reasonable to assume that water will have an important role in protein–protein interaction and in general in any activity. As formulated by Ball (2008b), proteins extend their range of influence via their hydration shells. Water molecules can guide a fully solvated protein to recognize another fully solvated protein by a gradual expulsion of water layers. Such a dynamic description has to be taken into account when addressing the desolvation mechanism of bringing two solvated proteins to form a specific assembly (Levy and Onuchic 2006).

6.2.3 Antifreeze and Ice Nucleating Proteins

Antifreeze proteins (AFP) enable the lowering of the water freezing temperature efficiently protecting cold-blooded animals at $T < 0°C$. During the winter season polar oceans can achieve $-1.9°C$ but in spite of such unfavorable conditions some fish are able to survive (DeVries 1983). AFPs are also found in plants, insects and in sea ice microorganisms. Fish AFP are divided in two families: antifreeze glycoproteins and AFP of different types (I, II, and III). Although they are structurally different,

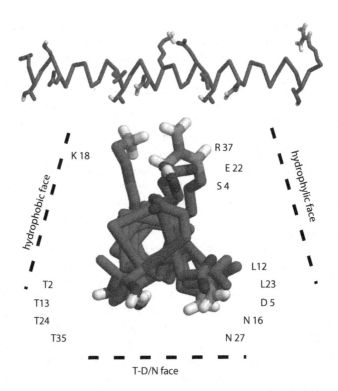

FIGURE 6.5 X-ray structure of the winter flounder AFP HPCL6 viewed (top) perpendicular and (bottom) along the α helical axis indicating the hydrophobic, hydrophilic and T-D/N faces. D = aspartic acid, E = glutamic acid, K = lysine, L = leucine, N = asparagine, R = arginine, S = serine, and T = threonine. (Reprinted from A. Jorov, B.S. Zhorov, and D.S.C. Yang. *Protein Sci.* 13:1524–1537, 2004. With permission.)

antifreeze glycoproteins and type I AFP are built from about 67% of alanine (Duman and DeVries 1976). AFP of type II (sea raven, *Hemitripterus americanus*) are rich in cysteine and type III (ocean pout *Macrozoarces americanus*) is neither alanine-rich nor contains any cysteine residues. The most extensively studied species of AFP type I comes from the winter flounder fish (*Pseudopleuronectus americanus*). Its major AFP, HPLC-6, consists of 37 amino acids with the following distribution:

DTASDAAAAAALTAANAKAAAELTAANAAAAAAATAR

where D = aspartic acid, N = asparagine, and E = glutamic acid (see Section 6.2.1 for the rest of the acronyms). The crystallographic structure shows a single α-helix secondary structure, as shown in Figure 6.5 (Yang et al. 1988). The α-helix is amphiphilic. When viewed along the α-helical axis, three faces may be distinguished: a hydrophobic face formed by alanines and methyl groups of threonines; a hydrophilic face formed by arginine, glutamic acid, serine, and asparagine residues; and finally a face formed by hydrophilic groups of threonine and asparagine/aspartic residues.

The mechanism leading to the lowering of the freezing temperature involves the binding of AFP to ice surfaces inhibiting the growth of ice crystals. The lowering of the freezing point is not a colligative effect in this case because of the observation of thermal hysteresis (different freezing and melting points). In a colligative system, no difference between the freezing and melting points is at hand. The most simplified mechanism for the inhibition of growth can be explained with the Gibbs–Thomson equation (3.13), introduced in Section 3.4.2, replacing r_{pore} by the radius of a spherical ice particle. When $\Delta\gamma > 0$ and/or the ice particle radius decreases, then $\Delta T > 0$, so that the local freezing temperature decreases. This equation describes the freezing temperature lowering, due to the presence of finite-dimensional crystals. The inhibition of the ice crystal growth derives from the local ice surface curvature effects induced by the adsorption of these proteins at the ice/solution interface. As formulated by Yeh and Feeney (1996), "Why do they wait until crystals have already been formed before adsorbing onto their surfaces and poisoning their growth sites? Why not simply prevent the nucleation of ice crystals completely?" Mother Nature's long-term experience shows that it simply works!

Sicheri and Yang (1995) showed, based on a geometrical model, that there is a match between AFP structures and the topology of the $(20\bar{2}1)$ ice plane along the equivalent $[01\bar{1}2]$ directions. This particular ice surface is corrugated, hence the proclivity to host biomolecules with protrusions. Threonine residues at the hydrophobic face form protrusions 16.5 Å apart, a distance large enough to prevent AFP dimerization but short enough to fit in the parallel grooves of the surface. In addition, the AFP hydrophobic face is not self-complementary (otherwise self-aggregation would occur) which explains why it binds preferably to rough surfaces of ice, such as $(20\bar{2}1)$. The AFP ice-binding surface needs hydrophobic groups that can provide large entropic gain to the ice-AFP binding energy and they should have a rather rigid backbone conformation to prevent hydrophobic collapse (Jorov, Zhorov, and Yang 2004). In addition, fish AFP I and AFP III are able to hinder the ice nucleating activity of AgI (Inada et al. 2012).

As opposed to AFP, a different kind of protein is able to induce the nucleation of ice, hence the term *ice nucleating proteins* (INP; Kawahara 2002). *Pseudomonas syringae* and a few other bacterial species are able to nucleate supercooled water to form ice (Arny, Lindow, and Upper 1976; Hirano and Upper 2000). *Pseudomonas syringae* have ice nucleation active genes that produce INP, which translocate to the outer bacterial cell wall on the surface of the bacteria where the INP act as nuclei for ice formation. INP assemble to form aggregates of various sizes. The larger the aggregates is, the more efficient is ice nucleation. *Pseudomonas syringae* is responsible for the surface frost damage in plants exposed to the environment because it can cause water to freeze at temperatures as high as -1.8 to -3.8 °C (Maki et al. 1974) but have a large economical impact in snow making; many ski resorts use a commercially available freeze-dried preparation to produce snow as required. The presence of biological INP in the atmosphere has remained elusive and only recently have their concentrations been determined. The outcome is that they are abundant in fresh snow and are ubiquitous worldwide in precipitation (Christner et al. 2008), thus they should be considered active actors in the water cycle (see Figure 5.1).

6.3 NUCLEIC ACIDS AND DNA

In their famous seminal work on the determination of the molecular structure of DNA, the macromolecule that encodes the genetic information, J. D. Watson and F. H. C. Crick (1953) pointed out that "the structure is an open one, and its water content is rather high." Both authors were awarded the Nobel Prize in Physiology or Medicine 1962 jointly to M.H.F. Wilkins "for their discoveries concerning the molecular structure of nucleic acids and its significance for information transfer in living material". The double-helix structure has become an icon of modern science and human knowledge and it would have certainly inspired A. Gaudí (1852–1926), who was fascinated by the geometries found in Nature. The building blocks of DNA are nucleotides, in analogy to amino acids for proteins. Nucleotides involve a phosphate group, a pentose (deoxyribose), and a base (adenine, guanine, thymine, and cytosine), as sketched in the inset to Figure 6.6. The phosphoribose group (phosphate and pentose) is the architectural backbone of DNA constituting the repeat unit of the chain. Each base is bound to the deoxyribose part. The four bases establish stable H-bond configurations but restricted to the adenine–thymine and guanine–cytosine couples that allow the formation of double strands of DNA that adopt the familiar helicoidal conformation. Despite the 3D complexity, the underlying rules are clear and simple and make life possible.

The stability of DNA is quite remarkable even in dry conditions, a property not shared by proteins, which become denatured in the absence of water. DNA owes such stability to the referred helicoidal double strand conformation; the interstrand H-bond configuration is not modified by the presence or absence of water molecules. The best way to convince ourselves of such stability is the fact that ancient DNA has been successfully isolated from several-thousand-year-old fossils (Oskam et al. 2010). However, the conformation and function can only be maintained in an aqueous medium. As for proteins, hydrophobic forces dictate the final conformation: the DNA interior is mainly hydrophobic and stabilized by the stacking interactions between the consecutive base pairs, and its surface is rich with hydrophilic groups from the phosphates and sugars. Whereas proteins can have hydrophilic residues in the core and hydrophobic residues at the surface, as previously discussed, the core of nucleic acids is composed of the aromatic bases of each nucleotide and is thus uniform (Levy and Onuchic 2006).

DNA exists in different forms. The B-form is the biologically more relevant with more than 20 water molecules per nucleotide on average. The double helix is right-handed, making a turn every 34 Å (Wing et al. 1980). Upon dehydration B-DNA transforms into A-DNA, with an hydration number of about 15 (Saenger, Hunter, and Kennard 1986). In the A-form, the DNA helix remains right-handed but becomes shorter and broader. Simulated hydration shells of both forms are shown in Figure 6.6 as 2D cuts through C–G and T–A base-planes (Feig and Pettitt 1999). The calculations reveal that the water density is increased up to a factor 6 in the first hydration layer near oxygen and nitrogen atoms and around 2 near carbon atoms as compared to bulk water density. Second and even third solvation shells can also be observed defining a locally enhanced density of water out to 0.8–1.0 nm from the molecular surface of DNA (Makarov, Pettitt, and Feig 2002). In the solid state the existence of a hydration spine has been proposed in which one water layer bridges the nitrogen and oxygen

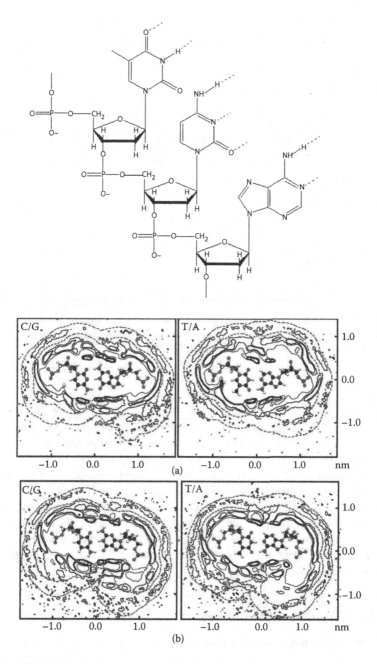

FIGURE 6.6 Simulated water oxygen density contours in the base-planes of C–G and T–A base-pairs in the (a) B- and (b) A-conformations, respectively. Density contours are shown in increasing continuous line thickness at 35.1, 45.2, and 116.3 water counts nm³ and broken lines indicate water counts below 15.5 nm³. The bulk water number density is 33 water counts nm³. The inset shows a portion of a single strand of DNA formed with three nucleotides and thymine, cytosine and guanine, where H-bonds are indicated by discontinuous bonds. (Reprinted from M. Feig and B.M. Pettitt. *J. Mol. Biol.* 286:1075–1095, 1999. With permission from Elsevier.)

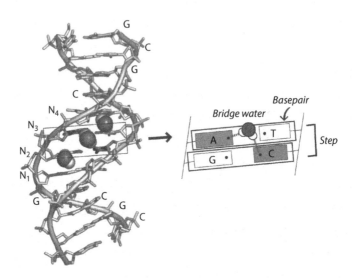

FIGURE 6.7 Snapshot from MD trajectories of the hydration of a 12-mer DNA observed in the minor groove. Water molecules, represented by large medium gray (oxygen) and light gray (hydrogen) balls, form H-bonding bridges between the bases of different strands as shown in the scheme in the right part of the figure. (Reprinted from Y. Yonetani and H. Kono. *Biophys. Chem.* 160: 54–61, 2012. With permission from Elsevier.)

atoms of bases in the minor groove (Kopka et al. 1983). The spine also persists in aqueous solution with water residence times in the minor groove above 1 ns (Liepinsh, Otting, and Wüthrich 1992), much larger than the residence times in the major groove and comparable to those of buried water molecules in globular proteins.

Figure 6.7 shows MD calculations of water molecules bridging bases in the minor groove (Yonetani and Kono 2012). The studied double-stranded B-form DNA is composed of a 12-mer, $CGCGN_1N_2N_3N_4CGCG$, where for $N_1N_2N_3N_4$ all possible arrangements of A, T, G, and C have been considered. The calculated water-bridge lifetime is in the $\sim$1 to 300 ps range and varies depending on the DNA sequence. The comparison of characteristic times with other works is sometimes not straightforward due to the particular definition chosen by the authors and, as found for proteins, the scattering of values is rather large.

6.4   BIOLOGICAL MEMBRANES

6.4.1   WATER/PHOSPHOLIPID INTERFACES

In Section 2.1.2 the binomial system formed by water and amphiphilic molecules was introduced highlighting how the surface of liquid water is able to structure such usually long molecules in layers, given their spatially separated opposed affinity to water, and in turn how such molecules are able to structure interfacial water. According to a particular theory on the origin of life, known as compartmentalistic, the combination of water and amphiphilic molecules has played an essential role, in that the confinement of molecules within vesicles enabled prebiotic metabolic pathways (Luisi 2006). Biological membranes are mainly formed by phospholipid bilayers that constitute the cell walls (Milhaud 2004). Phospholipids are amphiphilic, with a negatively charged phosphate group as polar head, and a lipidic chain tail. Membranes are physical barriers separating the inner part of the cell from the outer part but they can (and must) be permeable to the passage of water and ions in order to maintain the vital functionality of the cells. The doors are nothing other than embedded proteins, and in the next section the case of proteins able to allow the passage of individual water molecules is introduced. It is important to remark here that the architectural bilayer conformation is due to the presence of water: the hydrophobic/hydrophilic segregation is only possible in aqueous media. The bilayered self-assembly is thus driven by hydrophobic forces. The cooperative effect of liquid water, here meaning a large army of individual small molecules, enables the orientation and structuration of large phospholipid biomolecules.

In spite of the structuring power of water, the membrane surface is by no means flat and ideal (otherwise it would not be functional). We saw in Figure 1.9 that liquids become layered close to a flat surface but when surfaces are rough such oscillatory behavior is smeared out. The calculated water density as a function of the distance from the middle of the bilayer shows a smooth increase from zero to the bulk density value over a range of $\sim$2 nm (Pandit, Bostick, and Berkowitz 2003), larger that the $\sim$0.4 nm found for the (flat) vapor/liquid interface, as discussed in Section 2.1.1. This can be viewed as the convolution of oscillatory functions with different origins, because one can safely assume that locally surfaces can be considered as being flat. However, there is still some layering in spite of the broad character of the interface. When the nonplanar character of the membrane is considered, then the oscillatory behavior reflects different regions of hydrated water: inside the membrane, primary hydration shell, secondary hydration shell, and bulk water (Berkowitz, Bostick, and Pandit 2006). A similar situation was described in Figure 4.4 for the case of the (011) surface of L-alanine, that exhibits roughness at the molecular scale. In the case of ideal membranes, that is, prepared in the laboratory on flat surfaces such as mica, water layering can be experimentally observed. This has been achieved using AFM cantilever tips with carbon nanotubes attached to the tip apex and operated in the highly sensitive frequency modulation mode (Higgins et al. 2006). Force plots in the pN/nm range exhibit few maxima separated by about 0.3 nm, the mean water radius, from the membrane wall (see Section 3.3 for a discussion of water structuration induced by mica).

When considering membrane bilayers in saline aqueous solutions one has to take into account the influence of ions. The simulated arrangement of lipids and electrolytes is shown in Figure 6.8 for a dipalmitoylphosphatidylcholine (DPPC) bilayer in an NaCl solution. The addition of salt slightly decreases the area per headgroup and many Na^+ ions become tightly bound to the lipids, creating ion–lipid complexes containing on average two lipids per ion. However, Cl^- ions become only very slightly bound to lipids. Although the Na^+ ions penetrate into the bilayer and coordinate with carbonyl oxygens Cl^- ions remain in bulk water, building an electrical double layer. The cations penetrating the headgroups of phospholipid molecules confer mechanical stability, giving rise to a more packed phospholipid network and stronger phospholipid–phospholipid lateral interactions. This has been proved by means of

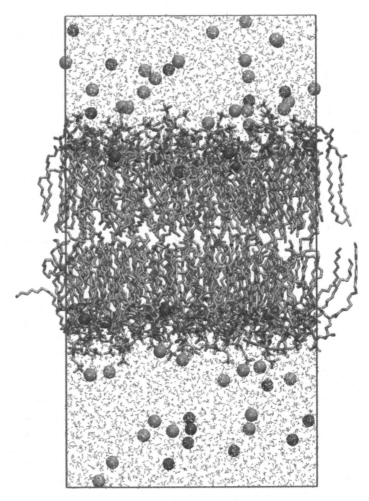

FIGURE 6.8 Snapshot from a computer simulation of a DPPC bilayer in an NaCl electrolyte solution. Na^+ and Cl^- ions are represented by dark and medium gray spheres, respectively. (Reprinted from M.L. Berkowitz, D.L. Bostick, and S. Pandit. *Chem. Rev.* 106:1527–1539, 2006, American Chemical Society. With permission.)

nanoindentation experiments with AFM, where it has been observed that the plastic yield increases for increasing ionic concentration: the higher the ionic strength, is the higher the force that must be applied to penetrate the bilayer (Garcia-Manyes, Oncins, and Sanz 2005). The charged character of the phospholipid can be used as an ideal system to study the ion distribution when submerged in an electrolyte solution. An example is given for supported 2.7-nm–thick lipid monolayers in a diluted $ZnCl_2$ solution (Bedzyk et al. 1990). By means of X-ray standing waves, the Debye length of the diffuse-double layer could be determined (see Figure 3.35), which varies in the 0.3–6 nm range, further validating the Gouy–Chapman–Stern model described in Section 3.5.2.

6.4.2 WATER CHANNELS: AQUAPORINS

The mechanisms of water transport across cell membranes was disputed until Agre and collaborators were able to isolate water channel proteins (Denker et al. 1988). Water channels exhibit a narrow selectivity permitting an extremely high water permeability ($\sim 3 \times 10^9$ molecules s^{-1}) but preventing the flow of H_3O^+. The protein known as aquaporine-1 (AQP1) was purified from red cell membranes (Agre, Sasaki, and Chrisp 1993). Aquaporins are fundamental to mammalian physiology, but they are also very important for microorganisms and plants. P. Agre was awarded the Nobel Prize in Chemistry 2003 "for the discovery of water channels". The determination of the structure has been essential in order to understand the mechanism of water transport (Murata et al. 2000). Such a structure can be imagined, as a first approximation, as an hourglass. Figure 6.9 shows the ribbon diagram of the structure of an AQP1 subunit (a) and the schematic architecture of the channel within the AQP1 subunit (b). The protein is a tetramer formed of four AQP1 monomers, each containing six tilted, bilayer-spanning α-helices surrounding the two asparagine–proline–alanine (NPA)-containing loops that enter the membrane from the opposite surfaces and are juxtaposed in the center, as shown in Figure 6.9a (Agre and Kozono 2003).

MD simulations of water transport by AQP1 have led to an advanced understanding of how water can be rapidly transported across membranes and hydronium ions are repelled (de Groot and Grubmüller 2001). An extremely specialized molecular architecture is needed to permit the passage of water but not of H_3O^+ because according to the Grotthus effect, columns of H-bonded water molecules are known to permit rapid

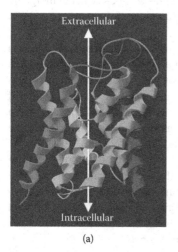
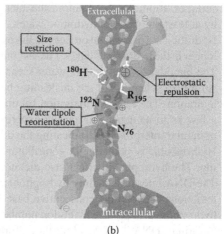

(a)   (b)

FIGURE 6.9 Structure of a AQP1 subunit and schematic of water transport. (a) Ribbon model of a AQP1 monomer showing six tilted domains and two pore-forming loops with short transmembrane α-helices entering the membrane. (b) Schematic diagram of a channel pore in same orientation as (a). The flow of water from extracellular to intracellular reservoirs occurs through the narrow column. (Republished from D. Kozono, et al. *J. Clin. Invest.* 109:1395–1399, 2002, American Society for Clinical Investigation. With permission conveyed through Copyright Clearance Center, Inc.)

conduction of protons. The hourglass structure of AQP1 has cone-shaped extracellular and intracellular vestibules that are filled with water. Such vestibules are separated by a narrow 2-nm–long channel through which the water molecules must pass in single file without H-bonding. The passage of molecules is controlled by both a geometrical restriction and electrostatic repulsion involving the R195 arginine and H180 histidine residues (see Figure 6.9b). The former is given by a 0.3-nm constriction, which is the mean molecular diameter of a single water molecule, whereas the latter is induced by the positive charge of both residues. Thus, R195 and H180 provide a size restriction, for molecules larger than water, and fixed positive charges to repel protons and other cations. It is interesting that the orientation of a water molecule changes as it passes through the channel. The oxygen faces down when the water molecule enters from the extracellular side. The water molecule flips when interacting with both asparagines of the NPA motifs, and moves farther down the channel with the oxygen facing upward as shown in Figure 6.9b.

6.5   SUMMARY

- The water molecule can be considered as a biomolecule: the 21st amino acid.
- Hydration shell: biological liquids are characterized by having a rather high concentration of large biomolecules, implying that water is mainly interfacial with a characteristic thickness of 1 nm, involving 2–3 water molecule diameters. Note that this layering effect is not exclusive of biomolecular surfaces because it has also been observed on liquid water close to flat inorganic surfaces. Biomolecules cannot exert their functions if they are not covered by at least one layer of hydration water.
- Hydration is a dynamical process on the ps timescale. Water molecules reside longer in the hydration shell (<1 ns) as compared to bulk liquid water (~1 ps). The bimodal *bound* and *free* water scenarios have to be considered only as simplified schemes but are misleading because of the dynamical nature of hydration. The range of hydration lifetimes is rather broad, spanning from a few ps to ns depending on the experimental technique used and on the particular interpretation, indicating the difficulty of the task.
- Hydrophobicity is the major contributor to protein stability inducing folding. The hydrophobic effect also dictates the final conformation of DNA. The DNA interior is mainly hydrophobic and stabilized by the stacking interactions between the consecutive base pairs, and its surface is rich with hydrophilic groups. The hydration number of the biologically more relevant B-form of DNA is >20 and for the A-form it is ~15.
- Antifreeze proteins are capable of lowering the water freezing temperature efficiently protecting fish, plants, insects, and microorganisms at temperatures below the freezing point. Such proteins attach to ice surfaces inhibiting the growth of ice crystals.
- Biological membranes are mainly formed by phospholipid bilayers that constitute the cell walls. The bilayered self-assembly is driven by hydrophobic forces. Interfacial water is also layered but the oscillatory distribution is

smeared out due to the molecular scale surface roughness. Water is transported across cell membranes through water channel proteins, which exhibit a narrow selectivity permitting an extremely high water permeability but preventing the flow of H_3O^+ thanks to size restriction (2.8 Å channel diameter) and electrostatic repulsion.

REFERENCES

1. Agre, P. and Kozono, D. 2003. Aquaporin water channels: Molecular mechanisms for human diseases. *FEBS Letters* 555:72–78.
2. Agre, P., Sasaki, S., and Chrispeels J. 1993. Aquaporins: A family of water channel proteins. *Am. J. Physiol.-Renal Physiol.* 265:F461.
3. Ansari, A., Berendzen, J., Bowne, S.F., Frauenfelder, H., Iben, I.E., Sauke, T.B., Shyamsunder, E., and Young, R.D. 1985. Protein states and proteinquakes. *Proc. Natl. Acad. Sci. USA* 82:5000–5004.
4. Arny, D.C., Lindow, S.E., and Upper, C.D. 1976. Frost sensitivity of *Zea mays* increased by application of *Pseudomonas syringae*. *Nature* 262:282–284.
5. Ball, P. 2008a. Water as a biomolecule. *Chem. Phys. Chem.* 9:2677–2685.
6. Ball, P. 2008b. Water as an active constituent in cell biology. *Chem. Rev.* 108:74–108.
7. Bedzyk, M.J., Bommarito, G.M., Caffrey, M., and Penner, T.L. 1990. Diffuse-double layer at a membrane-aqueous interface measured with X-ray standing waves. *Science* 248:52–56.
8. Berkowitz, M.L., Bostick, D.L., and Pandit, S. 2006. Aqueous solutions next to phospholipid membrane surfaces: Insights from simulations. *Chem. Rev.* 106:1527–1539.
9. Berne, B.J., Weeks, J.D., and Zhou, R. 2009. Dewetting and hydrophobic interaction in physical and biological systems. *Annu. Rev. Phys. Chem.* 60:85–103.
10. Blanco, S., Lesarri, A., López, J.C., and Alonso, J.L. 2004. The gas-phase structure of alanine. *J. Am. Chem. Soc.* 126:11675–11683.
11. Chandler, D. 2005. Interfaces and the driving force of hydrophobic assembly. *Nature* 437:640–647.
12. Cheung, M.S., García, A.E., and Onuchic, J.N. 2002. Protein folding mediated by solvation: Water expulsion and formation of the hydrophobic core occur after the structural collapse. *Proc. Natl. Acad. Sci. USA* 99:685–690.
13. Christner, B.C., Morris, C.E., Foreman, C.M., Cai, R., and Sands, D.C. 2008. Ubiquity of biological ice nucleators in snowfall. *Science* 319:1214.
14. Cruz, V., Ramos, J., and Martínez-Salazar, J. 2011. Water–mediated conformations of the alanine dipeptide as revealed by distributed umbrella sampling simulations, quantum mechanics based calculations, and experimental data. *J. Phys. Chem. B* 115:4880–4886.
15. Daggett, V. 2006. Protein folding-simulation. *Chem. Rev.* 106:1898–1916.
16. de Groot, B.L., and Grubmüller, H. 2001. Water permeation across biological membranes: Mechanism and dynamics of aquaporin-1 and GlpF. *Science* 294:2353–2357.
17. Degtyarenko, I., Jalkanen, K.J., Gurtovenko, A.A., and Nieminen, R.M. 2008. The aqueous and crystalline forms of L-alanine zwitterion. *J. Comput. Theor. Nanosci.* 5:277–285.
18. Denker, B.M., Smith, B.L., Kuhajda, F.P., and Agre, P. 1988. Identification, purification, and partial characterization of a novel Mr 28,000 integral membrane protein from erythrocytes and renal tubules. *J. Biol. Chem.* 263:15634–15642.

19. DeVries, A.L. 1983. Antifreeze peptides and glycopeptides in cold–water fishes. *Ann. Rev. Physiol.* 45:245–260.
20. Duman, J.G. and DeVries, A.L. 1976. Isolation, characterization, and physical properties of protein antifreezes from the winter flounder, *Pseudopleuronectes Americanus. Comp. Biochem. Physiol.* 54B:375–380.
21. Faraudo, J. 2011. The missing link between the hydration force and interfacial water: Evidence from computer simulations. *Curr. Opin. Colloid Interface Sci.* 16:557–560.
22. Feig, M., and Pettitt, B.M. 1999. Modeling high-resolution hydration patterns in correlation with DNA sequence and conformation. *J. Mol. Biol.* 286:1075–1095.
23. Frauenfelder, H., Chen, G., Berendzen, J., Fenimore, P.W., Jansson, H., McMahon, B.H., Stroe, I.R., Swenson, J., and Young, R.D. 2009. A unified model of protein dynamics. *Proc. Natl. Acad. Sci. USA* 106:5129–5134.
24. Garcia-Manyes, S., Oncins, G., and Sanz, F. 2005. Effect of ion-binding and chemical phospholipid structure on the nanomechanics of lipid bilayers studied by force spectroscopy. *Biophys. J.* 89:1812–1826.
25. Halle, B. 2004. Protein hydration dynamics in solution: A critical survey. *Phil. Trans. R. Soc. Lond. B* 359:1207–1224.
26. Higgins, M.J., Polcik, M., Fukuma, T., Sader, J.E., Nakayama, Y., and Jarvis, S.P. 2006. Structured water layers adjacent to biological membranes. *Biophys. J.* 91:2532–2542.
27. Hirano, S.S. and Upper, C.D. 2000. Bacteria in the leaf ecosystem with emphasis on Pseudomonas syringae–a pathogen, ice nucleus, and epiphyte. *Microbiol. Mol. Biol. Rev.* 64:624–653.
28. Inada, T., Koyama, T., Goto, F., and Seto, T. 2012. Inactivation of ice nucleating activity of silver iodide by antifreeze proteins and synthetic polymers. *J. Phys. Chem. B* 116:5364–5371.
29. Jorov, A., Zhorov, B.S., and Yang, D.S.C. 2004. Theoretical study of interaction of winter flounder antifreeze protein with ice. *Protein Sci.* 13:1524–1537.
30. Kauzmann, W. 1959. Some factors in the interpretation of protein denaturation. *Adv. Protein Chem.* 14:1–63.
31. Kawahara, H. 2002. The structures and functions of ice crystal-controlling proteins from bacteria. *J. Biosci. Bioeng.* 94:492–496.
32. Kendrew, J.C., Bodo, G., Dintzis, H.M., Parrish, R.G., Wyckoff, H., and Phillips, D.C. 1958. A three-dimensional model of the myoglobin molecule obtained by x-ray analysis. *Nature* 181:662–666.
33. Kopka, M.L., Fratini, A.V., Drew, H.R., and Dickerson, R.E. 1983. Ordered water structure around a B-DNA dodecamer: A quantitative study. *J. Mol. Biol.* 163:129–146.
34. Kuntz, I.D. and Kauzmann, W. 1974. Hydration of proteins and peptides. *Adv. Protein Chem.* 28:239–345.
35. Lehmann, M.S., Koetzle, T.F., and Hamilton, W.C. 1972. Precision neutron diffraction structure determination of protein and nucleic acid components. I. Crystal and molecular structure of the amino acid L-alanine. *J. Am. Chem. Soc.* 94:2657–2660.
36. Levy, Y. and Onuchic, J.N. 2006. Water mediation in protein folding and molecular recognition. *Annu. Rev. Biophys. Biomol. Struct.* 35:389–415.
37. Liepinsh, E., Otting, G., and Wüthrich, K. 1992. NMR observation of individual molecules of hydration water bound to DNA duplexes: Direct evidence for a spine of hydration water present in aqueous solution. *Nucl. Acids Res.* 20:6549–6553.

38. Liu, P., Huang, X., Zhou, R., and Berne, B.J. 2005. Observation of a dewetting transition in the collapse of the melittin tetramer. *Nature* 437:159–162.
39. Luisi, P.L. 2006. *The Emergence of Life, from Chemical Origins to Synthetic Biology.* Cambridge, UK: Cambridge University Press.
40. Lum, K., Chandler, D., and Weeks, J.D. 1999. Hydrophobicity at small and large length scales. *J. Phys. Chem. B* 103:4570–4577.
41. Makarov, V., Pettitt, B.M., and Feig, M. 2002. Solvation and hydration of proteins and nucleic acids: A theoretical view of simulation and experiment. *Acc. Chem. Res.* 35:376–384.
42. Maki, L.R., Galyan, E.L., Chang–Chien, M.M., and Caldwell, D. R. 1974. Ice nucleation induced by pseudomonas syringae. *Appl. Microbio.* 28:456–459.
43. Merzel, F. and Smith, J.C. 2002. Is the first hydration shell of lysozyme of higher density than bulk water? *Proc. Natl. Acad. Sci. USA* 99:5378–5383.
44. Milhaud, J. 2004. New insights into water–phospholipid model membrane interactions. *Biochim. Biophys. Acta* 1663:19–51.
45. Murata, K., Mitsuoka, K., Hiral, T., Walz, T., Agre, P., Heymann, J.B., Engel, A., and Fujiyoshi, Y. 2000. Structural determinants of water permeation through aquaporin-1. *Nature* 407:599–605.
46. Oberhauser, A.F., Marszalek, P.E., Erickson, H.P., and Fernandez, J.M. 1998. The molecular elasticity of the extracellular matrix protein tenascin. *Nature* 393:181–185.
47. Oskam, C.L., Haile, J., McLay, E., Rigby, P., Allentoft, M.E., Olsen, M.E. et al.. 2010. Fossil avian eggshell preserves ancient DNA. *Proc. R. Soc. B* 277:1991–2000.
48. Otting, G., Liepinsh, E., and Wüthrich, K. 1991. Protein hydration in aqueous solution. *Science* 254:974–980.
49. Pandit, S.A., Bostick, D., and Berkowitz, M.L. 2003. An algorithm to describe molecular scale rugged surfaces and its application to the study of a water/lipid bilayer interface. *J. Chem. Phys.* 119:2199–2205.
50. Prabhu, N. and Sharp, K. 2006. Protein–solvent interactions. *Chem. Rev.* 106:1616–1623.
51. Rhee, Y.M., Sorin, E.J., Jayachandran, G., Lindahl, E., and Pande, V.S. 2004. Simulations of the role of water in the protein-folding mechanism. *Proc. Natl. Acad. Sci. USA* 101:6456–6461.
52. Rief, M., Gautel, M., Oesterhelt, F., Fernandez, J.M., and Gaub, H.E. 1997. Reversible unfolding of individual titin immunoglobin domains by AFM. *Science* 276:1109–1112.
53. Saenger, W., Hunter, W.N., and Kennard, O. 1986. DNA conformation is determined by economics in the hydration of phosphate groups. *Nature* 324:385–388.
54. Sicheri, F.V. and Yang, D.S.C. 1995. Ice–binding structure and mechanism of an antifreeze protein from winter flounder. *Nature* 375:427–431.
55. Svergun, D.I., Richard, S., Koch, M.H.J., Sayers, Z., Kuprin, S., and Zaccai, G. 1998. Protein hydration in solution: Experimental observation by x-ray and neutron scattering. *Proc. Natl. Acad. Sci. USA* 95:2267–2272.
56. Takano, K., Funahashi, J., Yamagata, Y., Fujii, S., and Yutani, K. 1997. Contribution of water molecules in the interior of a protein to the conformational stability. *J. Mol. Biol.* 274:132–142.
57. Tanford, C. and Kirkwood, J.G. 1957. Theory of protein titration curves. I. General equations for impenetrable spheres. *J. Am. Chem. Soc.* 79:5333–5339.
58. Tobias, D.J. and Brooks, C.L. 1992. Conformational equilibrium in the alanine dipeptide in the gas phase and aqueous solution: A comparison of theoretical results. *J. Phys. Chem.* 96:3864–3870.

59. Watson, J.D. and Crick, F.H.C. 1953. Molecular structure of nucelic acids, a structure for deoxyribose nucleic acid. *Nature* 171:737–738.
60. Wing, R., Drew, H., Takano, T., Broka, C., Tanaka, S., Itakura, K., and Dickerson, R.E. 1980. Crystal structure analysis of a complete turn of B-DNA. *Nature* 287:755–758.
61. Yang, D.S.C., Sax, M., Chakrabartty, A., and Hew, C.L. 1988. Crystal structure of an antifreeze polypeptide and its mechanistic implications. *Nature* 333:232–237.
62. Yeh, Y. and Feeney, R.E. 1996. Antifreeze proteins: Structures and mechanisms of function. *Chem. Rev.* 96:601–618.
63. Yonetani, Y. and Kono, H. 2012. What determines water–bridge lifetimes at the surface of DNA? Insight from systematic molecular dynamics analysis of water kinetics for various DNA sequences. *Biophys. Chem.* 160:54–61.
64. Zhong, D., Pal, S.K., and Zewail, A.H. 2011. Biological water: A critique. *Chem. Phys. Lett.* 503:1–11.
65. Zhou, R., Huang, X., Margulis, C.J., and Berne, B.J. 2004. Hydrophobic collapse in multidomain protein folding. *Science* 305:1605–1609.

A   Buoyancy and Surface Tension

We are all familiar with Archimedes' principle, which states that an object immersed in a liquid (e.g., water on Earth) experiences an upward (buoyant) force equal to the weight of the displaced liquid. This millenarian principle is true provided the influence of surface tension, γ_{lv}, can be neglected, for example, for sufficiently large objects (think of people and ships). However, small dense objects can float at the water's surface even when their densities are much larger than the density of liquid water, violating Archimedes' principle in its initial classical formulation. If we consider small hydrophilic particles, capillarity pulls them deeper into the water, so that the mass of the displaced liquid exceeds the particle mass. In the case of small hydrophobic particles their mass is greater than that of the liquid they displace.

The force provided by γ_{lv} is precisely equal to the weight of the liquid that is displaced in the meniscus around the edge of the object. This elegant generalization of the Archimedes principle (already hypothesized by G. Galilei back in 1612) allows us to take γ_{lv} into account (Keller 1998). The generalization, which requires advanced mathematical tools, shows that the vertical γ_{lv} force on an object can dominate the buoyant force when the typical scale of the object is small compared to the length scale over which interfacial deformations decay. Interfacial deformations have an intrinsic length scale, the capillary length:

$$l_c = \sqrt{\frac{\gamma_{lv}}{\rho g}} \tag{A.1}$$

where g stands for the gravity constant. For a pure air/water interface $l_c = 2.7$ mm, so that dense objects may float at such interface provided that their typical size is $\leq l_c$.

In the case of a cylinder of density ρ_c, radius r_c, and contact angle θ_c at the air/water interface, the maximum density above which the cylinder cannot be in equilibrium and must sink follows the expression (Vella, Lee, and Kim 2006):

$$\frac{\rho_c}{\rho} \sim \frac{2}{\pi}\left[\frac{l_c}{r_c}\right]^2 \tag{A.2}$$

for $r_c \ll l_c$ provided that $\theta_c > \pi/2$, and is thus independent of θ_c. Thus, if r_c decreases ρ_c increases noticeably and applies, for example, to insects living at the air/water interface (Gao and Jiang 2004). However, for a sphere it turns out that:

$$\frac{\rho_c}{\rho} \sim \frac{3}{4}\left[\frac{l_c}{r_c}\right]^2 (1 - \cos\theta_c) \tag{A.3}$$

that is, it depends on the contact angle. In conclusion, for objects that are small compared to l_c, flotation at relatively large densities is possible because of the force of surface tension acting on the object.

REFERENCES

1. Gao, X. and Jiang, L. 2004. Water-repellent legs of water striders. *Nature* 432:36.
2. Keller, J.B. 1998. Surface tension force on a partly submerged body. *Phys. Fluids* 10:3009–3010.
3. Vella, D., Lee, D.-G. and Kim, H.-Y. 2006. The load supported by small floating objects. *Langmuir* 22:5979–5981.

B   Capillary Forces

Next we derive the simplest expression for the capillary force F_{cap} between a sphere of radius r and a flat surface separated by a distance D bound by a water meniscus with principal radii r_1 and r_2 describing the curvature κ through $1/\kappa = 1/r_1 + 1/r_2$ and exhibiting the same contact angle θ_c, as schematized in Figure B.1. A more detailed general analysis can be found in Israelachvili (2011).

F_{cap} can be calculated from the Young–Laplace expression of the pressure, $P_L = \gamma_{lv}/\kappa$, and the area $\pi r^2 \sin^2 \phi$, which can be approximated by $2\pi r d$ in the limit $r \gg d$. Under the assumption that $r_2 \gg r_1$, $P_L \simeq \gamma_{lv}/r_1$ and for small values of ϕ it can be shown that $d + D \simeq 2r_1 \cos \theta$. Thus, F_{cap} can be approximated by the expression:

$$F_{cap}(D) \simeq \frac{4\pi r \gamma_{lv} \cos \theta_c}{1 + D/d} \tag{B.1}$$

According to this expression, the maximum capillary force occurs for $D = 0$ (contact) and is given by $F_{cap}(0) = 4\pi r \gamma_{lv} \cos \theta_c$. Taking $\theta_c = 0°$ and $R = 10$ nm we obtain $F_{cap}(0) \simeq 9$ nN, of the order of the capillary force observed in Figure 2.2. For surfaces exhibiting different contact angles, θ_{c1} and θ_{c2}, $F_{cap}(0) = 2\pi r \gamma_{lv}(\cos \theta_{c1} + \cos \theta_{c2})$. The F_{cap} expressions corresponding to different radial symmetric geometries, including sphere–sphere, sphere–plane, cone–plane, cylinder–plane, and so on, can be found in Butt and Kappl (2009). The dependence of $F_{cap}(D)$ on RH could in principle be obtained by using the Kelvin equation (2.7) and those interested are referred to the work by Pakarinen et al. (2005).

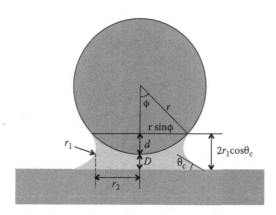

FIGURE B.1 Scheme of a meniscus formed between a sphere of radius r and a flat surface showing the relevant geometrical parameters. (Adapted from J.N. Israelachvili, *Intermolecular and Surface Forces*, 2011. Amsterdam: Elsevier. With permission.)

REFERENCES

1. Butt, H.J. and Kappl, M. 2009. Normal capillary forces. *Adv. Colloid Interface Sci.* 146:48–60.
2. Israelachvili, J.N. 2011. *Intermolecular and Surface Forces.* Amsterdam: Elsevier.
3. Pakarinen, O.H., Foster, A.S., Paajanen, M., Kalinainen, T., Katainen, J., Makkonen, I. et al. 2005. Towards an accurate description of the capillary force in nanoparticle-surface interactions. *Modelling Simul. Mater. Sci. Eng.* 13:1175–1186.

C Marangoni–Bénard Patterns

The Marangoni–Bénard effect is induced by thermal fluctuations of the surface tension at the liquid/air interface. Imagine a clean glass container with a flat bottom wall (e.g., a Petri dish), lying horizontally and filled with a thin layer of pure water with the condition that the thickness of the liquid is much smaller than the diameter of the dish. Two main interfaces are involved: liquid/air and liquid/glass. We consider the ideal conditions leading to a flat, smooth, and unperturbed free surface of water. If the bottom of the dish is heated, the liquid/air interface will undergo temperature fluctuations that will lead to gradients in surface tension. The surface tension decreases with increasing temperature (see Figure 2.4) and water will flow from warmer regions to nearby colder regions, so that a net fluid flow will be generated between regions of lower surface tension to regions of higher surface tension. This is the so-called Marangoni effect. Such a flow will be radial (parallel to the water/air interface) and convection lines will be generated because warm bulk water will ascend perpendicular to the water/air interface and the colder liquid will descend building cells. This is the Marangoni–Bénard effect. When the liquid is confined between two parallel rigid surfaces, cellular patterns are built due to buoyant convection, because of the thermally induced density gradient. This is the so-called Rayleigh–Bénard effect, which is also present for the Marangoni–Bénard case. Hence, the case of a free surface is also referred to as the Rayleigh–Bénard–Marangoni effect.

An example is illustrated in Figure C.1 with surface visualization by infrared imaging of pure water on a Petri dish (Ienna, Yoo, and Pollack 2012). The cellular structure becomes visible above about 30°C and the contrast between the inner part of the cells (warmer) and the boundaries (cooler) increases with temperature. Note that the patterns do not only exist at the air/water interface but that the domain walls penetrate in the liquid bulk.

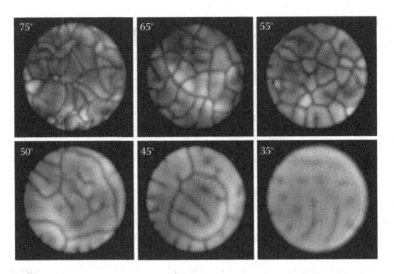

FIGURE C.1 Temperature dependence of surface images. The median surface temperature for each snapshot is noted in each frame in degrees Celsius. As the water cools, the cells stabilize, diminish in number, and eventually become less evident. (Reproduced from F. Ienna, H. Yoo, and G.H. Pollack, *Soft Matter* 8:11850–11856, 2012. With permission of The Royal Society of Chemistry.)

REFERENCES

1. Bénard, H. 1900. Les tourbillons cellulaires dans une nappe liquide. *Rev. Gén. Sci. Pures et Appl.* 11:1261–1271.
2. Ienna, F., Yoo, H., and Pollack, G.H. 2012. Spatially resolved evaporative patterns from water. *Soft Matter* 8:11850–11856.
3. Rayleigh, L. 1916. On convection currents in a horizontal layer of fluid when the higher temperature is on the under side. *Phil. Mag.* 32:529–546.

Index

Note: Page numbers ending in "e" refer to equations. Page numbers ending in "f" refer to figures. Page numbers ending in "t" refer to tables.

Printed in the United States
By Bookmasters